Springer Series in Statistics

Advisors:
J. Berger, S. Fienberg, J. Gani,
K. Krickeberg, I. Olkin, B. Singer

Springer Series in Statistics

Anderson: Continuous-Time Markov Chains: An Applications-Oriented Approach.
Andrews/Herzberg: Data: A Collection of Problems from Many Fields for the Student and Research Worker.
Anscombe: Computing in Statistical Science through APL.
Berger: Statistical Decision Theory and Bayesian Analysis, 2nd edition.
Bolfarine/Zacks: Prediction Theory for Finite Populations.
Brémaud: Point Processes and Queues: Martingale Dynamics.
Brockwell/Davis: Time Series: Theory and Methods, 2nd edition.
Daley/Vere-Jones: An Introduction to the Theory of Point Processes.
Dzhaparidze: Parameter Estimation and Hypothesis Testing in Spectral Analysis of Stationary Time Series.
Farrell: Multivariate Calculation.
Fienberg/Hoaglin/Kruskal/Tanur (Eds.): A Statistical Model: Frederick Mosteller's Contributions to Statistics, Science, and Public Policy.
Goodman/Kruskal: Measures of Association for Cross Classifications.
Grandell: Aspects of Risk Theory.
Hall: The Bootstrap and Edgeworth Expansion.
Härdle: Smoothing Techniques: With Implementation in S.
Hartigan: Bayes Theory.
Heyer: Theory of Statistical Experiments.
Jolliffe: Principal Component Analysis.
Kotz/Johnson (Eds.): Breakthroughs in Statistics Volume I.
Kotz/Johnson (Eds.): Breakthroughs in Statistics Volume II.
Kres: Statistical Tables for Multivariate Analysis.
Leadbetter/Lindgren/Rootzén: Extremes and Related Properties of Random Sequences and Processes.
Le Cam: Asymptotic Methods in Statistical Decision Theory.
Le Cam/Yang: Asymptotics in Statistics: Some Basic Concepts.
Manoukian: Modern Concepts and Theorems of Mathematical Statistics.
Miller, Jr.: Simultaneous Statistical Inference, 2nd edition.
Mosteller/Wallace: Applied Bayesian and Classical Inference: The Case of *The Federalist Papers.*
Pollard: Convergence of Stochastic Processes.
Pratt/Gibbons: Concepts of Nonparametric Theory.
Read/Cressie: Goodness-of-Fit Statistics for Discrete Multivariate Data.
Reiss: Approximate Distributions of Order Statistics: With Applications to Nonparametric Statistics.
Ross: Nonlinear Estimation.
Sachs: Applied Statistics: A Handbook of Techniques, 2nd edition.
Salsburg: The Use of Restricted Significance Tests in Clinical Trials.
Särndal/Swensson/Wretman: Model Assisted Survey Sampling.
Seneta: Non-Negative Matrices and Markov Chains.
Shedler: Regeneration and Networks of Queues.

(continued after index)

Heleno Bolfarine Shelemyahu Zacks

Prediction Theory for Finite Populations

Springer-Verlag
New York Berlin Heidelberg London Paris
Tokyo Hong Kong Barcelona Budapest

Heleno Bolfarine
Department of Statistics
Institute of Mathematics and Statistics
University of Sao Paulo
Sao Paulo
Brazil

Shelemyahu Zacks
Department of Mathematical Sciences
State University of New York
at Binghamton
Binghamton, NY 13902-6000
USA

Mathematics Subject Classification (1991): 62D05

Library of Congress Cataloging-in-Publication Data
Bolfarine, Heleno.
Prediction theory for finite populations/Heleno Bolfarine, Shelemyahu Zacks.
p. cm. – (Springer series in statistics)
Includes bibliographical references and index.
ISBN-13:978-1-4612-7713-2
1. Prediction theory. I. Zacks, Shelemyahu. II. Title. III. Series.
QA279.2.B65 1992
519.5′4 – dc20 91-45891

Printed on acid-free paper.

Production managed by Natalie Johnson; manufacturing supervised by Robert Paella.
Photocomposed copy prepared from the authors' $\mathcal{A}_{\mathcal{M}}\mathcal{S}$-$\TeX$ files.

9 8 7 6 5 4 3 2 1

ISBN-13:978-1-4612-7713-2 e-ISBN-13:978-1-4612-2904-9
DOI:10.1007/978-1-4612-2904-9

To our families

Preface

A large number of papers on predicting finite population quantities have appeared in the literature during the last two decades. This book is designed to present current theory in a systematic and consistent manner. An overview of the material in the book is presented in the Synopsis, which guides the reader to the various topics treated. Theoretical examples, which illustrate concepts and provide formulae of predictors for special cases of common interest, are given in each chapter. In addition, a large number of problems for solution accompany each chapter. Many of these problems are based on material found in research papers. Thus, the book can be used in a graduate course and can well prepare students for research in the field. The first draft was written while H. Bolfarine spent his sabbatical year at the Center for Statistics at the State University of New York at Binghamton, New York. We started by translating a short monograph (in Portuguese) written by Rodrigues and Bolfarine (1984). Soon we realized that much more work was required. This monograph turned out to be completely different from the short Portuguese version. We would like to acknowledge the advice and suggestions offered generously by Professor J. Rodrigues. H. Bolfarine would like to acknowledge the financial support of Conselho National de Pesquisas (CNPq–Brasil) and the University of São Paulo, which made the writing of this book possible. Special thanks are due to Mrs. Marge Pratt for her excellent TEX typing of the book. We would like to thank also the State University of New York at Binghamton for providing facilities and our families for their understanding and support.

Heleno Bolfarine
Shelemyahu Zacks

Binghamton, NY
October, 1991

Contents

Synopsis

Prediction theory for sampling surveys can be considered as a general framework for statistical inference on the characteristics of finite populations. Well–known estimators of population totals or population variances encountered in the classical theory, as expansion, ratio, regression, and other estimators, can be obtained as predictors in a general prediction theory, under some special (sometimes degenerate) models. The general prediction theory is based on superpopulation models, which consider the values of the population elements as random variables having joint distributions. These joint distributions might be specified completely or partially.

The stochastic structure for inferential purposes is introduced via the superpopulation model. Hence, prediction theory is focused on the inference aspects, based on the given data, rather than on the sampling design for data collection. As will be seen in different parts of the text, various properties of predictors, like model–unbiasedness or model–mean squared error, may depend on the values of certain auxiliary variables (regressors, for example) of the elements selected to the sample. This fact implies often, from the model point of view, which elements should be sampled. The emphasis is, however, on the analysis of the data rather than on the design. The classical theory, known also as the fixed population model, is actually a distribution–free component of the general theory. As in the case of randomization tests or bootstrapping, in inference from infinite populations, the classical theory attains its stochastic properties from the random selection procedures of the sample elements.

A strong drive for the development of the general prediction theory was the inadequacy of the classical theory, in providing reasonable predictors in various situations, as described by several authors. In particular, see the paper of Basu (1971, p. 212, Example 3). Thomsen and Tesfu (1988) presented some practical applications of superpopulation models in sampling from finite populations.

A large number of papers have been written on this subject during the last twenty years. In this monograph, we attempt to provide a consistent exposition of this theory and fill in several gaps.

The book consists of nine chapters. In Chapter 1, we present the basic models and frameworks of the classical and modern approaches. Six basic superpopulation models are introduced, as special cases of the general regression model. The concepts of model–unbiasedness and model–mean squared error are introduced. The general structure of predictors of population quantities (as the total or the variance) is discussed and illustrated. The combination of sampling design models with superpopulation models for the purpose of attaining the robustness of predictors is discussed too.

Chapter 2 is devoted to the development of optimal predictors for various superpopulation models. We start with the theory of best linear unbiased predictor (BLUP) of the population total T, for the general regression model. We prove also that the simple projection predictor (SPP) of T is BLUP if, and only if, the error covariance matrix $\mathbf{V}$ satisfies the so–called *Condition L.* This condition requires the existence of some linear combination of the column vectors of $\mathbf{V}$, which belongs to the space spanned by the column vectors of the matrix of regressors, $\mathbf{X}$. The derivation of BLUP requires only the modeling of $\mathbf{X}$ and $\mathbf{V}$, irrespective of the actual joint distribution of the population vector of interest $\mathbf{y}$. If the model specifies also the family of joint distribution of $\mathbf{y}$, and if this family possesses complete predictive–sufficient statistics (Definition 2.2.1), then, the best unbiased predictor (BUP), which minimizes the prediction MSE (mean squared error), exists and is essentially unique (analogous to the Lehmann–Scheffé–Blackwell–Rao Theorem). The structure of BUP is studied in Section 2.2. We derive the BUP of T of the population variance S_y^2 and of the finite population regression coefficients β_N, for the normal regression model. The theory of optimal equivariant predictors is presented in Section 2.3, for location, scale, and location–scale superpopulation models. This section provides minimum–risk equivariant predictors (MREP) for T and S_y^2. Section 2.4 studies Stein type shrinkage predictors.

Chapter 3 is devoted to Bayes and minimax predictors. We start in Section 3.1 with Bayes predictors of T for the normal regression model, with a normal prior distribution of the regression parameters $\boldsymbol{\beta}$. The Bayes predictor of S_y^2 is derived too. Section 3.2 presents Bayes linear predictors of T, which are of the form of BLUP, with a Bayes estimator of $\boldsymbol{\beta}$. Several models of such predictors of T are discussed. Section 3.3 discusses minimax and admissible predictors. The general theory of minimaxity and admissibility is embedded in prediction theory. It is shown that the BUP of T under the normal regression model is minimax. The minimax predictor of S_y^2 is derived under the normal location model. Section 3.4 develops the theory of dynamic prediction of T and S_y^2 at a sequence of time epochs. The theory of Kalman–Bucy filter is applied there to the prediction framework. Section 3.5 presents the theory of empirical bayes predictors (EBP).

Chapter 4 presents maximum–likelihood predictors (MLP). The question of an appropriate definition of MLP is crucial one, since in the classical theory (design based), the likelihood function attains a constant value for all vectors $\mathbf{y}$ in E^N, such that the coordinates y_i, with $i \in \mathbf{s}$, are equal to the corresponding observed sample values. Accordingly, classical theory cannot develop useful maximum–likelihood estimators. On the other hand, in prediction theory, which is based on superpopulation models, one can discuss various notions of predictive likelihoods and develop corresponding maximum–likelihood predictors. Four types of predictive likelihoods are discussed in Section 4.1: estimative, profile, Lauritzen–Hinkley, and the Royall predictive likelihoods. The MLP of the population total T, under a normal superpopulation model, is derived in Section 4.2, for each one of the four types of predictive likelihoods. In Section 4.3, we derive the MLP of the population variance S_y^2.

Chapter 5 develops the theory of classical and Bayesian prediction intervals. We start in Section 5.1 with confidence prediction intervals (CPI) for the population total. Section 5.2 is devoted to tolerance prediction intervals (TPI) for the population total.

In Chapter 6, we study the question of model robustness of predictors. Predictors, which are BUP under one model, might be biased under another model. This type of model dependence of predictors could lead to undesirable consequences, if the applied model differs from the one that correctly reflects reality. In Section 6.1, we present conditions on two alternative regression models, under which a predictor of T preserves its BLUP property. We illustrate in examples that these conditions for robustness could be satisfied if the selected sample is *balanced* with respect to several regressors simultaneously. In Section 6.2, we discuss the question of robust estimation of the prediction variance of predictors of T. Section 6.3 provides a numerical example based on simulations. Section 6.4 discusses Bayesian robustness of predictors of T. Finally, we present, in Section 6.5, an attempt to overcome the problem of the lack of robustness by considering several models simultaneously within a Bayesian framework.

Chapter 7 is devoted to superpopulation models with measurement errors. Both the values of the regressor $\mathbf{x}$ and the predicted variable $\mathbf{y}$ are not directly observable due to measurement errors. In Section 7.1, we develop model–unbiased regression–type predictors and derive their prediction variances. Section 7.2 presents a Bayesian model with errors of measurement and Bayes predictors of T and S_y^2.

Chapter 8 formulates the framework for asymptotic analysis, as the population size $N \to \infty$ and the sample size $n \to \infty$. The chapter contains three sections, in which conditions are established for the asymptotic (model) consistency and normality of $\hat{T}_{BLU}$ and of the estimator $\hat{\boldsymbol{\beta}}_s$, under regression models with and without measurement errors.

Finally, in Chapter 9, we study the characteristics of predictors under different sampling design models. We introduce a general class of QR predictors, and study conditions under which these predictors are asymptotically design unbiased (ADU). If these conditions are satisfied, optimal predictors that are model dependent (unrobust) become asymptotically unbiased (weakly robust). We establish also the Godambe Joshi (1965) variance lower bound for unbiased predictors (estimators), and thus define the notion of design optimal ADU predictor.

Each chapter contains examples, a section with exercises and over one hundred references are listed at the end.

1
Basic Ideas and Principles

Sampling theory develops the principles upon which the selection of a sample from a finite population and the inference about characteristics of the population are based. As emphasized by Cassel et al. (1977), the sources of randomness that provide the stochastic structure for statistical inference depend upon:

(i) the method used to select the sample units;
(ii) the method used to measure the values of the characteristics of interest of the sample units;
(iii) the knowledge of some random process that might have generated the values of the characteristics of interest associated with each unit of the population.

Category (i) pertains to possible random selection of the sample from the given population. As explained later, the various sampling designs or strategies fall under this category. Notice that the selection of the sample could, according to a given strategy, be nonrandomized (purposive). The classical approach to survey sampling considers mainly category (i).

Category (ii) relates to sources of randomness, which are due to measurement errors.

Category (iii) encompasses stochastic components or processes, which randomly determine the (true) values of the characteristic of interest of each unit in the population. Explicit modeling of these stochastic components provides the "superpopulation" or Bayesian framework for inference on the quantities of interest of the finite population. In this chapter, we present the theoretical models for the different approaches to inference and prediction in finite populations.

1.1. The Fixed Finite Population Model

Let $\mathcal{P} = \{1, \ldots, N\}$ be the set of labels of the units of a finite population of size N, where N is known. Associated with each unit of $\mathcal{P}$, there is a fixed value y_k, $k = 1, \ldots, N$. Let $\mathbf{y} = (y_1, \ldots, y_N)$ denote the vector of fixed values (parameters) associated with $\mathcal{P}$. In order to gain information about a function of $\mathbf{y}$, denoted by $\theta = \theta(\mathbf{y})$, a sample $\mathbf{s}$ of size $n(\leq N)$ is selected from $\mathcal{P}$. Examples of functions of $\mathbf{y}$ that might be of interest are: linear functions $\theta_L = \theta_L(\mathbf{y}) = \mathbf{l}'\mathbf{y}$, where $\mathbf{l}$ is an $N \times 1$ known vector, and quadratic functions $\theta_Q = \theta_Q(\mathbf{y}) = \mathbf{y}'\mathbf{A}\mathbf{y}$, where $\mathbf{A}$ is an $N \times N$ known matrix. Special cases of linear functions are the population total, $T = \sum_{i=1}^{N} y_i$, and the population mean $\bar{y} = T/N$. A well–known special case of a quadratic function is the population variance $S_y^2 = \sum_{i=1}^{N}(y_i - \bar{y})^2/N$.

The sampling plan (or the sampling design) is represented by a probability function, defined on the set $\mathcal{S}$ (the sample space) of all possible samples $\mathbf{s}$ of size n. This function, $p(\cdot)$, satisfies

(i) $p(s) \geq 0$, for all $\mathbf{s} \in \mathcal{S}$, and
(ii) $\sum_{\mathbf{s} \in \mathcal{S}} p(s) = 1$.

A detailed and general discussion of the properties of the function p is given by Cassel et al. (1977), where important examples are also presented. As pointed out by Royall (1988), the sampling designs serve many useful purposes. They might provide protection against unconscious bias due to selection procedures which are based on convenience or informal judgement. Also, as will be seen later, certain sampling designs may provide some protection against model misspecification.

After $\mathbf{s}$ has been observed, we may denote by $\mathbf{y}_s$ and $\mathbf{y}_r$ the observed and unobserved parts of $\mathbf{y}$, respectively. An estimator of θ is a function $\hat{\theta} = \hat{\theta}(\mathbf{y}_s)$ of $\mathbf{y}_s$. The properties of the estimator $\hat{\theta}$ are often expressed in terms of its expected value and its mean squared error with respect to the sampling design p. These quantities are defined, respectively, by

$$E_p[\hat{\theta}] = \sum_{\mathbf{s} \in \mathcal{S}} p(\mathbf{s})\hat{\theta}(\mathbf{y}_s) \tag{1.1.1}$$

and

$$E_p[\hat{\theta} - \theta]^2 = \sum_{\mathbf{s} \in \mathcal{S}} p(\mathbf{s})[\hat{\theta}(\mathbf{y}_s) - \theta(\mathbf{y})]^2, \tag{1.1.2}$$

where $\sum_{\mathbf{s} \in \mathcal{S}}$ denotes summation over all sample $\mathbf{s}$ in $\mathcal{S}$. Notice that the expected value $E_p[.]$ in Eqs. (1.1.1) and (1.1.2) depends on the sampling distribution of $\hat{\theta}$ induced by the function $p(.)$.

We introduce now the definition of p-unbiasedness.

Definition 1.1.1. *An estimator $\hat{\theta}$ is said to be p-unbiased if*

$$E_p[\hat{\theta}] = \theta,$$

for all $\mathbf{y}$.

If $\hat{\theta}$ is a p-unbiased estimator of θ, then, from Eq. (1.1.2),

$$E_p[\hat{\theta} - \theta]^2 = \mathrm{Var}_p[\hat{\theta}],$$

where the right–hand side of the above expression is the p-variance of θ. The following are examples of p-unbiased estimators.

Example 1.1.1. Let

$$I_i(\mathbf{s}) = \begin{cases} 1, & \text{if } i \in \mathbf{s}, \\ 0, & \text{otherwise.} \end{cases}$$

Let π_i, $i = 1, \ldots, N$, denote the inclusion probability of the ith unit of $\mathcal{P}$. These inclusion probabilities depend on the sampling design and are given by

$$\begin{aligned} \pi_i &= E_p[I_i(\mathbf{s})] \\ &= \sum_{\mathbf{s} \in \mathcal{S}} p(\mathbf{s}) I_i(\mathbf{s}). \end{aligned} \tag{1.1.3}$$

Similarly, let

$$I_{i,j}(\mathbf{s}) = \begin{cases} 1, & \text{if } i \in \mathbf{s} \text{ and } j \in \mathbf{s}, \\ 0, & \text{otherwise.} \end{cases}$$

The second–order inclusion probabilities are given by

$$\begin{aligned} \pi_{ij} =& E_p[I_{ij}(\mathbf{s})] \\ =& \sum_{\mathbf{s} \in \mathcal{S}} p(\mathbf{s}) I_{ij}(\mathbf{s}). \end{aligned} \tag{1.1.4}$$

An important and widely used p–unbiased estimator of the population total T is the Horvitz–Thompson (H–T) estimator

$$\hat{T} = \sum_{i \in \mathbf{s}} \frac{y_i}{\pi_i}. \tag{1.1.5}$$

To prove that $\hat{T}$ is p unbiased, write

$$\hat{T} = \sum_{i=1}^{N} I_i(\mathbf{s}) \frac{y_i}{\pi_i}.$$

Thus,

$$E_p[\hat{T}] = \sum_{i=1}^{N} E_p[I_i(\mathbf{s})]\frac{y_i}{\pi_i}$$
$$= \sum_{i=1}^{N} y_i.$$

It is a straightforward matter to show that the variance under p of $\hat{T}$ is (Exercise 1.1)

$$\text{(1.1.6)} \qquad \mathrm{Var}_p[\hat{T}] = \sum_{j=1}^{N} \frac{y_i^2}{\pi_i} + \sum_{i=1}^{N}\sum_{i \neq j} \frac{y_i}{\pi_i}\frac{y_j}{\pi_j}\pi_{ij} - T^2.$$

Notice that the so called *expansion estimator* (see Cochran, 1977, p. 21)

$$\hat{T}_E = N\bar{y}_s,$$

where $\bar{y}_s$ is the sample mean, is a special case of the H–T estimator for simple random sampling, in which $\pi_i = n/N$, $i = 1, \ldots, N$.

Example 1.1.2. Unbiased sampling estimators of the population variance are considered in Cochran (1977). Most of the estimators are based on the sample variance

$$\text{(1.1.7)} \qquad s_y^2 = \frac{1}{n-1}\sum_{j \in \mathbf{s}}(y_j - \bar{y}_s)^2.$$

It is simple to verify that, under simple random sampling, the estimator (Exercise 1.2)

$$\text{(1.1.8)} \qquad \hat{S}_y^2 = \frac{(N-1)}{N} s_y^2$$

is p-unbiased for S_y^2.

1.2. The Superpopulation Model

The superpopulation model assumes that the value of the variable of interest, associated with the ith unit of the population , y_i, $i = 1, \ldots, N$, is comprised of a deterministic element η_i and a random element e_i; that is,

$$y_i = \eta_i + e_i,$$

$i = 1, \ldots, N$. The random vector $\mathbf{e} = (e_1, \ldots, e_N)$ is assumed to have zero mean and a positive definite covariance matrix, $\mathbf{V}$. In the case of uncorrelated errors with equal variances, $\mathbf{V} = \sigma^2 \mathbf{I}_N$, $\sigma^2 > 0$, where $\mathbf{I}_N$ is the identity matrix of dimension N. The model is however more general. We exclude models under which e_i do not have finite second moments, like the Cauchy distribution model.

1.2.1. The Regression Model

The regression model can be applied when p auxiliary variables, $x_1, \ldots, x_p$, are given for each unit of $\mathcal{P}$. In this case the deterministic component η_i, may be modeled as linear functions of the auxiliary variables; that is,

$$\eta_i = \sum_{l=1}^{p} \beta_l x_{il}, \tag{1.2.1}$$

$i = 1, \ldots, N$. Typically, it is assumed that the values of $x_1, \ldots, x_p$ are known for all units of $\mathcal{P}$. Let $\mathbf{X}$ denote the $N \times p$ matrix $\mathbf{X} = (x_{il}; i = 1, \ldots, N, l = 1, \ldots, p)$, and let $\boldsymbol{\eta}' = (\eta_1, \ldots, \eta_N)$. The linear regression model can be expressed as

$$\mathbf{y} = \mathbf{X}\boldsymbol{\beta} + \mathbf{e}, \tag{1.2.2}$$

in which the random vector $\mathbf{e}$ is as described above. The superpopulation linear regression model (1.2.2) is represented by the parameter $\boldsymbol{\psi} = (\boldsymbol{\beta}, \mathbf{V})$, and is called the $\boldsymbol{\psi}$-model.

A sampling design is called noninformative if $p(\mathbf{s})$, $\mathbf{s} \in \mathcal{S}$, is independent of $\boldsymbol{\psi}$. Accordingly, if the sampling design is noninformative, the conditionality principle (Basu, 1975) implies that the inference on $\boldsymbol{\psi}$ should be based only on the observed part of $\mathbf{y}$, $\mathbf{y}_s$, and the model $\boldsymbol{\psi}$, which is the link between $\mathbf{y}_s$ and $\mathbf{y}_r$. The following are examples of some of the most commonly used regression models.

Model SM1. Suppose that the elements of the $\boldsymbol{\psi}$-model are $\mathbf{X} = \mathbf{1}_N$ and $\mathbf{V} = \sigma^2 \mathbf{I}_N$, where $\mathbf{1}_N$ is a vector of ones of dimension N. In this case, the $\boldsymbol{\psi}$ model is usually known as the *simple location model*, with β as the location parameter. Therefore, given β, which is an unknown scalar, the random variables $y_1, \ldots, y_N$ are uncorrelated and exchangeable.

The following is an example of a case to which model SM1 fits. The finite population consists of N bags of flour, filled by a machine. All bags should contain a nominal amount of β [kg]. However, due to the filling process, the actual amount poured into each bag is $\beta + e$, where e is a random variable with zero mean and variance σ^2. The values of e corresponding to different bags are uncorrelated.

Model SM2. In this model $\mathbf{X} = \mathbf{x}_N$ and $\mathbf{V} = \sigma^2 \mathbf{W}$, where $\mathbf{x}'_N = (x_1, \ldots, x_N)$ and $\mathbf{W}$ is a diagonal matrix whose diagonal elements are the elements of $\mathbf{x}_N$. We denote this as $\mathbf{W} = \text{diag}\{x_1, \ldots, x_N\}$. This model is usually known in the literature as the *simple regression model through the origin.* In practical applications, the model is used in situations where the expected value of y is proportional to x, and the variance of y is also proportional to x. For example, if the population elements are farms, x is the number of acres on which wheat is planted. We expect β [ton] of wheat from each acre, and the variance of the yield in σ^2 [ton]^2 per acre. y is the total yield in a farm.

Model SM3. A generalization of the previous model is obtained by considering

$$\mathbf{X} = \begin{pmatrix} 1 & x_1 \\ \vdots & \vdots \\ 1 & x_N \end{pmatrix}, \quad \boldsymbol{\beta} = \begin{pmatrix} \beta_0 \\ \beta_1 \end{pmatrix}$$

and $\mathbf{W} = \text{diag}\{x_1^g, \ldots, x_N^g\}$, where $g = 0, 1, 2$. In the case of $g = 0$, we obtain the *simple regression model.* For possible applications of this regression model, see Cochran (1977, p. 189).

Model SM4. Consider again model SM2, for the case where the population is stratified into K disjoint subgroups (strata) of size N_h, $h = 1, \ldots, K$. Clearly, $N = N_1 + \cdots + N_K$. The double subscript hj denotes quantities associated with the jth unit in the hth stratum, where $j = 1, \ldots, N_h$ and $h = 1, \ldots, K$. The strata may be formed by partitioning the elements of $\mathcal{P}$ according to the values of x. The N_1 units in the first stratum correspond to the N_1 smallest values of x; the N_2 units that form the second stratum correspond to the N_2 next smallest values of x, and so on. From stratum h, a sample $\mathbf{s}_h$ of size n_h is selected, where $h = 1, \ldots, K$ and $n = n_1 + \cdots + n_K$ is the total sample size. A more general version of this model is obtained by considering $\mathbf{X}$ as a block diagonal matrix with the hth diagonal element $\mathbf{X}_h' = (x_{h1}, \ldots, x_{hN_h})$ and $\boldsymbol{\beta}' = (\beta_1, \ldots, \beta_K)$. Also, we may consider $\mathbf{V} = \text{diag}\{\mathbf{V}_1, \ldots, \mathbf{V}_K\}$, with the hth diagonal element $\mathbf{V}_h = \sigma_h^2 \mathbf{W}_h$, where $\mathbf{W}_h = \text{diag}\{x_{h1}, \ldots, x_{hN_h}\}$, $h = 1, \ldots, K$ (Pfeffermann, 1984). We call this model the *stratified regression model.*

Stratification may result in significant gains in precision since it makes it possible to divide a heterogeneous population into homogeneous strata. Thus, if the elements of a population are textile plants, stratification can be done according to the number of workers, x, in each plant. A more reliable covariate could be the number of looms in each plant. Stratification can be obviously performed according to different criteria.

Model SM5. Suppose that the elements of the ψ-model are

$$\mathbf{X} = \begin{pmatrix} 1 & x_1 & \dots & x_1^J \\ \vdots & \vdots & \vdots & \vdots \\ 1 & x_N & \dots & x_N^J \end{pmatrix}, \quad \boldsymbol{\beta} = \begin{pmatrix} \beta_0 \\ \beta_1 \\ \vdots \\ \beta_J \end{pmatrix},$$

where $J \geq 0$, $\mathbf{V} = \sigma^2 \mathrm{diag}\{f_1(\mathbf{X}), \dots, f_N(\mathbf{X})\}$, $f_j(\mathbf{X}) = \sum_{i=0}^J c_i x_j^i$, and c_i are specified constants. In this case, $E[y_j] = \sum_{i=0}^J \beta_i x_j^i$. This model is typically known as the *polynomial regression model.*

Model SM6. Suppose that the population is divided into K distinct subpopulations (clusters) with N_h units in the hth cluster. $N = \sum_{h=1}^K N_h$ is the population size. In the first stage, a sample $\mathbf{s}$ of $k(\leq K)$ clusters is selected. In the second stage, a sample $\mathbf{s}_h$ of size n_h is selected from the hth cluster, $h \in s$. $n = \sum_{h \in \mathbf{s}} n_h$ is the total sample size. The model assumes that

$$E[y_{hj}] = \mu, \tag{1.2.3}$$

for $h = 1, \dots, K$, $j = 1, \dots, N_h$, and

$$\mathrm{Cov}[y_{hj}; y_{lq}] = \begin{cases} \sigma_h^2 + \sigma_v^2; & h = l, j = q, \\ \sigma_v^2; & h = l, j \neq q, \\ 0; & h \neq l, \end{cases} \tag{1.2.4}$$

where $j = 1, \dots, N_h$, $q = 1, \dots, N_l$, and $h, l = 1, \dots, K$. According to this model, units in the same cluster are correlated, but units in different clusters are not. In the notation of the ψ-model, $\mathbf{X} = (\mathbf{1}_{N_1}, \dots, \mathbf{1}_{N_K})'$, where $\mathbf{1}_{N_h}$ is a vector of ones of dimension N_h and $\mathbf{V} = \mathrm{diag}\{\mathbf{V}_1, \dots, \mathbf{V}_K\}$, is a block diagonal matrix, with $\mathbf{V}_h$, an $N_h \times N_h$ matrix with diagonal and off–diagonal elements given, respectively, by $\sigma_h^2 + \sigma_v^2$ and σ_v^2; that is

$$\mathbf{V}_h = \sigma_h^2 \mathbf{I}_{N_h} + \sigma_v^2 \mathbf{J}_{N_h}.$$

The identity matrix of dimension is denoted by $\mathbf{I}_{N_h}$, N_h and $\mathbf{J}_{N_h}$ denotes the $N_h \times N_h$ matrix of ones. Model SM6 can be considered as a variance components model in the following sense. Let $\mu_1, \dots, \mu_K$ be independent identically distributed (i.i.d.) random variables, such that $E[\mu_h] = \mu$ and Var $[\mu_h] = \sigma_v^2$. Suppose that for each $h = 1, \dots, K$,

$$E[Y_{hj} \mid \mu_h] = \mu_h, \qquad j = 1, \dots, N_h,$$

and

$$\mathrm{Cov}[y_{hj}, y_{hj'} \mid \mu_h] = \begin{cases} \sigma_h^2, & \text{if } j = j', \\ 0, & \text{if } j \neq j'. \end{cases}$$

Then, Eqs. (1.2.3) and (1.2.4) are obtained as the nonconditional (or total) expectation and covariance of the variables y_{hj}. Accordingly, σ_h^2 is the conditional variance within cluster h, and σ_v^2 is the variance between conditional cluster means.

Some variations of the above basic models are also considered in the text.

1.3. Predictors of Population Quantities

It is often possible to express a quantity of interest θ as the sum

(1.3.1) $$\theta = \theta_s + \theta_{sr},$$

where θ_s is a function of $\mathbf{y}_s$ and θ_{sr} is a function of $(\mathbf{y}_s, \mathbf{y}_r)$. The problem is therefore to predict the quantity θ_{sr} from the sample data given by $\mathbf{y}_s$. As pointed out by Ericson (1988), little is available in the literature on such predictors. The reader is referred to the study of Rodrigues et al. (1985).

As mentioned earlier, two quantities of primary interest are linear functions, θ_L, and quadratic functions, θ_Q. Writing $\theta_L = \mathbf{l}'\mathbf{y}$, where $\mathbf{l} = (\mathbf{l}'_s, \mathbf{l}'_r)$, we have $\theta_{sr} = \mathbf{l}'_r\mathbf{y}_r$. In the case of a quadratic population quantity, let $\theta_Q = \mathbf{y}'\mathbf{A}\mathbf{y}$. Make the corresponding partitions $\mathbf{y}' = (\mathbf{y}'_s, \mathbf{y}'_r)$ and

$$\mathbf{A} = \begin{pmatrix} \mathbf{A}_s & \mathbf{A}_{sr} \\ \mathbf{A}_{rs} & \mathbf{A}_r \end{pmatrix}.$$

We obtain the presentation $\theta_Q = \theta_s + \theta_{sr}$, where $\theta_s = \mathbf{y}'_s\mathbf{A}_s\mathbf{y}_s$ and

$$\theta_{sr} = 2\mathbf{y}'_s\mathbf{A}_{sr}\mathbf{y}_r + \mathbf{y}'_r\mathbf{A}_r\mathbf{y}_r.$$

In the special case of $\theta_Q = S_y^2$, we have $\mathbf{A} = (\mathbf{I}_N - \mathbf{J}_N/N)/N$, and simple algebraic manipulations yield the formula

(1.3.2) $$S_y^2 = \frac{n}{N}s_y^2 + (1 - \frac{n}{N})[S_{ry}^2 + \frac{n}{N}(\bar{y}_s - \bar{y}_r)^2],$$

where $s_y^2 = \mathbf{y}'_s(\mathbf{I}_s - \mathbf{J}_s/n)\mathbf{y}_s/n$ is the sample variance; $\mathbf{I}_s$ is the identity matrix of dimension n; $\mathbf{J}_N = \mathbf{1}_N\mathbf{1}'_N$, $\mathbf{J}_s = \mathbf{1}_s\mathbf{1}'_s$; $\bar{y}_s$ is the sample mean; $\bar{y}_r$ is the mean of the unobserved portion of $\mathcal{P}$ and S_{ry}^2 is its variance. Thus, in the case of S_y^2,

$$\theta_s = \frac{n}{N}s_y^2 \text{ and } \theta_{sr} = (1 - \frac{n}{N})[S_{ry}^2 + \frac{n}{N}(\bar{y}_s - \bar{y}_r)^2].$$

According to expression (1.3.1), we define a predictor $\hat{\theta}$ for θ, as

(1.3.3) $$\hat{\theta} = \theta_s + \hat{\theta}_{sr},$$

where $\hat{\theta}_{sr}$ is a predictor of θ_{sr}.

Example 1.3.1. Some commonly used predictors of T follow:

(i) The expansion estimator $\hat{T}_E$ can be written as

$$\hat{T}_E = n\bar{y}_s + (N-n)\bar{y}_s. \tag{1.3.4}$$

In this case, $\hat{\theta}_{sr} = (N-n)\bar{y}_s$.

(ii) The *ratio estimator* of T is given by

$$\hat{T}_R = n\bar{y}_s + (N-n)\bar{x}_r(\frac{\bar{y}_s}{\bar{x}_s}), \tag{1.3.5}$$

where $x_1, \ldots, x_N$ are known auxiliary quantities, $\bar{x}_s = (1/n)\sum_{i\in \mathbf{s}} x_i$ and $\bar{x}_r = [1/(N-n)]\sum_{i\in \mathbf{r}} x_i$.

We define now the notion of model unbiasedness.

Definition 1.3.1. *A predictor $\hat{\theta}$ is said to be ψ-unbiased if*

$$E_\psi[\hat{\theta}-\theta] = 0, \ \textit{for all } \psi = (\boldsymbol{\beta}, \mathbf{V}).$$

Notice that the expectation with respect to the superpopulation model, $E_\psi[.]$, is different from the previously defined $E_p[.]$. Whereas $E_p[.]$ is determined only according to the properties of the sampling selection strategy and requires explicit knowledge of $(\mathcal{S}, p(\mathbf{s}))$, $E_\psi[.]$ depends only on the model ψ and on the indexes of the elements in $\mathbf{s}$. The actual selection strategy is irrelevant.

Example 1.3.2. The expansion estimator is unbiased for T under the superpopulation model SM1. Indeed, under this model, $\mathbf{X} = \mathbf{1}_N$, $y_i = \beta + e_i$, $i = 1, \ldots, N$, $\bar{x} = \bar{x}_s = 1$, $\bar{y}_s = \beta + \bar{e}_s$, $\hat{T}_E = N\bar{y}_s = N\beta + N\bar{e}_s$, $T = N\beta + N\bar{e}$, where $\bar{e}_s = \sum_{i\in \mathbf{s}} e_i/n$ and $\bar{e} = \sum_{i=1}^N e_i/N$. Hence,

$$E_\psi[\hat{T}_E - T] = E_\psi[N(\bar{e}_s - \bar{e})] = 0.$$

On the other hand, with respect to the model SM2, the expansion predictor might be biased. To show this we write $\bar{y}_s = \beta\bar{x}_s + \bar{e}_s$ and

$$\hat{T}_E - T = N\beta(\bar{x}_s - \bar{x}) + N(\bar{e}_s - \bar{e}).$$

Hence,

$$E_\psi[\hat{T}_E - T] = N\beta(\bar{x}_s - \bar{x}).$$

Note that, under this model, $\hat{T}_E$ is ψ-unbiased if and only if, $\bar{x}_s = \bar{x}$. A sample design for which $\bar{x}_s = \bar{x}$ is called a *balanced design* (Royall and Herson, 1973a, Pereira and Rodrigues, 1983).

Example 1.3.3. The reader can verify (Exercise 1.5) that the predictor $\hat{T}_R$ is unbiased for the population total with respect to the model SM2.

On the other hand, $\hat{T}_R$ might be biased with respect to the ψ-model SM3. Indeed, under this model, we may write

$$\hat{T}_R - T = N\beta_o(\frac{\bar{x}}{\bar{x}_s} - 1) + N(\frac{\bar{x}}{\bar{x}_s}\bar{e}_s - \bar{e}).$$

Thus,

$$E_\psi[\hat{T}_R - T] = N\beta_o(\frac{\bar{x}}{\bar{x}_s} - 1).$$

The estimator $\hat{T}_R$ is ψ-unbiased under SM3 if and only if, $\bar{x} = \bar{x}_s$.

Example 1.3.4. Consider the statistic

$$s_y^2 = \sum_{j\in s}(y_j - \bar{y}_s)^2/n$$

as a predictor of the population variance S_y^2, with respect to the ψ-model SM1. In this case,

$$E_\psi[s_y^2 - S_y^2] = \frac{\sigma^2}{n}(1-f),$$

where $f = n/N$, is the sample fraction. Thus, s_y^2 is a biased predictor of S_y^2 with respect to the model SM1.

A general characterization of ψ-unbiased predictors is provided in the following lemmas, for the linear and quadratic cases.

Lemma 1.3.1. *$\hat{\theta} = \theta_s + \hat{\theta}_{sr}$ is ψ-unbiased for θ_L if, and only if,*

$$E_\psi[\hat{\theta}_{sr}] = \mathbf{l}_r'\mathbf{X}_r\boldsymbol{\beta} \quad \textit{for all} \quad \boldsymbol{\beta}.$$

The proof of the lemma is left as an exercise.

Lemma 1.3.2. *Let*

$$\boldsymbol{\mu} = \mathbf{X}\boldsymbol{\beta} = \begin{pmatrix} \mathbf{X}_s \\ \mathbf{X}_r \end{pmatrix} \boldsymbol{\beta} \text{ and } \mathbf{V} = \begin{pmatrix} \mathbf{V}_s & \mathbf{V}_{sr} \\ \mathbf{V}_{rs} & \mathbf{V}_r \end{pmatrix} \tag{1.3.6}$$

be the partition of the mean vector and variance–covariance matrix of $\mathbf{y}$, according to $(\mathbf{y}_s, \mathbf{y}_r)$. Then, $\hat{\theta}_Q = \theta_s + \hat{\theta}_{sr}$ is a ψ-unbiased predictor of θ_Q if, and only if,

$$E_\psi[\hat{\theta}_{sr}] = \mathrm{tr}\{\mathbf{A}_r\mathbf{V}_r + 2\mathbf{A}_{rs}\mathbf{V}_{sr}\} + \boldsymbol{\beta}'(\mathbf{X}_r'\mathbf{A}_r\mathbf{X}_r + 2\mathbf{X}_s'\mathbf{A}_{sr}\mathbf{X}_r)\boldsymbol{\beta}.$$

Proof. From Eq. (1.3.1), it follows that $\hat{\theta}$ is a ψ-unbiased predictor of θ_Q if, and only if,

$$E_\psi[\hat{\theta}_{sr}] = E_\psi[\mathbf{y}_r'\mathbf{A}_r\mathbf{y}_r + 2\mathbf{y}_s'\mathbf{A}_{sr}\mathbf{y}_r].$$

It is well known (Seber, 1977, p. 13) that if $\mathbf{A}$ is a symmetric matrix, $E[\mathbf{z}] = \boldsymbol{\mu}$ and $\text{Var}[\mathbf{z}] = \boldsymbol{\Sigma}$, then

$$E[\mathbf{z}'\mathbf{A}\mathbf{z}] = \text{tr}\{\mathbf{A}\boldsymbol{\Sigma}\} + \boldsymbol{\mu}'\mathbf{A}\boldsymbol{\mu}.$$

Similarly, if $\mathbf{Z}_n$ and $\mathbf{y}_m$ have means $\boldsymbol{\xi}_n$ and $\boldsymbol{\eta}_m$ and the covariance matrix is $\mathbf{C}_{nm}$, then

$$E[\mathbf{Z}_n'\mathbf{A}_{nm}\mathbf{y}_m] = \text{tr}\{\mathbf{A}_{nm}\mathbf{C}_{mn}\} + \boldsymbol{\xi}_n'\mathbf{A}_{nm}\boldsymbol{\eta}_m.$$

The proof of the lemma follows by taking the expected value of Eq. (1.3.3). (Q.E.D.)

Definition 1.3.2. *The prediction mean squared error of a predictor* $\hat{\theta}$ *with respect to model* ψ *(ψ-MSE) is*

$$E_\psi[\hat{\theta} - \theta]^2 = \text{Var}_\psi[\hat{\theta} - \theta] + \{E_\psi[\hat{\theta} - \theta]\}^2. \tag{1.3.7}$$

Of course, if $\hat{\theta}$ is a ψ-unbiased predictor of θ, then, the MSE in Eq. (1.3.7) reduces to the prediction variance of $\hat{\theta}$ under ψ.

Definition 1.3.3. *A predictor* $\hat{\theta}^*$ *is said to be* ψ*-MSE optimal (or* ψ*–best) if it has a minimal* ψ*-MSE for all* ψ*; that is,*

$$E_\psi[\hat{\theta}^* - \theta]^2 \le E_\psi[\hat{\theta} - \theta]^2, \tag{1.3.8}$$

for any other predictor $\hat{\theta}$.

The problem of determining optimal predictors with respect to the above and more general criteria of optimality will be discussed in the following chapters, for various superpopulation models. In the next example, we illustrate this problem in the case of the simple location model SM1.

Example 1.3.5. Consider the simple location model SM1. Let $\hat{\theta} = \sum_{i\in\mathbf{s}} y_i + \hat{\theta}_{sr}$ be a predictor of T. Its MSE under ψ is

$$\begin{aligned}\text{MSE}_\psi[\hat{\theta}] =& E_\psi[(\hat{\theta}_{sr} - \mathbf{1}_r'\mathbf{y}_r)^2] \\ =& E_\psi[\{(\hat{\theta}_{sr} - (N-n)\beta) - T_{re}\}^2],\end{aligned}$$

where $T_{re} = \sum_{i\in\mathbf{r}} e_i$. Since $\text{Var}_\psi[\mathbf{e}] = \sigma^2\mathbf{I}_N$, where $\mathbf{e} = (e_1, \ldots, e_N)'$ and

$$\text{Cov}_\psi[T_{re}, \hat{\theta}_{sr}] = 0,$$

we obtain

$$\text{MSE}_\psi[\hat{\theta}] = (N-n)\sigma^2 + (N-n)^2 E_\psi[(\frac{\hat{\theta}_{sr}}{N-n} - \beta)^2].$$

It turns out that a minimum MSE predictor of T (if it exists) has the form

$$\hat{\theta} = \mathbf{1}'_s \mathbf{y}_s + (N-n)\hat{\beta}_s,$$

where $\hat{\beta}_s$ is a minimum MSE estimator of β, based on $\mathbf{y}_s$. If we restrict attention to linear unbiased predictors, then, by the Gauss–Markov Theorem, the ψ-best linear unbiased predictor (ψ-BLUP) of T is $\hat{T}_E$, given by Eq. (1.3.4).

Another type of a finite population quantity, which has been extensively studied in the literature (Konijn, 1962; Hidiroglou, 1974; Fuller, 1975a; Hartley and Sielken, 1975; Shah et al., 1977; Särndal, 1982; and others) is the vector $\boldsymbol{\beta}_N$ of weighted least–squares regression coefficients. In relation to the superpopulation model $\psi = (\boldsymbol{\beta}, \mathbf{V})$, $\boldsymbol{\beta}_N$ is given by

$$\boldsymbol{\beta}_N = (\mathbf{X}'\mathbf{V}^{-1}\mathbf{X})^{-1}\mathbf{X}'\mathbf{V}^{-1}\mathbf{y}. \tag{1.3.9}$$

In cases where

$$\mathbf{V} = \begin{pmatrix} \mathbf{V}_s & \mathbf{0} \\ \mathbf{0} & \mathbf{V}_r \end{pmatrix},$$

the above formula of $\boldsymbol{\beta}_N$ is reduced to

$$\boldsymbol{\beta}_N = (\mathbf{X}'_s\mathbf{V}_s^{-1}\mathbf{X}_s + \mathbf{X}'_r\mathbf{V}_r^{-1}\mathbf{X}_r)^{-1}(\mathbf{X}'_s\mathbf{V}_s^{-1}\mathbf{y}_s + \mathbf{X}_r\mathbf{V}_r^{-1}\mathbf{y}_r). \tag{1.3.10}$$

Define the $p \times p$ matrices

$$\mathbf{E}_s = (\mathbf{X}'_s\mathbf{V}_s^{-1}\mathbf{X}_s + \mathbf{X}'_r\mathbf{V}_r^{-1}\mathbf{X}_r)^{-1}\mathbf{X}'_s\mathbf{V}_s^{-1}\mathbf{X}_s \tag{1.3.11}$$

and

$$\mathbf{E}_r = (\mathbf{X}'_s\mathbf{V}_s^{-1}\mathbf{X}'_s + \mathbf{X}'_r\mathbf{V}_r^{-1}\mathbf{X}_r)^{-1}\mathbf{X}'_r\mathbf{V}_r^{-1}\mathbf{X}_r. \tag{1.3.12}$$

One can express $\boldsymbol{\beta}_N$ in the form

$$\boldsymbol{\beta}_N = \mathbf{E}_s\hat{\boldsymbol{\beta}}_s + \mathbf{E}_r\boldsymbol{\beta}_r, \tag{1.3.13}$$

where

$$\hat{\boldsymbol{\beta}}_s = (\mathbf{X}'_s\mathbf{V}_s^{-1}\mathbf{X}_s)^{-1}\mathbf{X}'_s\mathbf{V}_s^{-1}\mathbf{y}_s \tag{1.3.14}$$

is the weighted least–squares estimator of $\boldsymbol{\beta}$ and $\boldsymbol{\beta}_r$ is a weighted least–squares "estimator" of $\boldsymbol{\beta}$ based on $\mathbf{X}_r$, $\mathbf{V}_r$, and $\mathbf{y}_r$. Since $\mathbf{y}_r$ has not been observed, $\boldsymbol{\beta}_r$ is treated as an unknown vector–valued quantity. A predictor of $\boldsymbol{\beta}_N$ can be written, according to Eq. (1.3.13), in the form

$$\hat{\boldsymbol{\beta}}_N = \mathbf{E}_s\hat{\boldsymbol{\beta}}_s + \mathbf{E}_r\hat{\boldsymbol{\beta}}_r, \tag{1.3.15}$$

where $\hat{\boldsymbol{\beta}}_r$ is a predictor of $\boldsymbol{\beta}_r$ based on $\mathbf{y}_s$. Definition 1.3.1 of ψ-unbiasedness holds also for vector–valued predictors.

Lemma 1.3.3. *Let* $\mathbf{C}_{ps}$ *be a* $p \times n$ *matrix* ($n \geq p$). $\hat{\boldsymbol{\beta}}_r = \mathbf{C}_{ps}\mathbf{y}_s$ *is a* ψ*-unbiased predictor of* $\boldsymbol{\beta}_r$ *if, and only if,*

$$\mathbf{C}_{ps}\mathbf{X}_s = \mathbf{I}_p. \tag{1.3.16}$$

Proof. Since $E_\psi[\mathbf{y}_s] = \mathbf{X}_s\boldsymbol{\beta}$ and $E_\psi[\boldsymbol{\beta}_r] = \boldsymbol{\beta}$,

$$E_\psi[\mathbf{C}_{ps}\mathbf{y}_s - \boldsymbol{\beta}_r] = (\mathbf{C}_{ps}\mathbf{X}_s - \mathbf{I}_p)\boldsymbol{\beta}. \tag{1.3.17}$$

Thus, $E_\psi[\hat{\boldsymbol{\beta}}_r - \boldsymbol{\beta}_r] = \mathbf{0}$, for all $\boldsymbol{\beta}$ if, and only if, $\mathbf{C}_{ps}\mathbf{X}_s - \mathbf{I}_p = \mathbf{0}$. This proves Eq. (1.3.16).

(Q.E.D.)

Notice that $\hat{\boldsymbol{\beta}}_s$ is a ψ-unbiased predictor of $\boldsymbol{\beta}_r$. Moreover, since $\mathbf{E}_s + \mathbf{E}_r = \mathbf{I}_p$, Eq. (1.3.15) yields that $\hat{\boldsymbol{\beta}}_s$ is a ψ-unbiased predictor of $\boldsymbol{\beta}_N$.

Definition 1.3.4. *The generalized prediction MSE of a vector–valued predictor* $\hat{\boldsymbol{\theta}}$ *of a vector valued quantity* $\boldsymbol{\theta}$ *is*

$$GMSE_\psi[\hat{\boldsymbol{\theta}}] = \mathbf{1}_p' E_\psi[(\hat{\boldsymbol{\theta}} - \boldsymbol{\theta})(\hat{\boldsymbol{\theta}} - \boldsymbol{\theta})']\mathbf{1}_p. \tag{1.3.18}$$

Definition 1.3.5. *A predictor* $\hat{\boldsymbol{\theta}}^*$ *of a vector valued quantity* $\boldsymbol{\theta}$ *is said to be* ψ*-best if it minimizes the* ψ*-GMSE, that is*

$$GMSE_\psi[\hat{\boldsymbol{\theta}}^*] \leq GMSE_\psi[\hat{\boldsymbol{\theta}}],$$

for any other predictor $\hat{\boldsymbol{\theta}}$.

1.4. The Model–Based Design–Based Approach

Most of the ψ-unbiased predictors derived according to the theory discussed in Sections 1.2 and 1.3 are not p-unbiased. In this respect, there is a conflict between the two types of modeling. Furthermore, as will be shown in later chapters, the optimality of predictors generally depends on the validity of the assumed ψ model (see Chapter 5 for the problem of robustness). In order to overcome some of these difficulties, combining the sampling with the superpopulation approach has been suggested (Godambe and Joshi, 1965; Särndal, 1980; Wright, 1983). A superpopulation model ψ is assumed. The sample is selected however according to a randomization strategy $[\mathcal{S}, p(\mathbf{s})]$. A predictor of $\theta(\mathbf{y})$ is selected to minimize the sampling anticipated ψ-MSE; that is,

$$E_p\{E_\psi[\hat{\theta}(\mathbf{y}_s) - \theta(\mathbf{y})]^2\}. \tag{1.4.1}$$

Godambe and Joshi (1965) considered expression (1.4.1) for estimating the population total T. They showed that if $\hat{T} = n\bar{y}_s + \hat{\theta}_{sr}$ is a p-unbiased estimator of T, where $[S, p(\mathbf{s})]$ is a design for which $\pi_i > 0$ for all $i = 1, \ldots, N$, and if, furthermore, $\psi = (\beta, \sigma^2\mathbf{W})$ where $\mathbf{W} = \text{diag}\{v_1, \ldots, v_N\}$, then

$$(1.4.2) \qquad E_p E_\psi [\hat{T} - T]^2 \geq \sigma^2 [\frac{1}{n} (\sum_{i=1}^{N} v_i^{1/2})^2 - \sum_{i=1}^{N} v_i].$$

They defined a p-unbiased estimator to be "best" if equality is attained in Eq. (1.4.2) for all σ^2. They also show that the Horvitz–Thompson estimator (1.1.5) with $\pi_i = n v_i^{1/2} / \sum_{i=1}^{N} v_i^{1/2}$, is best in this sense. The approach of Godambe and Joshi is, however, very restrictive. Wright (1983) showed that a class of predictors that are asymptotically p-unbiased (ADU predictors) attain asymptotically an anticipated ψ-MSE which is equal to the right-hand side of Eq. (1.4.2). We discuss this class of predictors and prove Eq. (1.4.2) in Chapter 9.

1.5. Exercises

[1.1] Show that the p-variance of the H–T estimator in Eq. (1.1.5) is Eq. (1.1.6), which can be also written as

$$\text{Var}_p[\hat{T}] = \frac{1}{2} \sum_{i=1}^{N} \sum_{j \neq i}^{N} (\pi_i \pi_j - \pi_{ij}) (\frac{e_i}{\pi_i} - \frac{e_j}{\pi_j})^2.$$

[1.2] Show that under simple random sampling, $\hat{S}_y^2$ in Eq. (1.1.7) is p-unbiased for S_y^2.

[1.3] Find the bias of predictors $\hat{T}_E$ and $\hat{T}_R$ with respect to the model SM5. Under what conditions are they unbiased?

[1.4] Prove Lemma 1.3.1.

[1.5] Show that the ratio predictor $\hat{T}_R$ is a ψ-unbiased predictor of T with respect to the model SM2. Furthermore, prove that its prediction variance is

$$(1.5.1) \qquad v_R = E_\psi [\hat{T}_R - T]^2 = \sigma^2 \frac{\sum_{j \in \mathbf{r}} x_j}{\sum_{j \in \mathbf{s}} x_j} \sum_{j=1}^{N} x_j.$$

[1.6] Let $\hat{\theta}_L = \theta_s + \hat{\theta}_{sr}$ be any predictor of $\theta_L = \mathbf{l}'\mathbf{y}$. Show that
(i) $\hat{\theta}_L$ is a linear predictor if, and only if, $\hat{\theta}_{sr}$ is linear.
(ii) $\hat{\theta}$ is an unbiased predictor of θ_L if, and only if, $\hat{\theta}_{sr}$ is an unbiased estimator of $E_\psi[\mathbf{l}_r'\mathbf{y}_r]$.
(iii) $\hat{\theta}^* = \theta + \hat{\theta}_{sr}^*$ is the best unbiased predictor of θ_L if, and only if, $\hat{\theta}_{sr}^*$ is the best unbiased estimator of $E_\psi[\mathbf{l}_r'\mathbf{y}_r]$.

(iv) If $\hat{\theta} = n\bar{y}_s + \mathbf{h}_s'\mathbf{y}_s$, where $\mathbf{h}_s$ is known, then $\hat{\theta}$ is ψ-unbiased for θ_L if, and only if,

$$\mathbf{h}_s'\mathbf{X}_s = \mathbf{l}_r'\mathbf{X}_r, \tag{1.5.2}$$

where $\mathbf{X}_s$ and $\mathbf{X}_r$ are as in Eq. (1.3.6).

[1.7] (Pereira and Rodrigues, 1983) Consider the ψ model given by Eq. (1.2.1), where $\mathbf{V} = \sigma^2\mathbf{W}$, and $\mathbf{W}$ is known and diagonal. Using the Gauss–Markov Theorem (Rao, 1977), and Exercise 1.6 show that

$$\hat{T}_{\text{BLU}} = \mathbf{1}_s'\mathbf{y}_s + \mathbf{1}_r'\mathbf{X}_r\hat{\boldsymbol{\beta}}_s \tag{1.5.3}$$

is the ψ-BLUP of T, according to Definition 1.2, where

$$\hat{\boldsymbol{\beta}}_s = (\mathbf{X}_s'\mathbf{W}_s^{-1}\mathbf{X}_s)^{-1}\mathbf{X}_s'\mathbf{W}_s^{-1}\mathbf{y}_s$$

is the weighted least–squares estimator of $\boldsymbol{\beta}$ and $\mathbf{1}_s$ and $\mathbf{1}_r$ are, respectively, vectors of ones of dimensions n and $N-n$. Show also that

$$E_\psi[\hat{T}_{BLU} - T]^2 = \mathbf{1}_r'\mathbf{V}_r\mathbf{1}_r + \mathbf{1}_r'\mathbf{X}_r(\mathbf{X}_s'\mathbf{V}_s^{-1}\mathbf{X}_s)^{-1}\mathbf{X}_r'\mathbf{1}_r. \tag{1.5.4}$$

[1.8] Using Exercise 1.7, show that $\hat{T}_E$ is the ψ-BLUP of T, with respect to the model SM1.

[1.9] (Royall, 1970) Consider the superpopulation model SM2. As a predictor of T consider

$$\hat{T}^{(j)} = n\bar{y}_s + \hat{\beta}_j \sum_{i\in\mathbf{r}} x_i, \tag{1.5.5}$$

where $\hat{\beta}_j$ is an estimator of β, $j = 1,2$. Show that

$$E_\psi[\hat{T}^{(1)} - T]^2 \le E_\psi[\hat{T}^{(2)} - T]^2,$$

if, and only if,

$$E_\psi[\hat{\beta}_1 - \beta]^2 \le E_\psi[\hat{\beta}_2 - \beta]^2.$$

[1.10] (Bellhouse, 1987) Assume that y_j, $j = 1,\ldots,N$, are independent and $E_\psi[y_j] = \mu_j$ and $\text{Var}[y_j] = \sigma_j^2 < \infty$, $j = 1,\ldots,N$. Let p be any sampling design of fixed size n. Let $\mu_r = \sum_{i\in r}\mu_i/(N-n)$. Let $\hat{T}^{(j)} = n\bar{y}_s + (N-n)\hat{\mu}_{ns}^{(j)}$, $j = 1,2$, be two predictors of the population total $T = n\bar{y}_s + (N-n)\sum_{i\in\mathbf{r}} y_i$, where $\hat{\mu}_{ns}^{(j)}$ is an estimator of μ_r. Then,

$$E_\psi E_p[\hat{T}^{(1)} - T]^2 \le E_\psi E_p[\hat{T}^{(2)} - T]^2,$$

if, and only if,

$$E_\psi[\hat{\mu}_{ns}^{(1)} - \mu_r]^2 \leq E_\psi[\hat{\mu}_{ns}^{(2)} - \mu_r]^2.$$

[1.11] (Royall and Herson, 1973b) Consider the stratified superpopulation model SM4, with $E[y_{hj}] = x_{hj}\beta$ and $\text{Var}[y_{hj}] = \sigma^2 x_{hj}$, $j = 1, \ldots, N_h$, and $h = 1, \ldots, K$. Consider the stratified ratio predictor

$$\hat{T}_{SR} = \sum_{h=1}^{K} \frac{\sum_{i \in s_h} y_{hi}}{\sum_{i \in s_h} x_{hi}} \sum_{i=1}^{N_h} x_{hi}. \tag{1.5.6}$$

Show the following.
(i) $\hat{T}_{SR}$ is ψ-unbiased;
(ii) the ψ-variance of $\hat{T}_{SR}$ is

$$E_\psi[\hat{T}_{SR} - T]^2 = \sigma^2 \sum_{h=1}^{K} \left(\sum_{i \in r_h} x_{hi} / \sum_{i \in s_h} x_{hi} \right) \sum_{i=1}^{N_n} x_{hi}; \tag{1.5.7}$$

(iii) Suppose that another model (model ψ^*) specifies that $E_{\psi^*}[y_{hj}] = \alpha + \beta x_{hj}$ and $\text{Var}_{\psi^*}[y_{hj}] = \sigma^2 x_{hj}$, for all h and j. Compute the bias of the predictor $\hat{T}_{SR}$ with respect to the model ψ^*. Which conditions should be imposed on the selected sample for $\hat{T}_{RS}$ to be unbiased with respect to ψ^*?
(iv) If a stratified balanced sample is selected from stratum h (i.e., $\bar{x}_h = \bar{x}_{sh}$, $h = 1, \ldots, K$), then, the prediction variance in (ii) may be written as

$$E_\psi[\hat{T}_{SR} - T]^2 = \sigma^2 \sum_{h=1}^{K} (\frac{N_h^2}{n_h})(1 - \frac{n_h}{N_h})\bar{x}_h.$$

(v) Suppose that the cost of sampling is given by a fixed amount c_0 plus a cost c_h for each unit sampled in stratum h. Then the total cost is $C = c_0 + \sum_{h=1}^{K} c_h n_h$. Using the Cauchy–Schwarz inequality (Cochran, 1977, pp. 97), show that the prediction variance in (iv) is minimized, for a fixed sample of size n, by

$$n_h = n \frac{N_h \bar{x}_h^{1/2} / c_h^{1/2}}{\sum_{h=1}^{K} N_h \bar{x}_h^{1/2} / c_h^{1/2}},$$

$h = 1, \ldots, K$. This is an optimal allocation formula.

[1.12] Suppose that $y_1, \ldots, y_N$ are independent and have a common normal distribution $N(\mu, \sigma^2)$, with known mean μ and unknown variance σ^2. Denote by $N(\boldsymbol{\mu}, \mathbf{V})$ a random vector having a multivariate

normal distribution with mean vector $\boldsymbol{\mu}$ and covariance matrix $\mathbf{V}$, show that:
(i)

$$\begin{pmatrix} \bar{y}_s \\ \bar{y} \end{pmatrix} \sim N\left(\mu\mathbf{1}, \sigma^2 \begin{bmatrix} n^{-1} & N^{-1} \\ N^{-1} & N^{-1} \end{bmatrix}\right).$$

(ii) The best predictor of $\bar{y}$ is

$$E_\psi[\bar{y}|\bar{y}_s] = \frac{m}{N}\bar{y}_s + \left(1 - \frac{n}{N}\right)\mu.$$

(iii)

$$\bar{y}_s - \bar{y} \sim N\left[0, \frac{(N-n)}{Nn}\sigma^2\right].$$

(iv)

$$(\bar{y}_s - \bar{y}) / \left(\frac{N-n}{Nn}\hat{S}^2\right)^{1/2} \sim t[n-1],$$

where $t[n-1]$ denotes the t distribution with $n-1$ degrees of freedom and $\hat{S}^2$ is as in Eq. (1.1.7).
(v) Using (iv) construct a 95% level prediction interval for $\bar{y}$ (see Chapter 5).

[1.13] Consider a population of size N divided into K strata of sizes N_h, $h = 1, \ldots, K$, according to some auxiliary information. Assume that, with unit j of stratum h, there is a random quantity y_{hj} such that

$$y_{hj} = \mu_h + e_{hj},$$

where $E[e_{hj}] = 0$ and

$$\text{Cov}[e_{hi}, e_{lj}] = \begin{cases} \sigma_h^2, & l = h, i = j, \\ 0, & \text{otherwise.} \end{cases}$$

From the hth stratum, a sample $\mathbf{s}_h$ of size n_h is selected. The total sample size is $n = \sum_{h=1}^K n_h$.
(i) Find the ψ-BLUP of $T = \sum_{h=1}^K \sum_{j=1}^{N_h} y_{hj}$ and derive its prediction variance.
(ii) Assuming a cost c_h for stratum h and a total cost $C = c_o + \sum_{h=1}^K c_h n_h$, use the Cauchy–Schwarz inequality (Cochran, 1977) to find the optimal allocation.

[1.14] The finite population distribution function $F_N(t)$ is defined as

$$F_N(t) = \frac{1}{N}\sum_{i=1}^N \Delta(t - y_i),$$

where

$$\Delta(u) = \begin{cases} 1, & u \geq 0, \\ 0, & u < 0. \end{cases}$$

(i) Show that we may express $\theta = F_N(t)$ as $\theta = \theta_s + \theta_{sr}$.
(ii) Show that under model SM1,

$$F_s(t) = \frac{1}{n} \sum_{i \in s} \Delta(t - y_i)$$

is an unbiased predictor of $F_N(t)$. Find the prediction variance of $F_s(t)$.

2
Optimal Predictors of Population Quantities

In the previous chapter, we presented the structure of predictors of population quantities for various superpopulation models (SM1–SM6). In particular, we considered unbiased predictors and introduced the criterion of minimum mean squared error predictors. In this chapter, we present the theory by which optimal predictors can be derived. This theory is an adaptation of optimal estimation theory for infinite populations to prediction theory for finite populations under the superpopulation model.

The superpopulation models of Chapter 1 assumed only the structure of the first two moments of the joint distribution of $\mathbf{y}$. In this generality, optimal linear unbiased predictors of the population total T can be derived using the Gauss–Markov theory. The form of these best linear unbiased predictors (BLUP) of T, $\hat{T}_{BLU}$, will be derived in Section 2.1.

If the statistician is willing to add distributional assumptions to the superpopulation model, then further results from optimal estimation theory could often be applied. For example, if it is assumed that the joint distribution of $(\mathbf{y}_s, \mathbf{y}_r)$ belongs to a parametric family for which complete and sufficient statistics exist, best unbiased predictors of quantities like T and S_y^2 can be derived by extending the notion of sufficiency and adapting the Blackwell–Rao–Lehmann–Scheffé Theorem. This will be presented in Section 2.2. Section 2.3 presents the theory of optimal equivariant prediction. Finally, Section 2.4 presents Stein–type and shrinkage predictors.

2.1. Best Linear Unbiased Predictors

We derive in the present section the minimum variance linear unbiased predictor of T, $\hat{T}_{BLU}$. This is performed under the general model $\psi = (\boldsymbol{\beta}, \mathbf{V})$. After the sample $\mathbf{s}$ has been selected, we may reorder the elements of $\mathbf{y}$ so that we have the corresponding partitions

$$\begin{pmatrix} \mathbf{y}_s \\ \mathbf{y}_r \end{pmatrix}, \quad \begin{pmatrix} \mathbf{X}_s \\ \mathbf{X}_r \end{pmatrix}, \text{ and } \begin{pmatrix} \mathbf{V}_s & \mathbf{V}_{sr} \\ \mathbf{V}_{rs} & \mathbf{V}_r \end{pmatrix}.$$

We assume that the matrix $\mathbf{X}_s$ is of full rank p. This may impose restrictions on the choice of $\mathbf{s}$. According to Eq. (1.3.1), a linear predictor of T is

$$\hat{T}_L = \mathbf{1}'_s \mathbf{y}_s + \mathbf{l}'_{sr} \mathbf{y}_s, \tag{2.1.1}$$

where $\mathbf{l}_{sr}$ is a specified n–dimensional vector of linear coefficients. A linear predictor $\hat{T}_L$ is ψ-unbiased if, and only if, (see Exercise 1.6),

$$\mathbf{l}'_{rs} \mathbf{X}_s = \mathbf{1}'_r \mathbf{X}_r. \tag{2.1.2}$$

The following lemma is instrumental for deriving the best linear unbiased predictor of T.

Lemma 2.1.1. *Under the superpopulation model $\psi = (\boldsymbol{\beta}, \mathbf{V})$, the ψ-MSE of a linear predictor $\hat{T}_L$ is given by*

$$\begin{aligned} E_\psi [\hat{T}_L - T]^2 &= E_\psi [(\mathbf{l}_{sr} - \mathbf{V}_s^{-1} \mathbf{V}_{sr} \mathbf{1}_r)'(\mathbf{y}_s - \mathbf{X}_s \boldsymbol{\beta})]^2 \\ &\quad + \mathbf{1}'_r \mathbf{V}_r \mathbf{1}_r - \mathbf{1}'_r \mathbf{V}_{rs} \mathbf{V}_s^{-1} \mathbf{V}_{sr} \mathbf{1}_r + [(\mathbf{l}_{sr} \mathbf{X}_s - \mathbf{1}'_r \mathbf{X}_r) \boldsymbol{\beta}]^2. \end{aligned} \tag{2.1.3}$$

Proof. The ψ-MSE of $\hat{T}_L$ can be written as

$$\begin{aligned} E_\psi [\mathbf{l}'_{sr} \mathbf{y}_s - \mathbf{1}'_r \mathbf{y}_r]^2 &= E_\psi [(\mathbf{l}'_{sr} \mathbf{y}_s - \mathbf{l}'_{sr} \mathbf{X}_s \boldsymbol{\beta} \\ &\quad + \mathbf{l}'_{sr} \mathbf{X}_s \boldsymbol{\beta} - \mathbf{1}'_r \mathbf{y}_r - \mathbf{1}'_r \mathbf{V}_{rs} \mathbf{V}_s^{-1} \mathbf{y}_s + \mathbf{1}'_r \mathbf{V}_{rs} \mathbf{V}_s^{-1} \mathbf{y}_s)^2] \\ &= E_\psi \{[(\mathbf{l}'_{sr} - \mathbf{1}'_r \mathbf{V}_{rs} \mathbf{V}_s^{-1})(\mathbf{y}_s - \mathbf{X}_s \boldsymbol{\beta}) \\ &\quad + \mathbf{1}'_r \mathbf{V}_{rs} \mathbf{V}_s^{-1} (\mathbf{y}_s - \mathbf{X}_s \boldsymbol{\beta}) + (\mathbf{l}'_{sr} \mathbf{X}_s \boldsymbol{\beta} - \mathbf{1}'_r \mathbf{y}_r)]^2\} \\ &= E_\psi \{[(\mathbf{l}'_{sr} - \mathbf{1}'_r \mathbf{V}_{rs} \mathbf{V}_s^{-1})(\mathbf{y}_s - \mathbf{X}_s \boldsymbol{\beta})]^2\} - \mathbf{1}'_r \mathbf{V}_{rs} \mathbf{V}_s^{-1} \mathbf{V}_{sr} \mathbf{1}_r \\ &\quad + E_\psi [(\mathbf{l}_{sr} \mathbf{X}_s \boldsymbol{\beta} - \mathbf{1}'_r \mathbf{y}_r)^2] \\ &= E_\psi \{[(\mathbf{l}_{sr} - \mathbf{V}_s^{-1} \mathbf{V}_{sr} \mathbf{1}_r)'(\mathbf{y}_s - \mathbf{X}_s \boldsymbol{\beta})]^2\} - \mathbf{1}'_r \mathbf{V}_{rs} \mathbf{V}_s^{-1} \mathbf{V}_{sr} \mathbf{1}_r \\ &\quad + \mathbf{1}'_r \mathbf{V}_r \mathbf{1}_r + [(\mathbf{l}_{rs} \mathbf{X}_s - \mathbf{1}'_r \mathbf{X}_r) \boldsymbol{\beta}]^2. \end{aligned}$$

(Q.E.D.)

Corollary 2.1.1. *If $\hat{T}$ is an unbiased linear predictor of T then its ψ-MSE is*

$$E_\psi [\hat{T} - T]^2 = \mathbf{l}'_{sr} \mathbf{V}_s \mathbf{l}_{sr} - 2\mathbf{l}'_{sr} \mathbf{V}_{sr} \mathbf{1}_r + \mathbf{1}'_r \mathbf{V}_r \mathbf{1}_r. \tag{2.1.4}$$

The following theorem (Royall, 1976a) provides the form of the $\hat{T}_{BLU}$.

Theorem 2.1.1. *The best unbiased linear predictor of T is*

(2.1.5)
$$\hat{T}_{BLU} = \mathbf{1}'_s\mathbf{y}_s + \mathbf{1}'_r[\mathbf{X}_r\hat{\boldsymbol{\beta}}_s + \mathbf{V}_{rs}\mathbf{V}_s^{-1}(\mathbf{y}_s - \mathbf{X}_s\hat{\boldsymbol{\beta}}_s)],$$

where $\hat{\boldsymbol{\beta}}_s$ is the best unbiased linear estimator of $\boldsymbol{\beta}$ (weighted least–squares estimator, WLSE) given in Eq. (1.3.14). The corresponding prediction variance of $\hat{T}_{BLU}$ is

(2.1.6)
$$E_\psi[\hat{T}_{BLU} - T]^2 = \mathbf{1}'_r\mathbf{V}_r\mathbf{1}_r - \mathbf{1}'_r\mathbf{V}_{rs}\mathbf{V}_s^{-1}\mathbf{V}_{sr}\mathbf{1}_r$$
$$+\mathbf{1}'_r(\mathbf{X}_r - \mathbf{V}_{rs}\mathbf{V}_s^{-1}\mathbf{X}_s)(\mathbf{X}'_s\mathbf{V}_s^{-1}\mathbf{X}_s)^{-1}(\mathbf{X}_r - \mathbf{V}_{rs}\mathbf{V}_s^{-1}\mathbf{X}_s)'\mathbf{1}_r.$$

Proof. Applying formula (2.1.4), the ψ-BLUP of T can be obtained by the method of the Lagrange multipliers by minimizing the Lagrangian

$$LG(\mathbf{l}_{sr}, \boldsymbol{\lambda}) = \mathbf{l}'_{rs}\mathbf{V}_s\mathbf{l}_{sr} - 2\mathbf{l}'_{sr}\mathbf{V}_{sr}\mathbf{1}_r + (\mathbf{l}'_{sr}\mathbf{X}_s - \mathbf{1}'_r\mathbf{X}_r)\boldsymbol{\lambda},$$

where $\boldsymbol{\lambda}$ is a $p\times 1$ vector of Lagrange multipliers. The gradients of $LG(\mathbf{l}_{rs}, \boldsymbol{\lambda})$ with respect to $\mathbf{l}_{sr}$ and $\boldsymbol{\lambda}$ yield the following equations:

$$\mathbf{V}_s\mathbf{l}^o_{sr} = \mathbf{V}_{sr}\mathbf{1}_r - \frac{1}{2}\mathbf{X}_s\boldsymbol{\lambda}$$

and

$$\mathbf{l}^{o'}_{sr}\mathbf{X}_s = \mathbf{1}'_r\mathbf{X}_r.$$

A solution of the latter equations yields

$$\mathbf{l}^o_{sr} = \mathbf{V}_s^{-1}\mathbf{V}_{sr}\mathbf{1}_r - \mathbf{X}_s(\mathbf{X}'_s\mathbf{V}_s^{-1}\mathbf{X}_s)^{-1}\mathbf{X}'_s\mathbf{V}_s^{-1}\mathbf{V}_{sr}\mathbf{1}_r + \mathbf{X}_s(\mathbf{X}'_s\mathbf{V}_s^{-1}\mathbf{X}_s)^{-1}\mathbf{X}'_r\mathbf{1}_r.$$

Taking the inner product $\mathbf{l}^{o'}_{sr}\mathbf{y}_s$ one obtains Eq. (2.1.5). The proof of Eq. (2.1.6) is obtained by substituting $\mathbf{l}^o_{sr}$ into Eq. (2.1.4).

(Q.E.D.)

In the particular case where $\mathbf{V} = \sigma^2\mathbf{W}$ and $\mathbf{W}$ is known, an unbiased estimator of Eq. (2.1.6) is considered in Exercise 2.4.

Example 2.1.1. Consider again model SM2 in which $\mathbf{X} = (x_1, \ldots, x_N)'$, $\sigma = 1$ and $\mathbf{W} = \text{diag}\{x_1, \ldots, x_N\}$. In this case,

$$\mathbf{X}'_s\mathbf{V}_s^{-1}\mathbf{X}_s = \sum_{i\in\mathbf{s}} x_i \quad \text{and} \quad \mathbf{X}'_s\mathbf{V}_s^{-1}\mathbf{y}_s = \sum_{i\in\mathbf{s}} y_i.$$

Thus, the WLSE of β is

$$\hat{\beta}_s = \frac{\sum_{i\in\mathbf{s}} y_i}{\sum_{i\in\mathbf{s}} x_i} = \frac{\bar{y}_s}{\bar{x}_s}.$$

Since $\mathbf{V}_{rs} = \mathbf{0}$, formula (2.1.4) yields the ratio predictor

$$\hat{T}_{BLU} = \sum_{i \in \mathbf{s}} y_i + \frac{\bar{y}_s}{\bar{x}_s} \sum_{i \in \mathbf{r}} x_i.$$

It can also be shown that the error variance of $\hat{T}_{BLU}$ is as given in Exercise 1.5. Therefore, the optimal sample to be used with $\hat{T}_R$ is the one that maximizes $\bar{x}_s$.

Example 2.1.2. Consider model SM6 of cluster sampling. In this model, there are K clusters from which we select k, $1 \leq k \leq K$, clusters at random. Let $\mathbf{s} = \{h_1, \ldots, h_k\}$ be the set of labels of the clusters selected in the first stage. From each cluster selected in the first stage we select a sample of units to observe.

Without loss of generality, we relabel the clusters and set $\mathbf{s} = \{\mathbf{s}_h : 1 \leq h \leq k\}$. We denote by $\mathbf{s}_h$ the sample from the hth cluster. The size of $\mathbf{s}_h$ is n_h. The total sample size is $n = \sum_{h=1}^{k} n_h$. We rearrange the rows and columns of $\mathbf{V}$ and $\mathbf{X}$. As before $\mathbf{X} = (\mathbf{1}'_s, \mathbf{1}'_r)'$ and

$$\mathbf{V} = \begin{pmatrix} \mathbf{V}_s & \mathbf{V}_{sr} \\ \mathbf{V}_{rs} & \mathbf{V}_{rr} \end{pmatrix}.$$

In the present case

$$\mathbf{V}_s = \text{diag}\{\mathbf{V}_{s_h}, \quad h = 1, \ldots, k\},$$

in which

$$\mathbf{V}_{s_h} = \sigma_h^2 \mathbf{I}_{n_h} + \sigma_v^2 \mathbf{J}_{n_h}, \qquad h = 1, \ldots, k.$$

Furthermore,

$$\mathbf{V}_{sr} = \left[\begin{array}{cccc:c} \mathbf{V}_{s_1 r_1} & & & 0 & \\ & \mathbf{V}_{s_2 r_2} & & & 0 \\ 0 & & \ddots & & \\ & & & \mathbf{V}_{s_k r_k} & \end{array} \right],$$

where

$$\mathbf{V}_{s_h r_h} = \sigma_v^2 \mathbf{1}_{n_h} \mathbf{1}'_{(N_h - n_h)}, \qquad h = 1, \ldots, k,$$
$$\mathbf{V}_{rs} = \mathbf{V}'_{sr},$$

and

$$\mathbf{V}_{rr} = \text{diag}\{\mathbf{V}_{rr}, \mathbf{V}^*_{rr}\},$$

where

$$\mathbf{V}_{rr} = \mathrm{diag}\{\mathbf{V}_{rh}, \quad h = 1, \ldots, k\},$$
$$\mathbf{V}^*_{rr} = \mathrm{diag}\{\mathbf{V}^*_{rh}, \quad h = k+1, \ldots, K\},$$
$$\mathbf{V}_{rh} = \sigma_h^2 \mathbf{I}_{N_h - n_h} + \sigma_v^2 \mathbf{J}_{N_h - n_h}, \quad h = 1, \ldots, k,$$
$$\mathbf{V}^*_{rh} = \sigma_h^2 \mathbf{I}_{N_h} + \sigma_v^2 \mathbf{J}_{N_h}, \quad h = k+1, \ldots, K.$$

Notice that

$$\mathbf{V}_{s_h}^{-1} = \frac{1}{\sigma_h^2}(\mathbf{I}_{n_h} - \frac{\sigma_v^2}{\sigma_h^2 + n_h \sigma_v^2} \mathbf{J}_{n_h}), \qquad h = 1, \ldots, k.$$

Hence,

$$\hat{\mu}_s = (\mathbf{1}_s' \mathbf{V}_s^{-1} \mathbf{1}_s)^{-1} \mathbf{1}_s' \mathbf{V}_s^{-1} \mathbf{y}_s$$

(2.1.7)
$$= \frac{\sum_{h=1}^k w_h \bar{y}_{s_h}}{\sum_{h=1}^k w_h},$$

where

(2.1.8)
$$w_h = \frac{n_h \sigma_v^2}{\sigma_h^2 + n_h \sigma_v^2}, \qquad h = 1, \ldots, k$$

and $\bar{y}_{s_h}$ is the mean of the y values in $\mathbf{s}_h$.

Furthermore,

$$\mathbf{1}_r' \mathbf{X}_r \hat{\mu}_s = (N - n)\hat{\mu}_s,$$
$$\mathbf{1}_r' \mathbf{V}_{rs} \mathbf{V}_s^{-1} \mathbf{y}_s = \sigma_v^2 \sum_{h=1}^k \mathbf{1}_{r_h}' \mathbf{1}_{r_h} \mathbf{1}_{n_h}' [\frac{1}{\sigma_n^2} \mathbf{I}_{n_h} - \frac{\sigma_v^2}{\sigma_h^2(\sigma_h^2 + n_h \sigma_v^2)} \mathbf{J}_{n_h}] \mathbf{y}_{s_h},$$
$$= \sum_{h=1}^k (N_h - n_h) w_h \bar{y}_{s_h}.$$

Similarly,

$$\mathbf{1}_r' \mathbf{V}_{rs} \mathbf{V}_s^{-1} \mathbf{X}_s \hat{\mu}_s = \hat{\mu}_s \sum_{h=1}^k (N_h - n_h) w_h.$$

Hence,

(2.1.9)
$$\hat{T}_{\mathrm{BLU}} = \sum_{h=1}^k n_h \bar{y}_{s_h} + \sum_{h=1}^k (N_h - n_h)[w_h \bar{y}_{s_h} + (1 - w_n)\hat{\mu}_s] + \sum_{n=k+1}^k N_h \hat{\mu}_s.$$

Notice that, if $N_h = M$ and $n_h = m$ for all $h = 1, \ldots, k$, then

$$\hat{T}_{\text{BLU}} = m \sum_{h=1}^{k} \bar{y}_{s_h} + (N - n)\hat{\mu}_s,$$

which is the simple projection predictor defined in Eq. (2.1.11). We derive now the prediction variance of Eq. (2.1.9). First,

$$\begin{aligned}
\mathbf{1}_r' \mathbf{V}_{rr} \mathbf{1}_r &= \sum_{h=1}^{k} \mathbf{1}'_{(N_h - n_h)} (\sigma_h^2 \mathbf{I}_{N_h - n_h} + \sigma_v^2 \mathbf{J}_{N_h - n_h}) \mathbf{1}_{(N_h - n_h)} \\
&\quad + \sum_{j=k+1}^{K} \mathbf{1}'_{N_j} (\sigma_h^2 \mathbf{I}_{N_j} + \sigma_v^2 \mathbf{J}_{N_j}) \mathbf{1}_{N_j}, \\
&= \sum_{h=1}^{k} (N_h - n_h)\sigma_h^2 + \sum_{h=k+1}^{K} \sigma_h^2 N_h \\
&\quad + \sigma_v^2 \left[\sum_{h=1}^{k} (N_h - n_h)^2 + \sum_{j=k+1}^{K} N_j^2 \right].
\end{aligned}$$

Second,

$$\begin{aligned}
\mathbf{1}_r' \mathbf{V}_{rs} \mathbf{V}_s^{-1} \mathbf{V}_{sr} \mathbf{1}_r &= \sigma_v^2 \sum_{h=1}^{k} (N_h - n_h)^2 \frac{\sigma_v^2 n_h}{\sigma_h^2 + n_h \sigma_v^2} \\
&= \sigma_v^2 \sum_{h=1}^{k} (N_h - n_h)^2 w_h.
\end{aligned}$$

Third,

$$\mathbf{X}_s \mathbf{V}_s^{-1} \mathbf{X}_s = \frac{1}{\sigma_v^2} \sum_{h=1}^{k} w_h,$$

$$\mathbf{1}_r' (\mathbf{X}_r - \mathbf{V}_{rs} \mathbf{V}_s^{-1} \mathbf{X}_s) = (N - n) - \sum_{h=1}^{k} (N_h - n_h) w_n.$$

Substituting these results into Eq. (2.1.6), we obtain

$$\begin{aligned}
\text{Var}_\psi [\hat{T}_{\text{BLU}} - T] &= \sum_{h=k+1}^{K} \sigma_h^2 N_h \\
&\quad + \sum_{h=1}^{k} (N_h - n_h)\sigma_h^2 + \sigma_v^2 \sum_{h=1}^{k} (N_h - n_h)^2 (1 - w_h) \qquad (2.1.10) \\
&\quad + \sigma_v^2 \sum_{j=k+1}^{K} N_j^2 + \frac{\sigma_v^2}{\sum_{h=1}^{k} w_h} [\sum_{h=1}^{k} (N_h - n_h)(1 - w_h)]^2.
\end{aligned}$$

Notice that, if $\sigma_h^2 = 0$, then $w_h = 1$, $h = 1, \ldots, k$, and the prediction variance reduces to

$$\mathrm{Var}_\psi[\hat{T}_{\mathrm{BLU}} - T] = \sigma_v^2 \sum_{j=k+1}^{N} N_j^2.$$

As pointed out by Särndal and Wright (1984), it is important that predictors should have good properties as well as simple and intuitive forms. An intuitive and simple predictor of the population total T is the *simple projection predictor* (SPP), which is defined as

$$\hat{T}_{SP} = \mathbf{1}_N'\mathbf{X}\hat{\boldsymbol{\beta}}_s, \tag{2.1.11}$$

with $\hat{\beta}_s$ as in Eq. (1.3.14). The following theorem, which appears in Bolfarine and Rodrigues (1988), presents conditions under which the SPP and the BLUP of T coincide. The superpopulation model ψ is considered, with $\mathbf{V} = \sigma^2\mathbf{W}$, where $\mathbf{W}$ is known and general. In this theorem, let $\mathcal{M}(\mathbf{X}_s)$ be the vector space spanned by the columns of $\mathbf{X}_s$. Let $\mathbf{K} = [\mathbf{W}_s, \mathbf{W}_{sr}]$.

Theorem 2.1.2. *$\hat{T}_{SP} = \hat{T}_{BLU}$ if, and only if, $\mathbf{K}\mathbf{1}_N \in \mathcal{M}(\mathbf{X}_s)$.*

Proof. The projection predictor $\hat{T}_{SP}$ is equal to $\hat{T}_{BLU}$ if, and only if,

$$\mathbf{1}_s'\mathbf{X}_s\hat{\boldsymbol{\beta}}_s + \mathbf{1}_r'\mathbf{X}_r\hat{\boldsymbol{\beta}}_s = \mathbf{1}_s'\mathbf{y}_s + \mathbf{1}_r'[\mathbf{X}_r\hat{\boldsymbol{\beta}}_s + \mathbf{W}_{rs}\mathbf{W}_s^{-1}(\mathbf{y}_s - \mathbf{X}_s\hat{\boldsymbol{\beta}}_s)].$$

Or,

$$\mathbf{1}_s'(\mathbf{y}_s - \mathbf{X}_s\hat{\boldsymbol{\beta}}_s) + \mathbf{1}_r'\mathbf{W}_{rs}\mathbf{W}_s^{-1}(\mathbf{y}_s - \mathbf{X}_s\hat{\boldsymbol{\beta}}_s) = 0. \tag{2.1.12}$$

Since

$$(\mathbf{y}_s - \mathbf{X}_s\hat{\boldsymbol{\beta}}_s) \in \mathcal{M}^\perp(\mathbf{X}_s),$$

where $\mathcal{M}^\perp(\mathbf{X}_s)$ is the vector space orthogonal to $\mathcal{M}(\mathbf{X}_s)$, Eq. (2.1.12) holds if, and only if,

$$\mathbf{W}_s\mathbf{1}_s + \mathbf{W}_{sr}\mathbf{1}_r \in \mathcal{M}(\mathbf{X}_s).$$

Finally, $\mathbf{K} = [\mathbf{W}_s, \mathbf{W}_{sr}]$, and $\mathbf{K}\mathbf{1}_N = \mathbf{W}_s\mathbf{1}_s + \mathbf{W}_{sr}\mathbf{1}_r$. Thus, the SPP is equivalent to the BLUP of T if, and only if,

$$\mathbf{K}\mathbf{1}_N \in \mathcal{M}(\mathbf{X}_s).$$

(Q.E.D.)

As direct consequences of Theorem 2.1.2, we have the following corollary.

Corollary 2.1.2.
(i) If $\mathbf{W}_s\mathbf{1}_s$ *and* $\mathbf{W}_{sr}\mathbf{1}_r \in \mathcal{M}(\mathbf{X}_s)$, *then*

$$\hat{T}_{SP} = \hat{T}_{BLU}.$$

(ii) If $\mathbf{W}\mathbf{1}_N \in \mathcal{M}(\mathbf{X})$, *then*

$$\hat{T}_{SP} = \hat{T}_{BLU}.$$

(iii) If $\mathbf{W}$ *is diagonal, then* $\hat{T}_{SP} = \hat{T}_{BLU}$ *if, and only if,*

$$\mathbf{W}_s\mathbf{1}_s \in \mathcal{M}(\mathbf{X}_s).$$

Condition $\mathbf{W}\mathbf{1}_N \in \mathcal{M}(\mathbf{X})$, which appears in Corollary 2.1.2(ii), and which may be written as

$$\mathbf{W}\mathbf{1}_N = \mathbf{X}\boldsymbol{\delta}, \tag{2.1.13}$$

for some vector $\boldsymbol{\delta}$, was termed by Royall (1976b), *condition* L.

Example 2.1.3. Suppose that superpopulation model ψ is such that $\mathbf{X} = \mathbf{1}_N$ and $\mathbf{W} = (1-\rho)\mathbf{I}_N + \rho\mathbf{1}_N\mathbf{1}_N'$. Since $\mathbf{W}\mathbf{1}_N = \mathbf{1}_N[1+(N-1)\rho]$, it follows that $\mathbf{W}\mathbf{1}_N \in \mathcal{M}(\mathbf{1}_N)$. Then, from Corollary 2.3.3(ii), we obtain

$$\hat{T}_{BLU} = \hat{T}_{SP} = \mathbf{1}_N'\mathbf{1}_N\hat{\beta}_s = N\hat{\beta}_s,$$

where $\hat{\beta}_s = \bar{y}_s$. Accordingly, $\hat{T}_{BLU} = N\bar{y}_s$ is the usual expansion estimator of T.

Example 2.1.4. Suppose that $\mathbf{X} = (x_1, \ldots, x_N)'$. The necessary and sufficient condition of Theorem 2.1.2 is satisfied whenever $\mathbf{W}$ is diagonal, with diagonal elements $w_{ii} \propto x_i$, $i = 1, \ldots, N$ as in SM2. In this case,

$$\hat{T}_{SP} = \sum_{i=1}^{N} x_i \sum_{i\in s} y_i \Big/ \sum_{i\in s} x_i,$$

is the ratio predictor, which is the ψ-BLUP of T.

Using the results of Example 2.1.2, one can show that condition L is generally not satisfied under model SM6.

2.2. Best Unbiased Predictors

As explained at the beginning of the chapter, if the superpopulation model assumes that the joint distribution of $\mathbf{y}$ belongs to a specified family, which admits predictive complete sufficient statistics (as defined bellow), then best unbiased predictors (BUP) can be derived. The method employed is an extension of the celebrated Blackwell–Rao–Lehmann–Scheffé Theorem.

Definition 2.2.1. *Let $\mathcal{F} = \{F_\psi; \psi \in \Psi\}$ be a family of joint distributions of $\mathbf{y}' = (\mathbf{y}'_s, \mathbf{y}'_r)$ indexed by a parameter ψ in a space Ψ. A statistic $S = S(\mathbf{y}_s)$ is called predictive sufficient with respect to $\mathcal{F}$ if:*

(i) the conditional distribution of $\mathbf{y}_s$ given S is independent of ψ, and

(ii) $\mathbf{y}_s$ and $\mathbf{y}_r$ are conditionally independent, given S.

A predictive sufficient statistic (p.s.s.) is said to be complete if the induced family $\mathcal{F}^s$ of sampling distributions of S is complete (Lehmann 1983, pp. 46). Notice that condition (ii) of the above definition means that S contains all the information on $\mathbf{y}_r$ given in $\mathbf{y}_s$ (through the model). In cases where $\mathbf{y}_s$ and $\mathbf{y}_r$ are independent condition (ii) holds. The following theorem is an extension of the Blackwell–Rao–Lehmann– Scheffé Theorem (see Skiener, 1983; Rodrigues et al, 1985).

Theorem 2.2.1. *Let S be a complete predictive sufficient statistic with respect to $\mathcal{F}$. Let $\theta = \theta_s + \theta_{sr}(S; \mathbf{y}_r)$ be a decomposition of a population quantity of interest as in Eq. (1.3.1), in which $\theta_{sr}(S; \mathbf{y}_r)$ depends on $\mathbf{y}_s$ only through S. If $\hat{\theta}_u = \theta_s + \hat{\theta}_{sr}$ is any unbiased predictor of θ then,*

$$\hat{\theta}_{BU} = \theta_s + \hat{\theta}_{sr}(S), \tag{2.2.1}$$

where $\hat{\theta}_{sr}(S) = E_\psi[\hat{\theta}_{sr}|S]$, is the essentially unique minimum variance unbiased predictor of θ_{sr}.

Proof. We may write

$$\mathrm{Var}_\psi[\hat{\theta}_u - \theta] \geq \mathrm{Var}_\psi\{E_\psi[\hat{\theta}_{sr} - \theta_{sr}(S; \mathbf{y}_r)|S, \mathbf{y}_r]\}.$$

Now, $\hat{\theta}_{sr}$ is a statistic (depending only on $\mathbf{y}_s$). Hence, since S is predictive sufficient with respect to $\mathcal{F}$,

$$E_\psi[\hat{\theta}_{sr}|S, \mathbf{y}_r] = E_\psi[\hat{\theta}_{sr}|S] = \hat{\theta}_{sr}(S). \tag{2.2.2}$$

Thus, the ψ-variance of $(\hat{\theta}_{BU} - \theta)$ is minimal in the class of all unbiased predictors. The essential uniqueness of $\hat{\theta}_{BU}$ follows from the completeness of S.

(Q.E.D.)

The notation ψ-BUP is used since $\hat{\theta}_{BU}$ is the *ψ-best* predictor in the class of all unbiased (not just linear) predictors of θ. According to Theorem 2.2.1, θ_{sr} and $\hat{\theta}_{sr}$ must depend on $\mathbf{y}_s$ only through S. If $\theta = \theta_L$, then $\theta_{sr} = \mathbf{l}'_r\mathbf{y}_r$. Hence, to construct the ψ-BUP of θ_L we have just to check if $\hat{\theta}_{sr}(S)$ is ψ-unbiased for θ_{sr}. On the other hand, if $\theta = \theta_Q$, then,

$$\theta_{sr} = 2\mathbf{y}'_s\mathbf{A}_{sr}\mathbf{y}_r + \mathbf{y}'_r\mathbf{A}_r\mathbf{y}_r,$$

which depends also on $\mathbf{y}_s$. Thus, in order to apply Theorem 2.2.1, we have first to check, for each $\mathbf{A}$, if θ_{sr} depends on $\mathbf{y}_s$ only through S.

The ψ-BUP is now characterized for linear and quadratic forms under normality.

Corollary 2.2.1. *Let $\theta = \theta_L$. Consider the ψ-model, where $\mathbf{y}$ has a multinormal distribution with mean vector $\mathbf{X}\boldsymbol{\beta}$ and covariance matrix $\mathbf{V}$; that is,*

$$\mathbf{y} \sim N(\mathbf{X}\boldsymbol{\beta}, \mathbf{V}), \tag{2.2.3}$$

where $\mathbf{V} = \sigma^2\mathbf{W}$, $\mathbf{W}$ is known, and $\mathbf{W}_{sr} = \mathbf{0}$. In this model, $(\hat{\boldsymbol{\beta}}_s, \hat{\sigma}_s^2)$ is a complete and sufficient statistic, and hence

$$\hat{\theta}_{BU} = \mathbf{1}_s'\mathbf{y}_s + \mathbf{1}_r'\mathbf{X}_r\hat{\boldsymbol{\beta}}_s, \tag{2.2.4}$$

is the ψ-BUP of θ_L. $\hat{\boldsymbol{\beta}}_s$ is given in (1.3.14) and

$$\hat{\sigma}_s^2 = \frac{1}{n-p}\mathbf{y}_s'\mathbf{W}_s^{-1}[\mathbf{W}_s - \mathbf{X}_s(\mathbf{X}_s'\mathbf{W}_s^{-1}\mathbf{X}_s)^{-1}\mathbf{X}_s']\mathbf{W}_s^{-1}\mathbf{y}_s. \tag{2.2.5}$$

Furthermore,

$$E_\psi[\hat{\theta}_{BU} - \theta]^2 = \sigma^2\left[\mathbf{1}_r'\mathbf{W}_r\mathbf{1}_r + \mathbf{1}_r'\mathbf{X}_r(\mathbf{X}_s'\mathbf{W}_s^{-1}\mathbf{X}_s)^{-1}\mathbf{X}_r'\mathbf{1}_r\right]. \tag{2.2.6}$$

If the normality assumption is dropped in Corollary 2.1.1, then, $\hat{\theta}_{BU}$ coincides with $\hat{\theta}_{BLU}$ as given in Eq. (2.1.5). An extension of Corollary 2.1.1 to the case where the matrix $\mathbf{W}$ is a general symmetric and positive–definite matrix and $\theta_L = T$ is provided by Tam (1987a). It is shown there that the model based predictor derived in Theorem 2.1.1 is indeed the ψ-BUP of T under normality. (See also Exercise 2.8.) Moreover, it is easy to check that Theorem 2.1.2 holds also under normality.

Corollary 2.2.2. *Under model (2.2.3) the BUP of S_y^2 is*

$$\begin{aligned}\hat{S}^2_{\mathrm{BU}} = {} & \frac{n}{N}s_y^2 + (1-\frac{n}{N})\hat{\sigma}_s^2[tr\{\mathbf{A}_r\mathbf{W}_r\} - tr\{\mathbf{B}_s'(\mathbf{X}_r'\mathbf{A}_r\mathbf{X}_r + \\ & \frac{n}{N}\mathbf{D}_{rs}\mathbf{D}_{rs}')\mathbf{B}_s\mathbf{W}_s\} + \frac{n}{N}(\frac{\mathbf{1}_s'\mathbf{W}_s\mathbf{1}_s}{n^2} + \frac{\mathbf{1}_r'\mathbf{W}_r\mathbf{1}_r}{(N-n)^2})] \\ & + (1-\frac{n}{N})\hat{\boldsymbol{\beta}}_s'(\mathbf{X}_r'\mathbf{A}_r\mathbf{X}_r + \frac{n}{N}\mathbf{D}_{rs}\mathbf{D}_{rs}')\hat{\boldsymbol{\beta}}_s,\end{aligned} \tag{2.2.7}$$

where $\hat{\sigma}_s^2$ is given in Eq. (2.2.5);

$$\mathbf{D}_{rs}' = [\bar{x}_r^{(1)} - \bar{x}_s^{(1)}, \ldots, \bar{x}_r^{(p)} - \bar{x}_s^{(p)}],$$

$\bar{x}_s^{(i)}$ and $\bar{x}_r^{(i)}$ are the respective means of the ith columns of $\mathbf{X}_s$ and $\mathbf{X}_r$, and $\mathbf{B}_s = (\mathbf{X}_s'\mathbf{W}_s^{-1}\mathbf{X}_s)^{-1}\mathbf{X}_s'\mathbf{W}_s^{-1}$.

Proof. Let $S_r^2 = S_{ry}^2 + \frac{n}{N}(\bar{y}_s - \bar{y}_r)^2$. Since $\mathbf{y}_s$ and $\mathbf{y}_r$ are independent,

$$\begin{aligned}E_\psi[S_r^2] = {} & \sigma^2[\mathrm{tr}\{\mathbf{A}_r\mathbf{W}_r\} + \frac{n}{N}(\frac{\mathbf{1}_s'\mathbf{W}_s\mathbf{1}_s}{n^2} + \frac{\mathbf{1}_r'\mathbf{W}_r\mathbf{1}_r}{(N-n)^2})] \\ & + \boldsymbol{\beta}'(\mathbf{X}_r'\mathbf{A}_r\mathbf{X}_r + \frac{n}{N}\mathbf{D}_{rs}\mathbf{D}_{rs}')\boldsymbol{\beta}.\end{aligned}$$

We develop now an unbiased predictor of S_r^2. Let $Q_{sr} = \hat{\boldsymbol{\beta}}_s'\mathbf{X}_r'\mathbf{A}_r\mathbf{X}_r\hat{\boldsymbol{\beta}}_s$. Hence, since $\mathbf{B}_s\mathbf{X}_s = \mathbf{I}_p$,

$$E_\psi[Q_{sr}] = \sigma^2\text{tr}\{\mathbf{B}_s'\mathbf{X}_r'\mathbf{A}_r\mathbf{X}_r\mathbf{B}_s\mathbf{W}_s\} + \boldsymbol{\beta}'\mathbf{X}_r'\mathbf{A}_r\mathbf{X}_r\boldsymbol{\beta}.$$

Hence, since $E_\psi[\hat{\sigma}_s^2] = \sigma^2$,

$$\hat{S}_{ry}^2 = \hat{\sigma}_s^2\text{tr}\{\mathbf{A}_r\mathbf{W}_r - \mathbf{B}_s'\mathbf{X}_r'\mathbf{A}_r\mathbf{X}_r\mathbf{B}_s\mathbf{W}_s\} + \boldsymbol{\beta}_s'\mathbf{X}_r'\mathbf{A}_r\mathbf{X}_r\boldsymbol{\beta}_s$$

is an unbiased predictor of S_{ry}^2.

Finally,

$$E_\psi[\boldsymbol{\beta}_s'\mathbf{D}_{rs}\mathbf{D}_{rs}'\boldsymbol{\beta}_s] = \sigma^2\text{tr}\{\mathbf{B}_s'\mathbf{D}_{rs}\mathbf{D}_{rs}'\mathbf{B}_s\mathbf{W}_s\} + \boldsymbol{\beta}'\mathbf{D}_{rs}\mathbf{D}_{rs}'\boldsymbol{\beta}.$$

This proves that Eq. (2.2.7) is an unbiased predictor. Moreover, since the unbiased predictor of S_r^2 is a function of the complete predictive sufficient statistics $(\hat{\boldsymbol{\beta}}_s, \hat{\sigma}_s^2)$, $\hat{S}_{\text{BU}}^2$ is BUP.

(Q.E.D.)

Several interesting aspects of the theory are highlighted in special examples. First we consider the problem of predicting the population total and population variance under particular versions of the ψ-model and normality. A very special case of linear prediction in the discrete case, which cannot be treated by the Gauss–Markov Theorem is also discussed.

Example 2.2.1. Consider model SM1 under normality. The quantities of interest are the population total T and the population variance S_y^2. If σ^2 is known, the statistic

$$S = \mathbf{1}_s'\mathbf{y}_s$$

is complete and predictive sufficient, since $y_1, \ldots, y_N$ are conditionally independent given β. On the other hand, if σ^2 is unknown,

$$S = (\sum_{j\in s} y_j, \sum_{j\in s} y_j^2)$$

is the complete predictive sufficient statistics. It is easy to check that θ_{sr} (where θ is either T or S_y^2) depends on $\mathbf{y}_s$ only through S. Hence, the ψ-BUP of T follows from Corollary 2.1.1 and is given by

$$\hat{T}_{BU} = N\bar{y}_s,$$

which is the usual expansion predictor.

If σ^2 is known, Corollary 2.1.3 implies that the ψ-BUP of S_y^2 is

$$\hat{S}_{BU}^2 = \frac{n}{N}s_y^2 + (1 - \frac{n}{N})\sigma^2. \tag{2.2.8}$$

Notice that $\hat{S}^2_{BU}$ is a weighted average of s^2_y and the model variance σ^2, which represents the variability in the unobserved portion of the population. If σ^2 is unknown, and if we replace it by the usual unbiased estimator $\hat{\sigma}^2 = \frac{n}{n-1}s^2_y$, formula (2.2.8) reduces to

$$\hat{S}^2_{BU} = (1 - \frac{1}{N})\hat{\sigma}^2. \tag{2.2.9}$$

Prediction variances of the ψ-BUP of S^2_y in Eqs. (2.2.8) and (2.2.9) are considered in Exercises 2.15 and 2.16.

Example 2.2.2. Consider the superpopulation model SM2, with the additional assumption that the error vector is normally distributed, that is,

$$\mathbf{y} \sim N(\mathbf{X}\beta, \sigma^2 \text{diag}\{x_1, \ldots, x_N\}).$$

Given $\mathbf{y}_s$ and $\mathbf{X}_s$, the likelihood function of (β, σ^2) is

$$\text{Lik}(\beta, \sigma^2; \mathbf{y}_s) = \frac{1}{(2\pi)^{n/2}\sigma^n(\prod_{i\in \mathbf{s}} x_i)^{1/2}}$$
$$\exp\left\{-\frac{1}{2\sigma^2}\left[\sum_{i\in \mathbf{s}} \frac{(y_i - \hat{\beta}_s x_i)^2}{x_i} + n\bar{x}_s(\hat{\beta}_s - \beta)^2\right]\right\},$$

where $\hat{\beta}_s = \bar{y}_s/\bar{x}_s$ is the ratio estimator of β. Thus, the complete sufficient statistic for (β, σ^2) is

$$S(\hat{\beta}_s, Q_s) = \begin{cases} \hat{\beta}_s, & \text{if } \sigma^2 \text{ is known,} \\ (\hat{\beta}_s, Q_s), & \text{if } \sigma^2 \text{ is unknown,} \end{cases}$$

where

$$Q_s = \sum_{i\in \mathbf{s}} \frac{1}{x_i}(y_i - \hat{\beta}_s x_i)^2.$$

Since, given $\psi = (\beta, \sigma^2)$, $\mathbf{y}_s$ is independent of $\mathbf{y}_r$, it follows that the best unbiased predictor of T is

$$\hat{T}_{BU} = n\bar{y}_s + \hat{\beta}_s \sum_{i\in \mathbf{r}} x_i,$$

which is the ratio predictor $\hat{T}_R$. It is easy to check with the aid of Corollary 2.2.1, that the prediction variance of $\hat{T}_{BU}$ is given by Eq. (1.5.1). Moreover, the prediction variance is minimized by a purposive sample maximizing $\bar{x}_s$.

Consider now the prediction of the population variance S^2_y. In the present model, $\mathbf{V}_s = \text{diag}\{x_i, i \in \mathbf{s}\}$, $\mathbf{V}_r = \text{diag}\{x_i, i \in \mathbf{r}\}$, and

$$\mathbf{X}'_r\mathbf{A}_r\mathbf{X}_r = S^2_{rx} = \sum_{i\in \mathbf{r}} (x_i - \bar{x}_r)^2/(N - n).$$

Hence, a direct application of Corollary 2.2.2 yields

$$\text{(2.2.10)}\quad \hat{S}^2_{BU} = \frac{n}{N}s^2_y + (1-\frac{n}{N})\{(\hat{\beta}^2_s - \frac{\sigma^2}{n\bar{x}_s})S^2_{r(x)} + \sigma^2[\bar{x}_r - \frac{1}{N}(\bar{x}_r - \bar{x}_s)]\},$$

where

$$S^2_{r(x)} = S^2_{rx} + \frac{n}{N}(\bar{x}_r - \bar{x}_s)^2.$$

If σ^2 is unknown, substitute in Eq. (2.2.10) the unbiased estimator

$$\text{(2.2.11)}\qquad \hat{\sigma}^2_s = \frac{1}{n-1}\sum_{i\in s}\frac{(y_i - \hat{\beta}_s x_i)^2}{x_i}.$$

We derive below the ψ-MSE of $\hat{S}^2_{BU}$ for the case of known σ^2.

(2.2.12)

$$\begin{aligned} E_\psi[\hat{S}^2_{BU} - S^2_y]^2 &= (1-\frac{n}{N})^2\mathrm{Var}_\psi\left\{S^2_{r(x)}\hat{\beta}^2_s - [S^2_{ry} + \frac{n}{N}(\bar{y}_r - \bar{y}_s)^2]\right\} \\ &= (1-\frac{n}{N})^2\left\{S^2_{r(x)}\mathrm{Var}_\psi[\hat{\beta}^2_s] + \mathrm{Var}_\psi[S^2_{ry} + \frac{n}{N}(\bar{y}_r - \bar{y}_s)^2]\right. \\ &\quad \left. - 2S^2_{r(x)}\frac{n}{N}\mathrm{Cov}_\psi[\hat{\beta}^2_s, (\bar{y}_s - \bar{y}_r)^2]\right\}, \end{aligned}$$

since $\hat{\beta}_s$ and S^2_{ry} are independent. Furthermore,

$$\mathrm{Var}_\psi[\hat{\beta}^2_s] = \frac{2\sigma^4}{n^2\bar{x}^2_s} + \frac{4\sigma^2\beta^2}{n\bar{x}_s}.$$

Let $\mathbf{Z}_r = \mathbf{W}_r^{-1/2}\mathbf{y}_r$, where $\mathbf{W}_r = \mathrm{diag}\{x_i, i \in \mathbf{r}\}$. It is not difficult to see that

$$\mathbf{Z}_r \sim N(\mathbf{Q}_r\beta, \sigma^2\mathbf{I}),$$

where $\mathbf{Q}_r = \mathbf{X}_r^{1/2}$. Hence, according to Seber (1977),

$$\mathrm{Var}_\psi[S^2_{ry}] = 2\sigma^4\mathrm{tr}\{\mathbf{B}^2_r\} + 4\sigma^2\beta^2\mathbf{Q}'_r\mathbf{B}^2_r\mathbf{Q}_r,$$

where

$$\mathbf{B}_r = \mathbf{W}_r^{1/2}\left(\mathbf{I}_r - \frac{1}{N-n}\mathbf{J}_r\right)\mathbf{W}_r^{1/2}.$$

After some algebraic manipulations, one obtains

$$\text{(2.2.13)}\qquad \begin{aligned} \mathrm{Var}_\psi[S^2_{ry}] &= \frac{2\sigma^4}{N-n}[(1-\frac{2}{N-n})m^{(2)}_{rx} + \frac{\bar{x}^2_r}{N-n}] \\ &\quad + \frac{4\beta^2\sigma^2}{N-n}[m^{(3)}_{rx} - \bar{x}_r m^{(2)}_{rx} - \bar{x}_r S^2_{rx}], \end{aligned}$$

where $m_{rx}^{(j)} = \sum_{i \in r} x_i^{(j)}/(N-n)$, $j = 2, 3$. Let $\chi^2[\nu, \lambda]$ denote a noncentral chi–squared random variable (r.v.), with ν degrees of freedom, and parameter of noncentrality λ. Since

$$(\bar{y}_s - \bar{y}_r)^2 \sim \sigma^2(\frac{\bar{x}_s}{n} + \frac{\bar{x}_r}{N-n})\chi^2[1, \lambda^*],$$

where

$$\lambda^* = \frac{\beta^2(\bar{x}_s - \bar{x}_r)^2}{2\sigma^2\left[\frac{\bar{x}_s}{n} + \frac{\bar{x}_r}{N-n}\right]},$$

we obtain

$$\mathrm{Var}_\psi[(\bar{y}_s - \bar{y}_r)^2] = 2\sigma^4(\frac{\bar{x}_s}{n} + \frac{\bar{x}_r}{N-n})^2\left[1 + \frac{2\beta^2(\bar{x}_s - \bar{x}_r)^2}{\sigma^2\left(\frac{\bar{x}_s}{n} + \frac{\bar{x}_r}{N-n}\right)}\right].$$

Similarly, it can be shown that

$$\mathrm{Cov}_\psi[S_{ry}^2, (\bar{y}_s - \bar{y}_r)^2] = \frac{2\sigma^4 S_{rx}^2}{(N-n)^2} + \frac{4\beta^2\sigma^2 S_{rx}^2}{N-n}(\bar{x}_r - \bar{x}_s) \tag{2.2.14}$$

and

$$\mathrm{Cov}_\psi[\hat{\beta}_s^2, (\bar{y}_s - \bar{y}_r)^2] = \frac{2\sigma^4}{n^2} - \frac{4\sigma^2\beta^2}{n}(\bar{x}_r - \bar{x}_s). \tag{2.2.15}$$

Substituting all the above results in Eq. (2.2.12), it follows that the prediction MSE of $\hat{S}_{BU}^2$ is

$$\begin{aligned} E_\psi[\hat{S}_{BU}^2 - S_y^2]^2 = {} & (1 - \frac{n}{N})^2 2\sigma^4\left\{\frac{S_{r(x)}^4}{n^2\bar{x}_s^2}\right. \\ & + \frac{1}{N-n}[(1 - \frac{2}{N-n})m_{rx}^{(2)} + \frac{\bar{x}_r^2}{N-n}] \\ & \left. + (\frac{n}{N})^2(\frac{\bar{x}_s}{n} + \frac{\bar{x}_r}{N-n})^2 + \frac{2n}{N}\frac{S_{rx}^2}{(N-n)^2} - \frac{2S_{r(x)}^2}{nN}\right\} \\ & + (1 - \frac{n}{N})^2 4\sigma^2\beta^2\left\{\frac{S_{r(x)}^4}{n\bar{x}_s} + \frac{1}{N-n}[m_{rx}^{(3)} - \bar{x}_r m_{rx}^{(2)} - \bar{x}_r S_{rx}^2]\right. \\ & + (\frac{n}{N})^2(\bar{x}_s - \bar{x}_r)^2(\frac{\bar{x}_s}{n} + \frac{\bar{x}_r}{N-n}) \\ & \left. + 2\frac{n}{N}\frac{S_{rx}^2}{N-n}(\bar{x}_r - \bar{x}_s) + 2S_{r(x)}^2\frac{1}{N}(\bar{x}_r - \bar{x}_s)\right\}. \end{aligned} \tag{2.2.16}$$

When N is sufficiently large relative to n, Eq. (2.2.16) reduces to

$$E_\psi[\hat{S}_{BU}^2 - S_y^2]^2 \approx 2\sigma^2\frac{S_{rx}^4}{n\bar{x}_s}(\frac{\sigma^2}{n\bar{x}_s} + 2\beta^2).$$

Thus, when N is large, the MSE of $\hat{S}^2_{BU}$ typically decreases as $\sum_{j\in\mathbf{s}} x_j$ increases. That is the ψ-optimal sample is the purposive sample maximizing $\sum_{j\in\mathbf{s}} x_j$.

Example 2.2.3. Consider the simple regression model SM3 with $g = 0$ and normal distribution for the error vector $\mathbf{e}$. It follows from Corollary 2.2.1 that the ψ-BUP of the population total T is

$$\hat{T}_{RE} = N\bar{y}_s + (N-n)\hat{\beta}_1(\bar{x}_r - \bar{x}_s), \tag{2.2.17}$$

which is the usual regression estimator of T (Cochran, 1977), where

$$\hat{\beta}_1 = \frac{\sum_{i\in\mathbf{s}}(x_i - \bar{x}_s)y_i}{\sum_{i\in\mathbf{s}}(x_i - \bar{x}_s)^2}$$

is the least-squares estimator of β_1. The prediction variance of $\hat{T}_{RE}$ is given in Exercise 2.7, formula (2.4.2). Notice that this prediction variance is minimized by a purposive sample satisfying $\bar{x}_s = \bar{x}_r$, namely, a balanced sample.

Example 2.2.4. In the present example the superpopulation model is binomial with covariates. More specifically, let $(x_{i1}, \ldots, x_{ik}, y_i)$, $i = 1, \ldots, N$ be binary variables associated with the ith element of the population. It is assumed that $y_i = 0, 1$ and $x_{ij} = 0, 1$ and, for each i, $\sum_{j=1}^k x_{ij} = 1$. The matrix $\mathbf{X} = (x_{ij};\ i = 1, \ldots, N,\ j = 1, \ldots, k)$ of covariates is known. The values of $y_1, \ldots, y_N$ are unknown. The objective is to predict $\bar{y} = \frac{1}{N}\sum_{j=1}^N y_j$ with the aid of n sample values $\{y_1, \ldots, y_n\}$ and the matrix X. The superpopulation model assumes that $y_1, \ldots, y_N$ are independent random variables, and that

$$E[y_i \mid x_{i1}, \ldots, x_{ik}] = \sum_{j=1}^k x_{ij}\beta_j, \qquad i = 1, \ldots, N$$

where

$$\beta_j = P[y_i = 1 \mid x_{ij} = 1], \qquad j = 1, \ldots, k.$$

Moreover, according to the model

$$\mathrm{Var}[y_i \mid x_1, \ldots, x_k] = w_i(1 - w_i), \qquad i = 1, \ldots, N$$

where $w_i = \sum_{j=1}^k x_{ij}\beta_j$. Thus, we consider the linear regression model $\mathbf{y} = \mathbf{X}\boldsymbol{\beta} + \mathbf{e}$, with parameters $\psi = (\boldsymbol{\beta}, \mathbf{V})$, where

$$\mathbf{V} = \mathrm{diag}\{w_i(1 - w_i),\ i = 1, \ldots, N\}.$$

The above binary superpopulation model was studied earlier by Thomsen (1981). We develop here the BUP of $\bar{y}$ and its prediction variance.

Without loss of generality we assume that $\mathbf{s}$ consists of the first n elements of the population; that is, $\mathbf{s} = \{1, \ldots, n\}$. The likelihood function of $\boldsymbol{\beta} = (\beta_1, \ldots, \beta_k)'$ given $\mathbf{y}_s$ and $\mathbf{X}_s$ is

$$\begin{aligned}\text{Lik}(\boldsymbol{\beta}; \mathbf{y}_s, \mathbf{X}_s) &= \prod_{i=1}^{n}\Big(\sum_{j=1}^{k} x_{ij}\beta_j\Big)^{y_i}\Big(1 - \sum_{j=1}^{k} x_{ij}\beta_j\Big)^{1-y_i} \\ &= \prod_{j=1}^{k} \beta_j^{f^{xy}_{s,j}}(1-\beta_j)^{n-f^{xy}_{s,j}},\end{aligned}$$

$0 < \beta_1, \ldots, \beta_k < 1$ where $f^{xy}_{s,j} = \sum_{i=1}^{n} x_{ij} y_i$, $j = 1, \ldots, k$, is the frequency (number) of sample elements for which $(x_{ij} = 1,\ y_i = 1)$. The right-hand side of the likelihood function is due to the fact that $\sum_{j=1}^{k} x_{ij} = 1$ for each $i = 1, \ldots, N$. Thus, $\mathbf{f}^{xy}_s = (f^{xy}_{s,1}, \ldots, f^{xy}_{s,k})'$ is a complete p.s.s., and the BUP of $\bar{y}$ is of the form

$$\hat{\bar{y}} = \frac{n}{N}\bar{y}_s + \Big(1 - \frac{n}{N}\Big)\hat{\theta}_{sr}(\mathbf{f}^{xy}_s),$$

where $\hat{\theta}_{sr}(\mathbf{f}^{xy}_s)$ is a BUP of $\bar{y}_r$. Here $\bar{y}_s = \dfrac{1}{N}\displaystyle\sum_{i=1}^{n} y_i$ and $\bar{y}_r = \dfrac{1}{N-n}\displaystyle\sum_{j=n+1}^{N} y_j$.

Notice that

$$\mathbf{f}^{xy}_s = \mathbf{X}'_s\mathbf{y}_s,$$

and that

$$E_\psi[\bar{y}_r] = \frac{1}{N-n}\mathbf{1}'_r\mathbf{X}_r\boldsymbol{\beta}.$$

Assume that the sample $\mathbf{s}$ was chosen so that $f^{x\cdot}_{sj} = \sum_{i=1}^{n} x_{ij} > 0$ for all $j = 1, \ldots, k$. Since $x_{ij}x_{ij'} = 0$ for all $j \neq j'$, $i = 1, \ldots, N$,

$$\mathbf{X}'_s\mathbf{X}_s = \text{diag}\{f^{x\cdot}_{sj},\ \ j = 1, \ldots, N\}$$

and the LSE of $\boldsymbol{\beta}$ is

$$\hat{\boldsymbol{\beta}}_s = \Big(\frac{f^{xy}_{s1}}{f^{x\cdot}_{s1}}, \ldots, \frac{f^{xy}_{sk}}{f^{x\cdot}_{sk}}\Big)'.$$

Thus, according to Theorem 2.2.1, since $E[\hat{\boldsymbol{\beta}}_s \mid \mathbf{X}_s] = \boldsymbol{\beta}$ and since $\hat{\boldsymbol{\beta}}_s$ is a function of the complete p.s.s., the BUP of $\bar{y}$ is

$$\hat{\bar{y}}_{BU} = \frac{n}{N}\bar{y}_s + \Big(1 - \frac{n}{N}\Big)\hat{\mathbf{p}}'_r\hat{\boldsymbol{\beta}}_s, \tag{2.2.18}$$

where $\hat{p}_{rj} = \dfrac{1}{N-n}\displaystyle\sum_{i=n+1}^{N} x_{ij}$, $j = 1, \ldots, k$, and $\hat{\mathbf{p}}_r = (\hat{p}_{r1}, \ldots, \hat{p}_{rk})'$.

We derive now the prediction risk of Eq. (2.2.18). Since $\hat{\boldsymbol{\beta}}_s$ is conditionally independent of $\mathbf{y}_r$, given $\mathbf{X}$,

$$\text{Var}_\psi[\hat{\bar{y}}_{BU} - \bar{y} \mid \mathbf{X}] = \Big(1 - \frac{n}{N}\Big)^2\{\text{Var}_\psi[\hat{\mathbf{p}}'_r\hat{\boldsymbol{\beta}}_s \mid \mathbf{X}] + \text{Var}_\psi[\bar{y}_r \mid \mathbf{X}]\}.$$

Starting with the conditional variance of $\bar{y}_r$ we have

$$\begin{aligned}\mathrm{Var}_\psi[\bar{y}_r \mid \mathbf{X}] &= \frac{1}{(N-n)^2}\mathbf{1}_r'\mathbf{V}_r\mathbf{1}_r \\ &= \frac{1}{(N-n)^2}\sum_{j=n+1}^{N} w_j(1-w_j) \\ &= \frac{1}{(N-n)^2}\sum_{j=n+1}^{N}(\sum_{l=1}^{k} x_{jl}\beta_l)(1-\sum_{l=1}^{k} x_{jl}\beta_l) \\ &= \frac{1}{(N-n)^2}\sum_{j=n+1}^{N}\sum_{l=1}^{k} x_{jl}\beta_l(1-\beta_l) \\ &= \frac{1}{N-n}\sum_{l=1}^{k}\hat{p}_{rl}\beta_l(1-\beta_l).\end{aligned}$$

In a similar fashion we show that

$$\mathrm{Var}_\psi[\hat{\beta}_{si} \mid \mathbf{X}] = \frac{\beta_i(1-\beta_i)}{f_{si}^{x\cdot}}, \qquad i = 1,\ldots,k,$$

and

$$\mathrm{Cov}_\psi[\hat{\beta}_{si}, \hat{\beta}_{si'} \mid \mathbf{X}] = 0$$

for all $i \neq i'$. Hence,

$$\mathrm{Var}_\psi[\hat{\mathbf{p}}_r'\hat{\boldsymbol{\beta}}_s \mid \mathbf{X}] = \sum_{i=1}^{k}\hat{\mathbf{p}}_{ri}^2\frac{\beta_i(1-\beta_i)}{f_{si}^{x\cdot}}.$$

Collecting all these expressions we obtain

$$\mathrm{Var}_\psi[\hat{\bar{y}}_{BU} - \bar{y} \mid \mathbf{X}] = \frac{1}{N-n}\sum_{l=1}^{k}\hat{p}_{rl}\beta_l(1-\beta_l)(1+\frac{f_{r_l}^{x\cdot}}{f_{s_l}^{x\cdot}}),$$

where $f_{r_l}^{x\cdot} = \sum_{j=n+1}^{N} x_{j_l}$. We can also write

$$\mathrm{Var}_\psi[\hat{\bar{y}}_{BU} - \bar{y} \mid \mathbf{X}] = \frac{N}{n(N-n)}\sum_{l=1}^{k}\frac{\bar{x}_l}{\bar{x}_{sl}}\hat{p}_{rl}\beta_l(1-\beta_l).$$

This expression is further simplified if the sample is balanced with respect to all covariates, that is, $\bar{x}_l = \bar{x}_{sl}$, $l = 1,\ldots,k$.

In Section 1.3, we discussed the structure of an unbiased predictor of the vector of regression coefficients $\boldsymbol{\beta}_N$. Under the restriction that $\mathbf{y}_s$ and $\mathbf{y}_r$ are ψ-uncorrelated, we have shown that

$$\boldsymbol{\beta}_N = \mathbf{E}_s\hat{\boldsymbol{\beta}}_s + \mathbf{E}_r\boldsymbol{\beta}_r,$$

and a predictor of $\boldsymbol{\beta}_N$ is

$$\hat{\boldsymbol{\beta}}_N = \mathbf{E}_s\hat{\boldsymbol{\beta}}_s + \mathbf{E}_r\hat{\boldsymbol{\beta}}_r,$$

where $\hat{\boldsymbol{\beta}}_r$ is a predictor of $\boldsymbol{\beta}_r$. According to Definition 1.3.4, the generalized prediction MSE of $\hat{\boldsymbol{\beta}}_N$ is

$$\text{GMSE}_\psi[\hat{\boldsymbol{\beta}}_N] = \mathbf{1}_p'\mathbf{E}_r E_\psi[(\hat{\boldsymbol{\beta}}_r - \boldsymbol{\beta}_r)(\hat{\boldsymbol{\beta}}_r - \boldsymbol{\beta}_r)']\mathbf{E}_r'\mathbf{1}_p. \tag{2.2.19}$$

The problem is that of finding a ψ–unbiased predictor $\hat{\boldsymbol{\beta}}_N$ that minimizes Eq. (2.2.19). As a direct consequence of Lemma 1.3.3, we have

Lemma 2.2.1. *$\hat{\boldsymbol{\beta}}_s$ is a ψ–unbiased predictor of $\boldsymbol{\beta}_N$.*

By extending Theorem 2.2.1, we prove that $\hat{\boldsymbol{\beta}}_s$ is indeed the ψ-BUP of $\boldsymbol{\beta}_N$.

Theorem 2.2.2. *Let*

$$\hat{\boldsymbol{\beta}}_{UN} = \mathbf{E}_s\hat{\boldsymbol{\beta}}_s + \mathbf{E}_r\hat{\boldsymbol{\beta}}_{Ur},$$

be any unbiased predictor of $\boldsymbol{\beta}_N$. Then, under the normal model (2.2.3) with $\mathbf{V}_{rs} = \mathbf{0}$, the ψ–unbiased predictor of $\boldsymbol{\beta}_N$ that minimizes (2.2.19) is

$$\hat{\boldsymbol{\beta}}_{BU} = \hat{\boldsymbol{\beta}}_s. \tag{2.2.20}$$

Furthermore,

$$GMSE_\psi[\hat{\boldsymbol{\beta}}_s] = \mathbf{1}_p'(\mathbf{X}'\mathbf{V}^{-1}\mathbf{X})^{-1}\mathbf{X}_r'\mathbf{V}_r^{-1}\mathbf{X}_r(\mathbf{X}_s'\mathbf{V}_s^{-1}\mathbf{X}_s)^{-1}\mathbf{1}_p. \tag{2.2.21}$$

Proof. According to Arnold (1981), if $\mathbf{V}$ is known, $\hat{\boldsymbol{\beta}}_s$ is a complete and sufficient statistic. Furthermore, since $\mathbf{V}_{sr} = \mathbf{0}$, $\mathbf{y}_s$ is independent of $\mathbf{y}_r$ under the ψ-model (2.2.3), $\hat{\boldsymbol{\beta}}_s$ is also predictive sufficient, according to Definition 2.2.1. We may then write

$$\begin{aligned} E_\psi[\mathbf{1}_p'(\hat{\boldsymbol{\beta}}_{uN} - \boldsymbol{\beta}_N)(\hat{\boldsymbol{\beta}}_{uN} - \boldsymbol{\beta}_N)'\mathbf{1}_p] &= \text{Var}_\psi[\mathbf{1}_p'(\hat{\boldsymbol{\beta}}_{uN} - \boldsymbol{\beta}_N)] \\ &\geq \text{Var}_\psi[\mathbf{1}_p' E_\psi[(\hat{\boldsymbol{\beta}}_{uN} - \boldsymbol{\beta}_N)|\mathbf{y}_r, \hat{\boldsymbol{\beta}}_s]] \\ &= \mathbf{1}_p'\mathbf{E}_r \text{Var}_\psi[E[\hat{\boldsymbol{\beta}}_{ur}|\mathbf{y}_r, \hat{\boldsymbol{\beta}}_s] - \boldsymbol{\beta}_r]\mathbf{E}_r'\mathbf{1}_p. \end{aligned}$$

Hence, since $\hat{\boldsymbol{\beta}}_s$ is predictive sufficient

$$E[\hat{\boldsymbol{\beta}}_{ur}|\mathbf{y}_r, \hat{\boldsymbol{\beta}}_s] = \hat{\boldsymbol{\beta}}_s.$$

This proves Eq. (2.2.20). Uniqueness follows from the completeness of $\hat{\boldsymbol{\beta}}_s$. To compute the ψ-generalized risk of $\hat{\boldsymbol{\beta}}_s$, notice that

$$\text{GMSE}_\psi[\hat{\boldsymbol{\beta}}_s] = \mathbf{1}_p'\mathbf{E}_r\text{Var}_\psi[\hat{\boldsymbol{\beta}}_s - \boldsymbol{\beta}_r]\mathbf{E}_r'\mathbf{1}_p.$$

Hence, Eq. (2.2.21) follows from the fact that

$$\text{Var}_\psi[\hat{\boldsymbol{\beta}}_s - \boldsymbol{\beta}_r] = (\mathbf{X}_s'\mathbf{V}_s^{-1}\mathbf{X}_s)^{-1} + (\mathbf{X}_r'\mathbf{V}_r^{-1}\mathbf{X}_r)^{-1}.$$

Bolfarine et al. (1991) extended the result of the above theorem to the general case, where the covariance matrix $\mathbf{V}$ is not necessarily block diagonal.

It can be shown that $\hat{\boldsymbol{\beta}}_s$ is also the BLUP of $\boldsymbol{\beta}_N$ under the Gauss–Markov setup of Section 2.1.1. To show this, let

$$\hat{\boldsymbol{\beta}}_{LN} = \mathbf{E}_s\hat{\boldsymbol{\beta}}_s + \mathbf{E}_r\hat{\boldsymbol{\beta}}_{Lr}, \tag{2.2.22}$$

where

$$\hat{\boldsymbol{\beta}}_{Lr} = \mathbf{C}_{ps}\mathbf{y}_s,$$

as considered in Lemma 1.3.3. The GMSE of this linear predictor is the same under both models (ψ and ψ with normality). The result follows from the fact that under normality, $\hat{\beta}_s$ minimizes Eq. (2.2.19) (see Exercise 2.26).

Example 2.2.5. In this example we illustrate a normal regression model, which is applied to educational data. Consider the data set of Cooley and Lohnes (1971, p. 349–351) collected under Project TALENT, which represents a "major psychometric sampling survey", supported in 1959 by the U.S. Office of Education. Using the first 150 records, we have a subpopulation of male students which constitutes for this example a finite population. Values of twenty variables are given for each record. We consider here only four variables: reading comprehension (x_1), abstract reasoning (x_2), physical science interest inventory (x_3), and mathematics test (y). We consider the problem of predicting the population mean of mathematics test (y), assuming that the values of the three predictors x_1, x_2, and x_3 are known for all the members of the population.

Regression analysis of y on (x_1, x_2, x_3) yields the population quantities shown in Table 2.2.1. The population mean of the y values is $\bar{y} = 27.667$. We consider two cases. In Case I, we have chosen the sample with the largest values of x_1. In Case II, we have chosen a random sample. In both cases, the sample is of size $n = 50$. We also assume that $y \mid \mathbf{X}, \boldsymbol{\beta} \sim N(\mathbf{X}\boldsymbol{\beta}, \sigma^2\mathbf{I})$.

Table 2.2.1. Population Quantities And Their Estimates.

		Case I		Case II	
	β_N	$\hat{\beta}_s$	$SE\{\hat{\beta}_s\}$	$\hat{\beta}_s$	$SE\{\hat{\beta}_s\}$
Constant	-9.3512	-20.1999	—	-6.4884	—
x_1	0.5116	0.5825	0.6567	0.5186	0.1390
x_2	1.0356	1.0843	0.5322	0.5809	0.5092
x_3	0.4040	0.6858	0.1589	0.4801	0.1578
σ	7.0920	7.9570	—	7.9400	—
R^2	0.5930	0.3540	—	0.5230	—

In Case I, we obtain that the BUP of $\bar{y}$ is $\hat{\bar{y}}_{BU} = 25.492$. On the other hand, the value of $\bar{y}_s = 36.63$. Indeed, $\bar{y}_s$ is generally a ψ–biased predictor of y, which is inefficient compared to $\hat{\bar{y}}_{BU}$ if $\bar{x}_s^{(i)}$ $(i = 1, 2, 3)$ are substantially different than the population means $\bar{x}^{(i)}$. In this case, the sample regressor means are quite different than the population regressor means, as seen in Table 2.2.2.

In Case II, the sample was chosen at random. As expected, the sample regression means are closer to the population regressor means. For this reason we obtained $\hat{\bar{y}}_{BU} = 27.662$ and $\bar{y}_s = 33.000$.

This example illustrates the fact that the regression predictor $\hat{\bar{y}}_{BU}$ yields better predictions, and that in the present case a random sample could yield better predictors than the sample of Case I.

Table 2.2.2. Regressor Means.

Case		x_1	x_2	x_3
I	Sample	43.82	11.70	27.22
	Population	34.63	11.10	21.93
II	Sample	40.06	10.08	25.06
	Population	34.63	11.10	21.93

2.3. Equivariant Predictors

A theory of equivariant prediction is developed for predicting population quantities. Minimum–risk equivariant predictors (MREP) are derived under the location, scale, and location–scale superpopulation models. Equivariant estimation in the infinite population context is considered in Lehmann (1984) and a more advanced treatment can be found in Zacks (1971). In the finite population context, very little has been done. Population variance prediction is considered in the works of Zacks and Solomon (1981) and Zacks (1981), using a Bayesian equivariant approach.

The methodology for determining the MREP of the population total T is to characterize first the class of all equivariant predictors and then, find the one within this class with minimum risk. This approach has been considered by Bolfarine (1989b). For the simple location model, the task of finding the MREP of T using the squared error loss function is made easier by exploring some relationships with the best (linear) unbiased predictor of T.

2.3.1. A General Formulation

Let $\mathbf{y} = (y_1, \ldots, y_N)'$ be the vector of $\mathbf{y}$ values associated with the elements of the population. We assume that $\mathbf{y}$ is a point in a sample space $\mathcal{Y}$. Let $F(\mathbf{y}; \psi)$ denote the joint distribution of $\mathbf{y}$ according to the superpopulation model. ψ is a parameter (vector valued) belonging to a parameter space Ψ. Let $\mathcal{F} = \{F(.; \psi); \psi \in \Psi\}$. Let g be a one to one transformation of $\mathcal{Y}$. Let

$$g\mathbf{y} = (g\mathbf{y}_s', g\mathbf{y}_r')'$$

be a representation of the transformed values associated with the observed and unobserved parts of $\mathbf{y}$. A model structure is said to be invariant under g if $F(g\mathbf{y}; \psi)$ is an element of $\mathcal{F}$. The distribution of $g\mathbf{y}$ is represented by the parametric point $\bar{g}\psi \in \Psi$. Thus, if the model is invariant with respect to g, then

$$F(g\mathbf{y}; \psi) = F(\mathbf{y}; \bar{g}\psi), \quad \text{for all } \psi \text{ in } \Psi. \tag{2.3.1}$$

Let $\mathcal{G}$ be a group of transformations on $\mathcal{Y}$ that leave the superpopulation model invariant. Let $\bar{\mathcal{G}}$ denote the group of transformations $\bar{g}$ on Ψ that are induced by $\mathcal{G}$. Let $\theta(\mathbf{y}_s)$ be a population quantity of interest (e.g., T, S_y^2, and so on). Let Θ be the range of $\theta(\mathbf{y})$, where $\mathbf{y}$ ranges over $\mathcal{Y}$. Applying g to $\mathbf{y}$ leads to

$$\theta(g\mathbf{y}) = \tilde{g}\theta(\mathbf{y}), \tag{2.3.2}$$

where $\tilde{g}$ is an element of an induced group $\tilde{\mathcal{G}}$, corresponding to $\mathcal{G}$. A predictor $\hat{\theta}(\mathbf{y}_s)$ is called equivariant if

$$\hat{\theta}(g\mathbf{y}_s) = \tilde{g}\hat{\theta}(\mathbf{y}_s), \quad \text{for all } g \in \mathcal{G}. \tag{2.3.3}$$

Let $L(.,.)$ be an invariant prediction loss function, that is,

$$L(\hat{\theta}(\mathbf{y}_s), \theta(\mathbf{y})) = L(\tilde{g}\hat{\theta}(\mathbf{y}_s), \tilde{g}\theta(\mathbf{y})), \tag{2.3.4}$$

for every $\mathbf{y} \in \mathcal{Y}$, every $g \in \mathcal{G}$, and every equivariant predictor $\hat{\theta}(\mathbf{y}_s)$. The prediction risk of a predictor $\hat{\theta}$ is defined as

$$R_\psi[\hat{\theta}; \theta] = E_\psi[L(\hat{\theta}, \theta)].$$

Notice that, for model invariant transformations, and equivariant predictors,

$$
\begin{aligned}
R_\psi[\hat{\theta};\theta] &= E_\psi[L(\hat{\theta}(\mathbf{y}_s),\theta(\mathbf{y}))] \\
&= E_\psi[L(\tilde{g}\hat{\theta}(\mathbf{y}_s),\tilde{g}\theta(\mathbf{y}))] \\
&= E_\psi[L(\hat{\theta}(g\mathbf{y}_s),\theta(g\mathbf{y}))] \qquad (2.3.5)\\
&= E_{\bar{g}\psi}[L(\hat{\theta}(\mathbf{y}_s),\theta(\mathbf{y}))] \\
&= R_{\bar{g}\psi}[\hat{\theta};\theta], \quad \text{for all } \ g \in \mathcal{G},\ \psi \in \mathbf{\Psi}.
\end{aligned}
$$

Let ψ be an element of $\mathbf{\Psi}$. The set of all points of $\mathbf{\Psi}$ that are images of ψ^o with respect to $\mathcal{G}$ is called an *orbit*, $\mathbf{\Psi}_{\mathcal{G}}(\psi^o)$, that is,

$$\mathbf{\Psi}_{\mathcal{G}}(\psi^o) = \{\psi; \psi \in \mathbf{\Psi}, \psi = \bar{g}\psi^o, \quad \text{for all } \ g \in \mathcal{G}\}.$$

Thus, $\mathcal{G}$ induces a partition of $\mathbf{\Psi}$ to orbits $\mathbf{\Psi}_{\mathcal{G}}(\cdot)$. If this partion contains a single orbit then $\mathcal{G}$ is called *transitive*. Using Eq. (2.3.5), we prove the following theorem

Theorem 2.3.1. *The risk function $R_\psi[\hat{\theta};\theta]$ of an equivariant predictor $\hat{\theta}$ assumes a constant value on each orbit of $\mathbf{\Psi}$ with respect to $\mathcal{G}$.*

Corollary 2.3.1. *Under the assumptions of Theorem 2.3.1, if the group of transformations is transitive then, for any equivariant predictor $\hat{\theta}$, $R_\psi[\hat{\theta};\theta]$ is constant over $\mathbf{\Psi}$.*

The result of Corollary 2.3.1 simplifies considerably the task of finding an MREP, since, if $\mathcal{G}$ is transitive, the risk of equivariant predictors could be minimized at any point ψ^o of $\mathbf{\Psi}$. Without loss of generality, we assume that $\mathbf{s} = \{1, \ldots, n\}$.

2.3.2. Location Equivariant Predictors of T Under Model SM1

The superpopulation simple location model assumes that $y_1, \ldots, y_N$ are independent and identically distributed having density

$$f(y_i;\eta) = f(y_i - \eta), \quad -\infty < \eta < \infty, \qquad (2.3.6)$$

$i = 1, \ldots, N$. Model (2.3.6) remains invariant under the group of translations,

$$y_i^* = y_i + c \text{ and } \eta^* = \eta + c, -\infty < c < \infty, \qquad (2.3.7)$$

$i = 1, \ldots, N$. The population total $T(\mathbf{y})$, when expressed in the new coordinate system, becomes $T(\mathbf{y}^*) = T(\mathbf{y}) + Nc$. Therefore, according to Eq. (2.3.3), a predictor $\widehat{T}(\mathbf{y}_s)$ is location equivariant if

$$\widehat{T}(\mathbf{y}_s^*) = \widehat{T}(\mathbf{y}_s) + Nc. \qquad (2.3.8)$$

Moreover, the loss function should be a function of the difference $\widehat{T}(\mathbf{y}_s) - T(y)$ only, that is

$$L(\widehat{T}(\mathbf{y}_s), T(\mathbf{y})) = \Delta(\widehat{T}(\mathbf{y}_s) - T(\mathbf{y})), \tag{2.3.9}$$

The squared error loss function

$$L(\widehat{T}(\mathbf{y}_s), T(\mathbf{y})) = (\widehat{T}(\mathbf{y}_s) - T(\mathbf{y}))^2, \tag{2.3.10}$$

satisfies condition (2.3.9).

Lemma 2.3.1. *Under the transformations (2.3.7), the bias, variance and risk function of any location–equivariant predictor are all independent of* η.

Lemma 2.3.1 follows directly from Corollary 2.3.1, since the group generated by transformations (2.3.7) is transitive.

The next theorem relates B(L)UP and MREP in finite populations.

Theorem 2.3.2. *Consider the location model (2.3.6) with a squared error loss function. Then,*

(i) an MREP of $T(\mathbf{y})$ is unbiased;

(ii) if the B(L)UP exists and is equivariant it is an MREP.

Proof. From Lemma 2.3.1 it follows that the risk function of any equivariant predictor is constant. Therefore, for any biased equivariant predictor, there is an unbiased equivariant predictor with smaller risk. This concludes (i). Part (ii) is a direct consequence of part (i).

(Q.E.D.)

Example 2.3.1. Let $y_1, \ldots, y_N$, be independent and $y_i \sim N(\eta, \sigma^2)$, where σ^2 is known, $i = 1, \ldots, N$. According to Example 2.1.1, $\widehat{T}_E = N\bar{y}_s$ (the usual expansion predictor), is the BUP of T. Since $\widehat{T}_E$ is a location–equivariant predictor, it is, according to Theorem 2.3.2, an MREP. If the normality assumption is dropped, $\hat{T}_E$ is, according to Example 1.3.5 and Theorem 2.3.2, an MREP within the class of all linear equivariant predictors.

The following result establishes a least favorable property of the normal distribution in finite populations.

Lemma 2.3.2. *Let $y_1, \ldots, y_N$, be independent and distributed according to $F(y_i - \eta)$, which belongs to the class of all univariate distributions having a density, where $E[y_i] = \eta$, and Var$[y_i] = \sigma^2$ is known, i=1, ... ,N. Let r_F be the risk function of the MREP of $T(\mathbf{y})$, with respect to the distribution F. Then, r_F takes on its maximum value when F is $N(\eta, \sigma^2)$.*

Proof. According to Example 2.3.1, the MREP of T under the $N(\eta, \sigma^2)$ is $\widehat{T}_E$, with risk

$$r_N = \sigma^2 N(N-n)/n.$$

Since the risk of $\widehat{T}_E$ is r_N, whatever F is, the MREP for any F must have risk $r_F \leq r_N$, and the result follows.

(Q.E.D.)

Example 2.3.2. Suppose now that $y_1, \ldots, y_N$ are independent and distributed according to the density

$$e^{-(y_i-\eta)}; \; y_i \;\geq\; \eta,$$

$i \;=\; 1, \ldots, N$. The predictive complete sufficient statistics is $y_{s(1)} = \min\{y_i, i \in \mathbf{s}\}$. It follows from Theorem 2.2.1 that the BUP of T is (see Exercise 2.5)

$$\widehat{T}_{BU}(\mathbf{y}_s) = n\bar{y}_s + (N-n)[y_{s(1)} - \frac{1}{n}].$$

Moreover, since $\widehat{T}_{BU}(\mathbf{y}_s)$ is equivariant, it is the MREP of T.

We derive an explicit and general expression for the MREP of T with respect to model (2.3.6).

Lemma 2.3.3. *Let $\widehat{T}^{(1)}(\mathbf{y}_s) = n\bar{y}_s + \widehat{T}_r^{(1)}(\mathbf{y}_s)$ be any equivariant predictor of T under model (2.3.7). Then, $\widehat{T}^{(2)}(\mathbf{y}_s)$ is equivariant if, and only if,*

$$\widehat{T}_r^{(2)}(\mathbf{y}) = \widehat{T}_r^{(1)}(\mathbf{y}_s) + w(\mathbf{z}_s),$$

where $\mathbf{z}_s = (z_1, \ldots, z_{n-1})$, and $z_i = y_i - y_n, i = 1, \ldots, n-1$.

The next theorem provides a general characterization of the MREP under model (2.3.6).

Theorem 2.3.3. *The MREP of T with respect to any loss function satisfying (2.3.9) is given by*

$$\widehat{T}_{MRE}(\mathbf{y}_s) = \widehat{T}(\mathbf{y}_s) + w^*(\mathbf{z}_s), \tag{2.3.11}$$

where $w^*(\mathbf{z}_s)$ minimizes

$$E_o\{\Delta[\widehat{T}_r(\mathbf{y}_s) + w(\mathbf{z}_s) - T_r]|\mathbf{z}_s\}, \tag{2.3.12}$$

and $\widehat{T}_r(\mathbf{y}_s)$ is any location–equivariant predictor of $T_r(\mathbf{y}) = 1_r'\mathbf{y}_r$ having a finite risk function.

Notice that the conditional distribution of $\mathbf{y}_s$ given $\mathbf{z}_s$ is independent of η.

The proof of Theorem 2.3.3 follows from the fact that if $\hat{T}_r$ is any equivariant predictor of T_r, its risk can be computed as

$$R_o[\hat{T}_r; T_r] = E_o\{E_o[\Delta(\hat{T}_r - T_r)|\mathbf{z}_s]\}.$$

Since $\Delta(\hat{T}_r - T_r) \geq 0$, predictor $\hat{T}_r^*$ which minimizes $E_o[\Delta(\hat{T}_r - T_r)|\mathbf{z}_s]$ is optimal.

Corollary 2.3.2. *Under the squared error loss function (2.3.10), the MREP of T is $\widehat{T}_{MRE}(\mathbf{y}_s)$ of Eq. (2.3.11), with*

$$w^*(\mathbf{z}_s) = -E_o[\widehat{T}_r(\mathbf{y}_s) - T_r(\mathbf{y})|\mathbf{z}_s], \tag{2.3.13}$$

where, $\widehat{T}(\mathbf{y}) = n\bar{y}_s + \widehat{T}_r(\mathbf{y}_s)$ is any equivariant predictor of T.

From Corollary 2.3.2, $\widehat{T}_E$ is the MREP under a distribution F if, and only if, $E_o[\bar{y}_s|\mathbf{z}_s] = 0$. This holds if, and only if, F is the normal distribution (see Kagan, Linik, and Rao, 1965). This result we summarize in the following lemma.

Lemma 2.3.4. *$\widehat{T}_E$ is an MREP only under the normal model of Example 2.3.1.*

The next theorem presents the Pitman–location predictor for finite populations.

Theorem 2.3.4. *The MREP of T with respect to the squared error loss function (2.3.10) for the location model (2.3.6) is*

$$\widehat{T}_{MRE}(\mathbf{y}_s) = n\bar{y}_s + (N-n)\frac{\int_{-\infty}^{\infty} uf(y_1 - u, \ldots, y_n - u)du}{\int_{-\infty}^{\infty} f(y_1 - u, \ldots, y_n - u)du}. \tag{2.3.14}$$

Proof. Obviously, $\widehat{T}(\mathbf{y}_s) = n\bar{y}_s + (N-n)y_n$ is equivariant. According to Corollary 2.3.2, under the squared–error loss function (2.3.10), the MREP of T is

$$\widehat{T}_{MRE}(\mathbf{y}_s) = \widehat{T}(\mathbf{y}_s) + w^*(\mathbf{z}_s),$$

where $\mathbf{z}_s = (y_1 - y_n, \ldots, y_{n-1} - y_n)$ is a maximal invariant statistic and

$$w^*(\mathbf{z}_s) = -(N-n)E_o[(y_n - \bar{y}_r)|\mathbf{z}_s] = -(N-n)E_o[y_n|\mathbf{z}_s],$$

since $\bar{y}_r$ and $\mathbf{z}_s$ are independent and $E_o[\bar{y}_r] = 0$. Formula (2.3.14) follows from the fact that

$$E_o[y_n|\mathbf{z}_s] = y_n - \frac{\int_{-\infty}^{\infty} uf(y_1 - u, \ldots, y_n - u)du}{\int_{-\infty}^{\infty} f(y_1 - u, \ldots, y_n - u)du},$$

a result that follows from Zacks (1971, Eq. 7.2.9).

(Q.E.D.)

Predictor (2.3.14) is called the Pitman–location predictor of the population total $T(\mathbf{y})$, since its structure is similar to that of the well-known Pitman estimator of a location parameter.

Example 2.3.3. Let $y_1, \ldots, y_N$ be independent and distributed uniformly over $(\theta - 1/2, \theta + 1/2)$; that is,

$$f(y_i - \theta) = \begin{cases} 1, & -1/2 \le y_i - \theta \le 1/2, \\ 0, & \text{elsewhere,} \end{cases}$$

where $i = 1, \ldots, N$. In this case, we obtain

$$E_o[y_n | \mathbf{z}_s] = y_n - (y_{s(1)} + y_{s(n)})/2,$$

where $y_{s(1)} = \min\{y_i, i \in s\}$, and $y_{s(n)} = \max\{y_i, i \in s\}$. It follows from Eq. (2.3.14) that the MREP of T is

$$\widehat{T}_{MRE}(\mathbf{y}_s) = n\bar{y}_s + (N - n)(y_{s(1)} + y_{s(n)})/2,$$

The main results of this section extend easily to the case of stratified populations (see Exercises 2.18 and 2.21).

2.3.3. Location Equivariant Predictors of T Under the Regression Model

As in Eq. (1.2.1), let $y_i = \eta_i + e_i$, where $\eta_i = \sum_{l=1}^{p} \beta_l x_{il}$, $i = 1, \ldots, N$. We consider the group $\mathcal{G}$ of translations of $\mathcal{Y}$, the sample space of y_i, $i = 1, \ldots, N$. The parameter space here is

$$\mathbf{\Pi}_p = \{\boldsymbol{\eta} = \mathbf{X}\boldsymbol{\beta}; \boldsymbol{\beta} \in \mathcal{R}^p\},$$

where $\boldsymbol{\eta}' = (\eta_1, \ldots, \eta_N)$. Accordingly, for each $g_c \in \mathcal{G}$,

$$\bar{g}_c \boldsymbol{\eta} = \boldsymbol{\eta} + \mathbf{c},$$

where $\mathbf{c} \in \mathbf{\Pi}_p$. The transformation g_c operates on each element of $\mathbf{y}$. Thus,

$$g_c y_i = y_i + c_i,$$

$i = 1, \ldots, N$. A predictor of T, $\hat{T}(\mathbf{y}_s)$, is called, location–equivariant, if

$$\hat{T}(g_c \mathbf{y}_s) = \hat{T}(\mathbf{y}_s) + \mathbf{c}'\mathbf{1}_N, \tag{2.3.15}$$

for all $\mathbf{c} \in \mathbf{\Pi}_p$. Since $\mathcal{G}_c$ is a transitive group, each equivariant predictor of T has a constant bias and a constant risk over $\mathbf{\Pi}_p$. Accordingly, the BLUP, $\hat{T}_{BLU}(\mathbf{y}_s) = n\bar{y}_s + \mathbf{1}_r'\mathbf{X}_r\hat{\boldsymbol{\beta}}_s$, is an MREP of T. The argument is the same as given in Section 2.3.2. Clearly, the expansion, ratio, and regression predictors are MREP for models SM1, SM2, and SM3 (g=0), respectively.

2.3.4. Scale Equivariant Predictors of T

Let $y_1, \ldots, y_N$ be independent and identically distributed random variables having a density that depends on a scale parameter τ; that is,

$$f(y_i;\tau) = \frac{1}{\tau} f(\frac{y_i}{\tau}),\ 0 < \tau < \infty, \tag{2.3.16}$$

$i = 1, \ldots, N$. Model (2.3.16) remains invariant under the scale change transformations

$$y_i^* = by_i \text{ and } \tau^* = b\tau, b \geq 0, \tag{2.3.17}$$

$i = 1, \ldots, N$. Accordingly, the condition of equivariance for scale transformations is

$$T(\mathbf{y}^*) = bT(\mathbf{y}).$$

Therefore, a predictor $\widehat{T}(\mathbf{y}_s)$ is equivariant under the transformations (2.3.17), if

$$\widehat{T}(b\mathbf{y}_s) = b\widehat{T}(\mathbf{y}_s), \text{ for all } 0 < b < \infty. \tag{2.3.18}$$

Such a predictor is called scale–equivariant. The problem of predicting T remains invariant under the transformations (2.3.17) provided the loss function is invariant; that is,

$$L[\widehat{T}(\mathbf{y}_s), T(\mathbf{y})] = \Delta \left[\frac{\widehat{T}(\mathbf{y}_s) - T(\mathbf{y})}{\tau} \right].$$

Notice that the quadratic loss function

$$L[\widehat{T}(\mathbf{y}_s), T(\mathbf{y})] = [\frac{\widehat{T}(\mathbf{y}_s) - T(\mathbf{y})}{\tau}]^2, \tag{2.3.19}$$

is invariant.

Since the group of transformations induced by (2.3.17) is transitive, it follows from Corollary 2.3.1 that the risk function of any equivariant predictor is independent of τ. The following result parallels that of Theorem 2.3.3 for the location model. Its proof is therefore omitted.

Theorem 2.3.5. *Let $\widehat{T}_1(\mathbf{y}) = n\bar{y}_s + \widehat{T}_r^{(1)}(\mathbf{y}_s)$ be any equivariant predictor of $T(\mathbf{y})$. A necessary and sufficient condition for $\widehat{T}_2(\mathbf{y}_s) = n\bar{y}_s + \widehat{T}_r^{(2)}(\mathbf{y}_s)$ to be scale–equivariant is that*

$$\widehat{T}_r^{(2)}(\mathbf{y}_s) = \frac{\widehat{T}_r^{(1)}(\mathbf{y}_s)}{w(\mathbf{z}_s)},$$

where $\mathbf{z}_s = (y_1/y_n, \ldots, y_{n-1}/y_n, |y_n|/y_n)$ *is a maximal invariant. Furthermore, if the conditional risk*

$$E_1\left\{\Delta\left[\frac{\widehat{T}_r^{(1)}(\mathbf{y}_s)}{w(\mathbf{z}_s)} - T_r(\mathbf{y})\right]|\mathbf{z}_s\right\}, \tag{2.3.20}$$

is finite for each $w(\mathbf{z}_s)$ *and* $w^*(\mathbf{z}_s)$ *minimizes expression (2.3.20), then*

$$\widehat{T}_{MRE}(\mathbf{y}_s) = n\bar{y}_s + \frac{\widehat{T}_r^{(1)}(\mathbf{y}_s)}{w^*(\mathbf{z}_s)} \tag{2.3.21}$$

is an MREP of T.

Notice that since the risk is independent of τ, the conditional expectation in expression (2.3.20) is performed with the distribution corresponding to $\tau = 1$.

Corollary 2.3.3. *Under the quadratic loss function (2.3.19), the MREP of* $T(\mathbf{y})$ *is given by Eq. (2.3.21), with*

$$w^*(\mathbf{z}_s) = \frac{E_1[(\widehat{T}_r^{(1)}(\mathbf{y}_s))^2|\mathbf{z}_s]}{E_1[\widehat{T}_r^{(1)}(\mathbf{y}_s)T_r(\mathbf{y}_r)|\mathbf{z}_s]}. \tag{2.3.22}$$

Example 2.3.4. Let $y_1, \ldots, y_N$ be independent and identically distributed random variables having a scale parameter exponential distribution; that is,

$$f(y_i;\tau) = \frac{1}{\tau}e^{-y_i/\tau}, \quad y_i > 0, \text{ and } 0 < \tau < \infty,$$

$i = 1, \ldots, N$. A complete and sufficient statistic for this family is the sample mean $\bar{y}_s$. According to Basu's theorem (Basu, 1955), $\bar{y}_s$ and $\mathbf{z}_s$ are independent. In addition, according to the model, $\mathbf{z}_s$ and $\bar{y}_r$ are independent. Consider the quadratic loss function (2.3.19) and the equivariant predictor $\hat{T}(\mathbf{y}_s) = n\bar{y}_s + (N-n)\bar{y}_s$. According to Eq. (2.3.21) and due to the independence of $\bar{y}_s$ and $\mathbf{z}_s$,

$$w^*(\mathbf{z}_s) = \frac{E_1[\bar{y}_s^2]}{E_1[\bar{y}_s]E_1[\bar{y}_r]} = \frac{n+1}{n},$$

Hence, the MREP of $T(\mathbf{y})$ is given by

$$\widehat{T}_{MRE}(\mathbf{y}_s) = n\bar{y}_s + (N-n)\frac{\sum_{i\in\mathbf{s}} y_i}{n+1}.$$

Example 2.3.5. Suppose that $y_1, \ldots, y_N$ are independent and follow a uniform distribution, that is, $y_i \sim U(0,\theta), i = 1, \ldots, N$. It is easy to see that predictor $\widehat{T}(\mathbf{y}_s) = n\bar{y}_s + (N-n)y_{s(n)}$, where $y_{s(n)} = \max\{y_i, i \in s\}$, is scale equivariant. Since $y_{s(n)}$ is a complete and sufficient statistics (Lehmann, 1984), it is independent of the maximal invariant $\mathbf{z}_s$. Thus, from Corollary 2.3.3, for the quadratic loss function,

$$w^*(\mathbf{z}) = \frac{E_1[y^2_{s(n)}]}{E_1[y_{s(n)}]E_1[\bar{y}_r]} = \frac{2(n+1)}{n+2}.$$

Hence, the MREP of $T(\mathbf{y})$ is

$$\widehat{T}_{MRE}(\mathbf{y}_s) = n\bar{y}_s + (N-n)\frac{n+2}{2(n+1)}y_{s(n)}.$$

Consider the model (2.3.16) for positive random variables $y_1, \ldots, y_N$. We derive now the general form of the MREP under the quadratic loss function. The predictor obtained is called the Pitman scale equivariant predictor. Starting with the simple scale equivariant predictor $\hat{T}(\mathbf{y}_s) = n\bar{y}_s + (N-n)y_n$, we apply formula (2.3.22) with a maximal invariant

$$\mathbf{z}_s = (\frac{y_1}{y_n}, \ldots, \frac{y_{n-1}}{y_n}).$$

By elementary transformation theory one obtains that the conditional density of y_n, given $\mathbf{z}_s$, under τ, can be written as

$$k_\tau(u|\mathbf{z}_s) = \frac{1}{\tau}(\frac{u}{\tau})^{n-1}f(\frac{u}{\tau})\prod_{i=1}^{n-1} f(\frac{uz_i}{\tau})/h(\mathbf{z}_s), \tag{2.3.23}$$

where $h(\mathbf{z}_s)$ is the joint density of $\mathbf{z}_s$. From Eqs. (2.3.22) and (2.3.23), we obtain

$$w^*(\mathbf{z}_s) = \frac{\int u^{n+1}f(u)\prod_{i=1}^{n-1} f(uz_i)du}{\mu\int u^n f(u)\prod_{i=1}^{n-1} f(uz_i)du},$$

where $\mu = E_1[\bar{y}_r]$.

According to Eq. (2.3.21), the MREP is

$$\begin{aligned}\hat{T}_{MRE}(\mathbf{y}_s) &= n\bar{y}_s + (N-n)\frac{y_n\mu\int u^n f(u)\prod_{i=1}^{n-1} f(uz_i)du}{\int u^{n+1}f(u)\prod_{i=1}^{n-1} f(uz_i)du} \\ &= n\bar{y}_s + (N-n)\frac{\mu\int u^n f(u)\prod_{i=1}^{n-1} f(uy_i)du}{\int u^{n+1}f(u)\prod_{i=1}^{n-1} f(uy_i)du}.\end{aligned} \tag{2.3.24}$$

2.3.5. Location–Scale Equivariant Predictors of T

Suppose now that $y_1, \ldots, y_N$ are independent and identically distributed with density

(2.3.25) $$f(y_i; \eta, \tau) = \frac{1}{\tau} f(\frac{y_i - \eta}{\tau}), \quad -\infty < \eta < \infty, \ \ 0 < \tau < \infty;$$

$i = 1, \ldots, N$. Model (2.4.25) remains invariant under the transformations

(2.3.26) $$y_i^* = by_i + a, \ \eta^* = b\eta + a, \ \text{and} \ \tau^* = b\tau,$$

where $b \geq 0$. Under the transformations (2.3.26), we obtain that $T(\mathbf{y}^*) = a + bT(\mathbf{y})$. Therefore, a predictor $\widehat{T}(\mathbf{y}_s)$ is equivariant under Eqs. (2.3.26) provided

(2.3.27) $$\widehat{T}(\mathbf{y}_s^*) = a + b\widehat{T}(\mathbf{y}_s).$$

A predictor satisfying Eqs. (2.3.26) is called a location–scale equivariant predictor. Moreover, the problem of predicting $T(\mathbf{y})$ remains invariant under transformations (2.3.26) if

(2.3.28) $$L[\widehat{T}(\mathbf{y}_s), T(\mathbf{y})] = \Delta \left[\frac{\widehat{T}(\mathbf{y}_s) - T(\mathbf{y})}{\tau} \right].$$

Note that the loss function (2.3.19) satisfies condition (2.3.28). Note also that the transformations (2.3.26) induce a transitive group. Therefore, according to Corollary 2.3.1, the risk function of any location–scale equivariant predictor is constant.

Lemma 2.3.5. *Consider the location–scale model (2.3.25). Suppose that $\hat{T}_{MRE}(\mathbf{y}_s)$ is the best location equivariant predictor of T, for a fixed τ. Then, if $\hat{T}_{MRE}(\mathbf{y}_s)$ is independent of τ, it is an MREP for the location–scale transformations (2.3.26).*

Example 2.3.6. Let $y_1, \ldots, y_N$ be independent and $y_i \sim N(\eta, \sigma^2)$, $i = 1, \ldots, N$, both parameters being unknown. It follows from Example 2.3.1 and Lemma 2.3.5 that $\widehat{T}_E$ is the MREP of T under the transformations (2.3.26).

Theorem 2.3.6. *Let $\hat{T}_1(\mathbf{y}_s) = n\bar{y}_s + T_r^{(1)}(\mathbf{y}_s)$, be a location–scale equivariant predictor of T. A predictor $\hat{T}_2(\mathbf{y}_s) = n\bar{y}_s + T_r^{(2)}(\mathbf{y}_s)$ is also location–scale equivariant if, and only if,*

(2.3.29) $$\hat{T}_r^{(2)}(\mathbf{y}_s) = \hat{T}_r^{(1)} + w(\mathbf{z}_s)\delta(\mathbf{u}_s),$$

where

$$\mathbf{z}_s = \left(\frac{y_1 - y_n}{y_{n-1} - y_n}, \ldots, \frac{|y_{n-1} - y_n|}{y_{n-1} - y_n}\right)$$

is a maximal location–scale invariant statistic, $\mathbf{u}_s = (y_1 - y_n, \ldots, y_{n-1} - y_n)$ *is a maximal location invariant statistic, and*

$$\delta(b\mathbf{u}_s) = b\delta(\mathbf{u}_s), \ \textit{for all } 0 < b < \infty.$$

In addition, the MREP of T is given by

$$(2.3.30) \qquad \widehat{T}_{MRE}(\mathbf{y}_s) = n\bar{y}_s + \widehat{T}_r^{(1)}(\mathbf{y}_s) + w^*(\mathbf{z}_s)\delta(\mathbf{y}_s),$$

where $w^*(\mathbf{z}_s)$ *minimizes the conditional risk*

$$(2.3.31) \qquad E_{o,1}\{\Delta[\widehat{T}_r^{(1)}(\mathbf{y}_s) + w(\mathbf{z}_s)\delta(\mathbf{u}_s) - T_r(\mathbf{y})]|\mathbf{z}_s\},$$

at $\eta = 0$ *and* $\tau = 1$, *provided Eq. (2.3.31) is positive and finite for each* $w(\mathbf{z}_s)$.

Corollary 2.3.4. *In the case of the quadratic loss function (2.3.19), if all expectations exist, then*

$$w^*(\mathbf{z}_s) = \frac{-E_{o,1}\{[\widehat{T}_r^{(1)})(\mathbf{y}_s) - T_r(\mathbf{y})]\delta(\mathbf{u}_s)|\mathbf{z}_s\}}{E_{o,1}[\delta^2(\mathbf{u}_s)|\mathbf{z}_s]}.$$

Example 2.3.7. Let $y_1, \ldots, y_N$ be independent and

$$f(y_i; \theta, \tau) = \frac{1}{\tau} e^{-(y_i - \theta)/\tau}; y_i \geq \theta,$$

$i = 1, \ldots, N$, where θ and τ are unknown. It is clear that $\widehat{T}(\mathbf{y}_s) = n\bar{y}_s + (N-n)y_{s(1)}$ is a location equivariant predictor, where $y_{s(1)} = \min\{y_i, i \in s\}$. Let

$$\delta(\mathbf{u}_s) = \sum_{i \in s} [y_i - y_{s(1)}].$$

Note that $\delta(\mathbf{u}_s)$ satisfies $\delta(b\mathbf{u}_s) = b\delta(\mathbf{u}_s)$. Basu's theorem (Basu, 1955), implies that $\widehat{T}_r^{(1)}(\mathbf{u}_s)$ and $\delta(\mathbf{u}_s)$ are independent since $y_{s(1)}$ is a function of the complete sufficient statistic $[y_{s(1)}, \bar{y}_s]$ and $\mathbf{u}_s$ is location invariant and also independent of $\mathbf{z}_s$. Therefore, from Corollary 2.3.4, we obtain

$$w^*(\mathbf{z}_s) = -\frac{(N-n)(n-1)}{n^2}.$$

Hence, the MREP of $T(\mathbf{y}_s)$, for the quadratic loss function is

$$\widehat{T}_{MRE}(\mathbf{y}_s) = n\bar{y}_s + (N-n)\left\{y_{s(1)} + \frac{(n-1)}{n^2}\sum_{i \in s}[y_i - y_{s(1)}]\right\}.$$

2.3.6. Location–Scale Equivariant Predictors of S_y^2

As in Section 2.3.5, consider the location–scale superpopulation model (2.3.25), and the group of transformations prescribed by Eqs. (2.3.26). We recall that S_y^2 may be written as in Eq. (1.3.2). If

$$\mathbf{y} \to \mathbf{y}^* = a\mathbf{1}_N + b\mathbf{y},$$

where $-\infty < a < \infty$ and $0 < b < \infty$, then

$$S_{y^*}^2 = b^2 S_y^2,$$

where $S_{y^*}^2$ denotes the population variance corresponding to the transformed vector $\mathbf{y}^*$. Accordingly, a predictor

$$\hat{S}_y^2 = \frac{n}{N} s_y^2 + (1 - \frac{n}{N}) \hat{S}_r^2(\mathbf{y}_s)$$

of S_y^2 is equivariant under the transformations (2.3.26) if, and only if, for any $\mathbf{y}_s^* = a\mathbf{1}_s + b\mathbf{y}_s$,

$$\hat{S}_r^2(\mathbf{y}_s^*) = b^2 \hat{S}_r^2(\mathbf{y}_s), \tag{2.3.32}$$

where $\hat{S}_r^2$ is a predictor of

$$S_r^2 = S_{ry}^2 + \frac{n}{N}(\bar{y}_s - \bar{y}_r)^2.$$

A predictor $\hat{S}_y^2$ satisfying Eq. (2.3.32) is called a location–scale equivariant predictor of S_y^2. Moreover, the loss function of any equivariant predictor should be invariant; that is, it should satisfy

$$L(b^2 \hat{S}_y^2, b^2 S_y^2) = L(\hat{S}_y^2, S_y^2). \tag{2.3.33}$$

When all the above conditions hold, we say that the problem of predicting S_y^2 remains invariant under the transformations (2.3.26). Any nonnegative function of $(\hat{S}_y^2 - S_y^2)/\tau^2$, say

$$\Delta(\frac{\hat{S}_y^2 - S_y^2}{\tau^2}), \tag{2.3.34}$$

is an invariant loss function. Notice, in particular, that the quadratic loss function

$$L(\hat{S}_y^2, S_y^2) = (\frac{\hat{S}_y^2 - S_y^2}{\tau^2})^2, \tag{2.3.35}$$

is invariant. The predictive risk of any predictor $\hat{S}_y^2$ is defined as

$$R_\psi[\hat{S}_y^2; S_y^2] = E_\psi[L(\hat{S}_y^2, S_y^2)],$$

where $\psi = (\theta, \tau)$. Moreover, since the group of transformations induced by Eqs. (2.3.26) is transitive, the risk function of any equivariant predictor with respect to any invariant loss function is, according to Corollary 2.3.1, constant over Ψ.

Let

$$\mathbf{z}_s = \frac{1}{s_y}(\mathbf{y}_s - \bar{y}_s \mathbf{1}_s).$$

It is easy to check that $\mathbf{z}_s$ is maximal invariant, and that any equivariant predictor may be represented as

$$\hat{S}_y^2 = \frac{n}{N} s_y^2 + (1 - \frac{n}{N}) w(\mathbf{z}_s) \hat{S}_{EQr}^2, \tag{2.3.36}$$

where $\hat{S}_{EQr}^2$ is a particular equivariant predictor of S_r^2. The following theorem provides the general structure for the MREP of S_y^2 under model (2.3.25), with respect to the transformations (2.3.26).

Theorem 2.3.7. *The MREP of S_y^2 with respect to any loss function satisfying function (2.3.34) is given by*

$$\hat{S}_{MRE} = \frac{n}{N} s_y^2 + (1 - \frac{n}{N}) w^*(\mathbf{z}_s) \hat{S}_{EQr}^2, \tag{2.3.37}$$

where $w^*(\mathbf{z}_s)$ minimizes

$$E_{0,1}\{\Delta[(1 - \frac{n}{N})(w(\mathbf{z}_s) \hat{S}_{EQr}^2 - S_r^2)] | \mathbf{z}_s\}, \tag{2.3.38}$$

and $\hat{S}_{EQr}^2$ is any equivariant predictor of S_{ry}^2.

Proof. According to Eq. (2.3.36), the MREP of S_y^2 may be obtained by minimizing

$$E_\psi\{\Delta[(1 - \frac{n}{N})(\frac{w(\mathbf{Z}_s)\hat{S}_{rEQ}^2 - S_r^2)}{\tau})]\}. \tag{2.3.39}$$

Now, since the group of transformations induced by Eqs. (2.3.26) is transitive, Corollary 2.3.1 implies that minimizing the expression (2.3.39) is equivalent to minimizing

$$E_{0,1}\{\Delta[(1 - \frac{n}{N})(w(\mathbf{z}_s) \hat{S}_{EQr}^2 - S_r^2)] | \mathbf{z}_s\},$$

from which the result follows.

(Q.E.D.)

In expression (2.3.38), $E_{0,1}[.]$ is the expectation operator under model (2.3.25) with $\eta = 0$ and $\tau = 1$. In particular, we have the following corollary.

Corollary 2.3.5. *For the particular case of the quadratic loss function (2.3.35), the MREP of S_y^2 is given by Eq. (2.3.37) with*

$$w^*(\mathbf{z}_s) = \frac{E_{0,1}[\hat{S}^2_{rEQ} S_r^2 | \mathbf{z}_s]}{E_{0,1}[\hat{S}^4_{rEQ} | \mathbf{z}_s]},$$

where $\hat{S}^2_{rEQ}$ is any equivariant predictor of S_r^2.

Notice that $w^*(\mathbf{z}_s)$ exists if, and only if, the conditional fourth moment of the distribution of y_i given $\mathbf{z}_s$ exists.

Example 2.3.8. Consider the normal location–scale superpopulation model of Example 2.3.8, that is,

$$\mathbf{y} \sim N(\eta \mathbf{1}_N, \sigma^2 \mathbf{I}_N), \quad -\infty < \eta < \infty, \ 0 < \sigma < \infty$$

Choose $\hat{S}^2_{rEQ} = s_y^2$. Notice first that S^2_{ry} is independent of $\mathbf{y}_s$; therefore, it is also independent of $\mathbf{z}_s$. Thus,

$$E_{0,1}[s_y^2 S^2_{ry} | \mathbf{z}_s] = E_{0,1}[S^2_{ry}] E_{0,1}[s_y^2 | \mathbf{z}_s].$$

It can be easily verified that

$$E_{0,1}[S^2_{ry}] = 1 - \frac{1}{N-n}.$$

Furthermore, since $(\bar{y}_s, s_y^2)$ is in the normal case a complete sufficient statistic and since $\mathbf{z}_s$ is invariant, Basu's theorem (Basu, 1955) implies that

$$E_{0,1}[s_y^2 | \mathbf{z}_s] = E_{0,1}[s_y^2] = 1 - \frac{1}{n}.$$

A similar argument yields

$$E_{0,1}[(\bar{y}_s - \bar{y}_r)^2 | \mathbf{z}_s] = E_{0,1}[(\bar{y}_s - \bar{y}_r)^2] = \frac{N}{n(N-n)}.$$

Finally, since, for $\eta = 0$ and $\tau = 1$,

$$s_y^2 \sim \frac{1}{n} \chi^2[n-1],$$

where $\chi^2[n-1]$ denotes the chi–squared distribution with $n-1$ degrees of freedom, it follows that

$$E_{0,1}[s_y^4 | \mathbf{z}_s] = E_{0,1}[s_y^4] = \frac{n^2-1}{n^2}.$$

Collecting all these results, we obtain that the MREP of S_y^2, for the quadratic loss (2.3.35) is

$$\begin{aligned}\hat{S}^2_{MRE} &= \frac{n}{N}s_y^2 + (1-\frac{n}{N})\frac{n}{n+1}s_y^2 \\ &= s_y^2\frac{n(N+1)}{N(n+1)}.\end{aligned} \tag{2.3.40}$$

The ψ-BUP of S_y^2, $\hat{S}^2_{BU}$ is given in Eq. (2.2.9). Note that both predictors may be considered as elements of the more general class of predictors

$$\hat{S}^2_{cy} = \frac{n}{N}s_y^2 + (1-\frac{n}{N})cs_y^2. \tag{2.3.41}$$

It can be shown after some algebraic manipulations that the risk function of predictor $\hat{S}^2_{cy}$ with respect to the quadratic loss function is

$$\frac{1}{\sigma^4}E_\psi[\hat{S}^2_{cy} - S_y^2]^2 = (1-\frac{n}{N})^2[c^2\frac{n^2-1}{n^2} - 2c\frac{n-1}{n} + \frac{N-n+2}{N-n}] \tag{2.3.42}$$

The risks of $\hat{S}^2_{\text{MRE}}$ and $\hat{S}^2_{\text{BU}}$ are easily computed from Eq. (2.3.42). It is clear that Eq. (2.3.42) is minimized for $c = n/(n+1)$. Moreover, $\hat{S}^2_{\text{MRE}}$ has uniformly smaller risk than $\hat{S}^2_{\text{BU}}$. Table 2.3.1 below provides an example of the efficiency (risks ratio),

$$E_\psi[\hat{S}^2_{\text{BU}} - S_y^2]^2/E_\psi[\hat{S}^2_{\text{MRE}} - S_y^2]^2,$$

for several values of N and $n = 10$.

Table 2.3.1. Prediction Risk Efficiency (PRE) of $\hat{S}^2_{\text{MRE}}$ Relative to $\hat{S}^2_{\text{BU}}$, $n = 10$.

N	50	100	200	250	300	∞
PRE	1.174	1.198	1.206	1.212	1.214	1.222

Example 2.3.9. Let $y_1, \ldots, y_N$ be independent and identically distributed with density

$$f(y;\theta,\tau) = \frac{1}{\tau}\exp(-\frac{y-\theta}{\tau}), \quad y \geq \theta.$$

Under this superpopulation model the statistic

$$\left\{y_{(1)}, \sum_{i\in\mathbf{s}}[y_i - y_{(1)}]\right\}$$

is complete and sufficient for $\psi = (\theta,\tau)'$. Since $y_1, \ldots, y_N$ are also independent, it follows that this statistic is also predictive sufficient. Moreover,

$$\frac{\sum_{i\in\mathbf{s}}(y_i - y_{s(1)})}{\tau} \sim G(1, n-1),$$

where $G(1,\nu)$ indicates the gamma r.v. with scale parameter 1 and shape parameter ν.

In order to derive the MREP of S_y^2 under quadratic loss function, we choose

$$\hat{S}_{r\ \mathrm{EQ}}^2 = (\sum_{i \in s}(y_i - y_{(1)}))^2.$$

Notice that $\hat{S}_{r\ \mathrm{EQ}}^2$ is independent of S_{ry}^2 and of $\bar{y}_r$. Also, according to Basu's theorem (Basu, 1955), it is independent of $\mathbf{Z}_s$. One can derive the following

$$\begin{aligned}
E_{0,1}[\hat{S}_{r\ \mathrm{EQ}}^2] &= (n-1)n, \\
E_{0,1}[\hat{S}_{r\ \mathrm{EQ}}^3] &= (n-1)n(n+1), \\
E_{0,1}[\hat{S}_{r\ \mathrm{EQ}}^4] &= (n-1)n(n+1)(n+2), \\
E_{0,1}[\bar{y}_r - y_{(1)}] &= 1 - \frac{1}{n}, \\
E_{0,1}[\bar{y}_r - y_{(1)}]^2 &= (1-\frac{1}{n})^2 + \frac{1}{N-n} + \frac{1}{n^2}.
\end{aligned}$$

Therefore,

$$\begin{aligned}
E_{0,1}[\hat{S}_{r\ \mathrm{EQ}}^2 S_{ry}^2 \mid \mathbf{Z}_s] &= E_{0,1}[\hat{S}_{r\ \mathrm{EQ}}^2] E_{0,1}[S_{ry}^2] \\
&= n(n-1)(1+\frac{6}{Nn}).
\end{aligned}$$

Thus,

$$w^*(\mathbf{z}_s) = \frac{1+6/nN}{(n+1)(n+2)},$$

and the MREP of S_y^2 is:

$$\hat{S}_{\mathrm{MREP}}^2 = \frac{n}{N}s_y^2 + (1-\frac{n}{N})\frac{1+6/nN}{(n+1)(n+2)}\left\{\sum_{i \in S}[y_{(i)} - y_{s(1)}]\right\}^2. \tag{2.3.43}$$

We consider in the sequel the predictor

$$\hat{S}_{\mathrm{EP}}^2 = \frac{n}{N}s_y^2 + (1-\frac{n}{N})\frac{1}{(n+1)(n+2)}\left[\sum_{i \in S} y_i - y_{s(1)}\right]^2, \tag{2.3.44}$$

which is close to Eq. (2.3.43). The prediction risk of $\hat{S}_{\mathrm{EP}}^2$ is

$$\begin{aligned}
\frac{1}{\tau^4}E[\hat{S}_{\mathrm{EP}}^2 - S_y^2]^2 = (1-\frac{n}{N})^2 E_{0,1}\{&\frac{\hat{S}_{r\ \mathrm{EQ}}^4}{(n+1)^2(n+2)^2} \\
&- \frac{2}{(n+1)(n+2)}\hat{S}_{r\ \mathrm{EQ}}^2 S_r^2 + S_r^4\}.
\end{aligned} \tag{2.3.45}$$

According to the previous derivations,

$$E_{0,1}[\frac{\hat{S}^4_{r\ \mathrm{EQ}}}{(n+1)^2(n+2)^2}] = \frac{(n-1)n}{(n+1)(n+2)},$$

and

$$-\frac{2}{(n+1)(n+2)}E_{0,1}[\hat{S}^2_{r\ \mathrm{EQ}}S^2_r] = -\frac{2n(n-1)}{(n+1)(n+2)}(1+\frac{6}{nN}).$$

It remains to derive $E[\hat{S}^4_r]$. Notice that

$$S^4_r = S^4_{ry} + 2\frac{n}{N}S^2_{ry}(\bar{y}_r - \bar{y}_s)^2 + (\frac{n}{N})^2(\bar{y}_s - \bar{y}_r)^4.$$

Under $\theta = 0$, $\tau = 1$, $\bar{y}_s \sim \frac{1}{n}G(1,n)$, $\bar{y}_r \sim \frac{1}{N-n}G(1,N-n)$. Thus, algebraic manipulations yield,

$$E_{0,1}[S^4_{ry}] = 1 + \frac{6}{N-n} - \frac{13}{(N-n)^2} + \frac{6}{(N-n)^3}.$$

Similar computations yield

$$E_{0,1}[S^2_{ry}(\bar{y}_r - \bar{y}_s)^2] = \frac{1}{N-n}[\frac{5}{N-n} - \frac{6}{(N-n)^2} + \frac{N-1}{n}].$$

Finally,

$$E_{0,1}[(\bar{y}_r - \bar{y}_s)^4] = 3\frac{N^2}{n^2(N-n)^2} + 6\frac{(N-n)^3 + n^3}{n^3(N-n)^3}.$$

Collecting all these results we obtain

$$\frac{1}{\tau^4}E[\hat{S}^2_{\mathrm{EP}} - S^2_y]^2 = (1-\frac{n}{N})^2[\frac{2(2n+1)}{(n+1)(n+2)}(1+\frac{12}{nN}) + \frac{8}{N-n} - \frac{6(2N-1)}{nN(N-n)}]. \tag{2.3.46}$$

One can readily verify that the best unbiased predictor of S^2_y is

$$\hat{S}^2_{\mathrm{BUP}} = \frac{n}{N}s^2_y + (1-\frac{n}{N})\frac{1}{n(n-1)}\left[\sum_{i\in S}(y_i - y_{s(1)})\right]^2. \tag{2.3.47}$$

The prediction risk of the BUP is

$$\frac{1}{\tau^4}E[\hat{S}^2_{\mathrm{BUP}} - S^2_y]^2 = (1-\frac{n}{N})^2[\frac{2(2n+1)}{n(n-1)} + \frac{8}{N-n} - \frac{6(2N-1)}{nN(N-n)}]. \tag{2.3.48}$$

In Table 2.3.2, we present the prediction risk efficiency of $\hat{S}^2_{\text{EP}}$ versus $\hat{S}^2_{\text{BUP}}$.

Table 2.3.2. Prediction Risk Efficiency (PRE) of $\hat{S}^2_{\text{EP}}$ Versus $\hat{S}^2_{\text{BUP}}$ for $n = 10$.

N	50	100	150	200	300	500	∞
PRE	1.284	1.294	1.395	1.412	1.429	1.443	1.466

2.4. Stein–Type Shrinkage Predictors

In the present section we derive Stein–type shrinkage predictors (STSP) for the superpopulation model. The idea is to shrink the BUP toward a point that is believed to be a reasonable value of the quantity under consideration. This point is called the "*natural origin*" of the shrinkage predictor.

The superpopulation model considered in the present section is the normal regression model, with $\mathbf{V} = \sigma^2\mathbf{W}$, where $\mathbf{W}$ is a known diagonal matrix. As we have proven before, if $\mathbf{W}$ satisfies condition L then $\hat{T}_{\text{BU}} = \hat{T}_{\text{SP}} = \mathbf{1}'_N\mathbf{X}\hat{\boldsymbol{\beta}}_s$.

We study the problem of estimating the parametric vector $\boldsymbol{\eta} = \mathbf{1}'_N\mathbf{X}\boldsymbol{\beta}$, by a shrinkage estimator, with a natural origin $\mathbf{1}'_N\mathbf{X}\boldsymbol{\beta}_0$. Following Arnold (1980, Ch. 11) and Berger (1980), we consider the estimator

$$\text{(2.4.1)}\quad \hat{\eta}_c(\boldsymbol{\beta}_0) = \left[1 - \frac{c\hat{\sigma}_s}{|\mathbf{1}'_N\mathbf{X}(\hat{\boldsymbol{\beta}}_s - \boldsymbol{\beta}_0)|}\right](\mathbf{1}'_N\mathbf{X}\hat{\boldsymbol{\beta}}_s - \mathbf{1}'_N\mathbf{X}\boldsymbol{\beta}_0) + \mathbf{1}'_N\mathbf{X}\boldsymbol{\beta}_0,$$

where c is a constant minimizing the quadratic risk

$$\text{(2.4.2)}\qquad R_\psi[\hat{\eta}_c(\boldsymbol{\beta}_0);\eta] = \frac{1}{\sigma^2}E_\psi[\hat{\eta}_c(\boldsymbol{\beta}_0) - \mathbf{1}'_N\mathbf{X}\boldsymbol{\beta}]^2.$$

Theorem 2.4.1. *The risk of* $\hat{\eta}_c \equiv \hat{\eta}_c(0)$ *is*

$$\text{(2.4.3)}\qquad R_\psi[\hat{\eta}_c;\eta] = R_\psi[\hat{\eta};\eta] - \{2c[K_N(p)]^{1/2}e^{-\delta_s^2/2} - c^2\},$$

where p *is the dimension of* $\boldsymbol{\beta}$ *and*

$$\begin{aligned}
\hat{\eta} &= 1'_N X\hat{\boldsymbol{\beta}}_s,\\
K_N(p) &= \frac{4\Gamma^2(\frac{n-p+1}{2})\xi_N^2}{(n-p)\pi\Gamma^2(\frac{n-p}{2})},\\
\xi_N^2 &= \mathbf{1}'_N\mathbf{X}(\mathbf{X}'_s\mathbf{W}_s^{-1}\mathbf{X}_s)^{-1}\mathbf{X}\mathbf{1}_N,\\
\delta_N^2 &= (\mathbf{1}'_N\mathbf{X}\boldsymbol{\beta})^2/\sigma^2\xi_N^2.
\end{aligned}$$

Proof. The following results can be obtained by standard computations

(i) $E_\psi[\hat{\sigma}_s] = \sigma\Gamma\left(\frac{n-p+1}{2}\right)2^{1/2}\Big/(n-p)^{1/2}\Gamma(\frac{n-p}{2})$;
(ii) $\hat{\eta} \sim N(\eta, \sigma^2\xi_N^2)$;
(iii) $E_\psi[\frac{\hat{\eta}}{|\hat{\eta}|}] = 1-2\Phi(-\delta_N)$, where $\Phi(\cdot)$ is the standard normal integral, and
(iv) $E_\psi[|\hat{\eta}|] = (\xi_N\sigma/\sqrt{2\pi})\{2e^{-\delta_N^2/2} + \sqrt{2\pi}\delta_N[1-2\Phi(-\delta_N)]\}$.

Furthermore, $\hat{\sigma}_s^2$ and $\hat{\eta}$ are independent, and $E_\psi[\hat{\sigma}_s^2] = \sigma^2$. Hence,

$$
\begin{aligned}
R_\psi[\hat{\eta}_c;\eta] &= \frac{1}{\sigma^2}E_\psi[(\hat{\eta}-\eta) - \frac{c\hat{\sigma}_s\hat{\eta}}{|\hat{\eta}|}]^2 \\
&= R_\psi[\hat{\eta};\eta] - 2\frac{c}{\sigma^2}E_\psi[\hat{\sigma}_s\frac{\hat{\eta}}{|\hat{\eta}|}(\hat{\eta}-\eta)] + c^2 \\
&= R_\psi[\hat{\eta};\eta] - 2\frac{c}{\sigma^2}E_\psi[\hat{\sigma}_s]E_\psi[|\hat{\eta}|] \\
&\quad + 2\frac{c}{\sigma^2}\eta E_\psi[\hat{\sigma}_s]E_\psi[\frac{\hat{\eta}}{|\hat{\eta}|}] + c^2.
\end{aligned}
\tag{2.4.4}
$$

Substituting into Eq. (2.4.4) results (i) – (iv), we obtain Eq. (2.4.3). (Q.E.D).

From the proof of Theorem 2.4.1 it is evident that, for any $\boldsymbol{\beta}_0$

$$
R_\psi[\hat{\eta}_0(\boldsymbol{\beta}_0);\eta] = R_\psi[\hat{\eta};\eta] - (2c(K_N(p))^{1/2}e^{-\delta_N^2(\boldsymbol{\beta}_0)/2} - c^2), \tag{2.4.5}
$$

where $\delta_N(\boldsymbol{\beta}_0) = \mathbf{1}_N'\mathbf{X}(\boldsymbol{\beta}-\boldsymbol{\beta}_0)/\sigma\xi_N$. This shows that Eq. (2.4.5) is minimized at $\boldsymbol{\beta}_0 = \boldsymbol{\beta}$. Thus, Eq. (2.4.5) is a decreasing continuous function of $||\boldsymbol{\beta}-\boldsymbol{\beta}_0||^2$. The value of c that minimizes Eq. (2.4.5) depends on $\boldsymbol{\beta}$ and is given by

$$
c^0(\boldsymbol{\beta}) = (K_N(p))^{1/2}e^{-\delta_N^2(\boldsymbol{\beta}_0)/2}. \tag{2.4.6}
$$

Thus, if we believe that $\boldsymbol{\beta}_0$ is close to $\boldsymbol{\beta}$, a good strategy will be to use the predictor $\hat{\eta}_{c^0}(\boldsymbol{\beta}_0)$, where $c^0 = (K_N(p))^{1/2}$.

Corollary 2.4.1. *The risk function $R_\psi[\hat{\eta}_c;\eta]$ of a shrinkage estimator of $\mathbf{1}_N'\mathbf{X}\boldsymbol{\beta}$ toward zero, when $c = \sqrt{K_N(p)}$ is,*

$$
R_\psi[\hat{\eta}_c;\eta] = R_\psi[\hat{\eta};\eta] - K_N(p)[2e^{-\delta_N^2/2} - 1].
$$

Furthermore, $R_\psi[\hat{\eta}_c;\eta] < R_\psi[\hat{\eta};\eta]$ if, and only if, $\delta_N^2 < \ln 4$.

The BUP of T is $\hat{T}_{\mathrm{BU}} = \mathbf{1}_s'\mathbf{y}_s + \mathbf{1}_r'\mathbf{X}_r\hat{\boldsymbol{\beta}}_s$, because $\mathbf{W}$ is diagonal. This motivates us to consider the shrinkage predictor (Rodrigues, 1987)

$$
\hat{T}_c = (1 - \frac{c\hat{\sigma}_s}{|\mathbf{1}_r'\mathbf{X}_r\hat{\boldsymbol{\beta}}_s|})(\hat{T}_{\mathrm{BU}} - \mathbf{1}_s'\mathbf{y}_s) + \mathbf{1}_s'\mathbf{y}_s. \tag{2.4.7}
$$

Theorem 2.4.2. *The prediction risk of $\hat{T}_{c^*}$ with $c^* = (K_r(p))^{1/2}$ is*

$$R_\psi[\hat{T}_{c^*};T] = R_\psi[\hat{T}_{\mathrm{BU}};T] - K_r(p)(2e^{-\delta_r^2/2} - 1), \tag{2.4.8}$$

where

$$K_r(p) = \frac{4\Gamma^2(\frac{n-p+1}{2})\xi_r^2}{(n-p)\Gamma^2(\frac{n-p}{2})\pi},$$

$$\xi_r^2 = \mathbf{1}_r'\mathbf{X}_r(\mathbf{X}_s'\mathbf{W}_s^{-1}\mathbf{X}_s)^{-1}\mathbf{X}_r\mathbf{1}_r,$$

and

$$\delta_r^2 = (\mathbf{1}_r'\mathbf{X}_r\boldsymbol{\beta})^2/\sigma^2\xi_r^2.$$

Moreover, $R_\psi[\hat{T}_{c^*};T] \le R_\psi[\hat{T}_{\mathrm{BU}};T]$ if, and only if, $\delta_r^2 \le \ln 4$.

Proof. Consider the population quantity $\eta_r = \mathbf{1}_r'\mathbf{X}_r\boldsymbol{\beta}$ and the shrinkage predictor $\hat{\eta}_{r,c^*}(\mathbf{0})$, which is obtained from Eq. (2.4.1) by substituting $\boldsymbol{\beta}_0 = \mathbf{0}$, $c = c^*$, and replacing $\mathbf{1}_N'\mathbf{X}$ by $\mathbf{1}_r'\mathbf{X}_r$. Notice that $\hat{\eta}_{r,0}(\mathbf{0}) = \mathbf{1}_r'\mathbf{X}_r\hat{\boldsymbol{\beta}}_s = \hat{T}_{\mathrm{BU}} - \mathbf{1}_s'\mathbf{y}_s$.

(Q.E.D.)

2.5. Exercises

[2.1] Prove Corollary 2.1.1.

[2.2] Derive the prediction variance (2.1.10).

[2.3] Prove Corollary 2.2.1.

[2.4] Consider the regression model $\psi = (\boldsymbol{\beta}, \mathbf{V})$, where $\mathbf{V} = \sigma^2\mathbf{W}$, with σ^2 unknown. Show that an unbiased estimator of the prediction variance (2.1.6) is obtained by replacing $\mathbf{V}$ by $\hat{\sigma}^2\mathbf{W}$ where

$$\hat{\sigma}_s^2 = \frac{1}{n-p}(\mathbf{y}_s - \mathbf{X}_s\hat{\boldsymbol{\beta}}_s)'\mathbf{W}_s^{-1}(\mathbf{y}_s - \mathbf{X}_s\hat{\boldsymbol{\beta}}_s). \tag{2.5.1}$$

[2.5] Suppose that $y_1, \ldots, y_N$ are independent and distributed according to a common exponential distribution with density

$$e^{(y-\theta)}, \ y \ge \theta.$$

Prove that the ψ-BUP of T is $\hat{T}_{\mathrm{BUP}} = n\bar{y}_s + (N-n)[y_{s(1)} + 1 - \frac{1}{n}]$.

[2.6] Suppose that $y_1, \ldots, y_N$ are independent and distributed according to the uniform distribution $U(0, \theta)$. Find the ψ-BUP of T.

[2.7] Consider the superpopulation model SM3 with $g = 0$ under normality. The ψ-BUP of T is derived in Example 2.2.3.

(i) Show that

$$\mathrm{Var}_\psi[\hat{T}_{RE} - T] = N\frac{(1-f)}{f}\sigma^2[1 + n(1-f)\frac{(\bar{x}_r - \bar{x}_s)^2}{\sum_{i\in s}(x_i - \bar{x}_s)^2}], \tag{2.5.2}$$

where $f = n/N$.

(ii) Show that an unbiased estimator of Eq. (2.5.2) is obtained by replacing σ^2 by

$$\hat{\sigma}_s^2 = \frac{1}{n-2}\sum_{i\in s}(y_i - \hat{\beta}_0 - \hat{\beta}_1 x_i)^2,$$

where $\hat{\beta}_o = \bar{y}_s - \bar{x}_s\hat{\beta}_1$ and $\hat{\beta}_1$ is given in Example 2.2.3.

[2.8] (Tam, 1987a,b) Consider the superpopulation model (2.2.3), with a general symmetric positive-definite covariance matrix. Show that the ψ-BUP of T is given by formula (2.1.5).

[2.9] (Rodrigues and Elian, 1989) Consider the regression model $\psi = (\boldsymbol{\beta}, \mathbf{V})$ where $\mathbf{V}$ is a known symmetric and positive–definite covariance matrix. Consider the linear predictors

$$\hat{T}_L^{(j)} = \mathbf{1}_s'\mathbf{y}_s + \mathbf{1}_{sr}^{(j)\prime}\mathbf{y}_s,$$

$j = 1, 2$ and any noninformative sampling plan p. Show that

$$E_\psi E_p[\hat{T}_L^{(1)} - T]^2 \le E_\psi E_p[\hat{T}_L^{(2)} - T]^2$$

if, and only if,

$$E_\psi\left[(\mathbf{1}_{sr}^{(1)} - \mathbf{V}_s^{-1}\mathbf{V}_{sr}\mathbf{1}_r)'\mathbf{y}_s - (\mathbf{1}_r - \mathbf{V}_s^{-1}\mathbf{V}_{sr}\mathbf{1}_r)'\mathbf{X}_s\boldsymbol{\beta}\right]^2$$
$$\le E_\psi\left[(\mathbf{1}_{sr}^{(2)} - \mathbf{V}_s^{-1}\mathbf{V}_{sr}\mathbf{1}_r)'\mathbf{y}_s - (\mathbf{1}_r - \mathbf{V}_s^{-1}\mathbf{V}_{sr}\mathbf{1}_r)'\mathbf{X}_s\boldsymbol{\beta}\right]^2.$$

[2.10] (Rodrigues and Elian, 1989) Use the Gauss–Markov theorem and Exercise 2.9 to prove that the ψ-BLUP of T and its prediction variance are the ones given in Eqs. (2.1.5) and (2.1.6). (This is an alternative proof to the one given in the text).

[2.11] Consider the superpopulation model ψ, where $\mathbf{X} = (x_1, \ldots, x_N)'$ and $\mathbf{V} = \sigma^2\mathbf{I}_N$, with normally distributed errors.
(i) Find the ψ-BUP of T and its prediction variance.
(ii) Find an unbiased estimator of the prediction variance in (i).
(iii) Find the ψ-BUP of S_y^2 and its prediction variance.
(iv) Find the ψ-BUP of β_N and its ψ-GMSE.

[2.12] Consider the superpopulation model SM4, with the additional assumption that

$$y_{hj}|\beta_h \sim N(x_{hj}\beta_h, \sigma_h^2 x_{hj})$$

for $j = 1, \ldots, N_h$ and $h = 1, \ldots, K$. All the random variables $\{y_{hj}\}$ are independent.
(i) Find the ψ-BUP of T and its prediction variance.
(ii) Find the ψ-BUP of S_y^2 and its prediction variance.

[2.13] Consider model SM6.
(i) What conditions should be imposed so that condition L would be satisfied?
(ii) What is the BLUP of T when condition L is satisfied?
(iii) What is the prediction variance of the predictor derived in (ii).

[2.14] Consider model SM2 with errors normally distributed. Find the ψ-BUP of β_N and its GMSE for $\lambda = 1$.

[2.15] Consider the superpopulation model SM1, with the errors normally distributed, where σ^2 is known. The ψ-BUP of S_y^2, namely $\hat{S}^2_{\text{BU}}$, is given in Eq. (2.2.8). Show that

$$E_\psi[\hat{S}^2_{BU} - S_y^2]^2 = \frac{2\sigma^4(N-n)}{N^2}.$$

[2.16] Consider the superpopulation model SM1, with errors normally-distributed and σ^2 unknown. The ψ-BUP of S_y^2 is given in Eq. (2.2.9). Find its prediction variance.

[2.17] (Skiener, 1983) Consider the stratified superpopulation model which is specified in Exercise 1.13, with normally distributed errors. Find the ψ-BUP of S_y^2 and its prediction variance.

[2.18] Consider the superpopulation model of Exercise 2.17, where σ_h^2 is known, $h = 1, \ldots, K$. Find the MREP of T.

[2.19] Prove Theorems 2.3.1, 2.3.3, 2.3.5, and 2.3.6.

[2.20] Derive the Pitman scale predictor of T given in Section 2.3.4.

[2.21] Consider the stratified location model of Exercise 1.13, with a distribution F for the error term, which depends on location parameters, μ_h, $h = 1, \ldots, k$, with $\sigma_h^2 = 1$ for all h. BLUP of T is considered in Exercise 1.13. Show that this predictor is an MREP if, and only if, F is a normal distribution. Derive the Pitman predictor of T with respect to the squared-error loss.

[2.22] Consider the normal model of Example 2.3.8. Show that the MSE of the predictor $\hat{S}^2_{cy}$, defined in Eq. (2.3.41), is given by Eq. (2.3.42). Show that this MSE is minimized by taking $c = n/(n+1)$.

[2.23] Consider the exponential location scale superpopulation model of Example 2.3.9. Derive the formula for S^2_{BUP} and find its prediction MSE.

[2.24] Consider model SM1 with $\sigma = 1$ and normally distributed errors. Show that the BUP of the empirical cumulative distribution function (CDF), $F_N(t)$ (Exercise 1.14), is

$$F_{BU}(t) = \frac{n}{N}F_s(t) + (1 - \frac{n}{N})\Phi\left(\sqrt{\frac{n}{n-1}}(t - \bar{y}_s)\right), \tag{2.5.3}$$

where $\Phi(\cdot)$ is the distribution function of $N(0,1)$.

[2.25] (Bolfarine and Sandoval, 1990) Consider model SM2 with normally distributed errors and σ^2 known. Show that the BUP of $F_N(t)$ is given by

$$\begin{aligned}\hat{F}_{BUN}(t) &= \frac{n}{N}F_s(t)\\ &+ \left(1-\frac{n}{N}\right)\sum_{i\notin s}\Phi\left(\frac{t-x_i\hat{\beta}_s}{\sigma\left[x_i\left(1-x_i\Big/\sum_{j\in s}x_j\right)\right]^{1/2}}\right)\end{aligned} \tag{2.5.4}$$

Derive an approximation for the prediction variance of $\hat{F}_{BUN}(t)$.

[2.26] Derive expressions (2.2.13), (2.2.14), and (2.2.15).

[2.27] Consider the regression model $\boldsymbol{\psi} = \boldsymbol{\psi}(\boldsymbol{\beta}, \mathbf{V})$, with $\mathbf{V}_{rs} = \mathbf{0}$. Prove that if condition L is satisfied, then $\mathbf{1}_s'(\mathbf{y}_s - \mathbf{X}_s\hat{\boldsymbol{\beta}}_s) = 0$.

[2.28] Derive the prediction risk of the shrinkage predictor of T, given by Eq. (2.4.7), in the case of models SM1 and SM2 under normality. Compute the relative prediction risk efficiency of these predictors compared to that of the BUP.

3
Bayes and Minimax Predictors

Various superpopulation models have been introduced in the previous two chapters. Most of the models considered assumed that the joint distribution of $\mathbf{y}$ belongs to a specified parametric family $\mathcal{F}$. We also considered distribution-free models, which assumed only the existence of finite first two moments, as in the case of models SM1–SM6.

A purely Bayesian model for a fixed finite population framework, assumes a specific (prior) distribution of $\mathbf{y}$ and can be considered as a superpopulation model with a single element of $\mathcal{F}$. In this book we deal with families $\mathcal{F}$ consisting of more than one element and focus attention on parametric models. Thus, generally, let $\mathcal{F} = \{F_{\boldsymbol{\psi}}; \boldsymbol{\psi} \in \boldsymbol{\Psi}\}$, where $\boldsymbol{\psi}$ is a parameter in a specified parametric space $\boldsymbol{\Psi}$. For example, in the linear regression models with normality assumption, $F_\theta = \text{Normal}_{\boldsymbol{\psi}}$, where $\boldsymbol{\psi} = (\boldsymbol{\beta}, \mathbf{V})$. The Bayes superpopulation model imposes on the family $\mathcal{F} = \{F_{\boldsymbol{\psi}}; \boldsymbol{\psi} \in \boldsymbol{\Psi}\}$ a prior distribution for $\boldsymbol{\psi}$, $\zeta(\boldsymbol{\psi}|\phi)$, where ϕ is known. Cassel et al. (1977, p. 135) call this model, "case B" of the Bayes model.

We start in Section 3.1 to develop Bayes predictors for the normal regression models. In Section 3.2, we consider Bayes linear predictors for the general regression models (not necessarily normal). Section 3.3 presents the theory of minimax and admissible predictors. Section 3.4 deals with (time) sequential dynamic models, including adaptations of the Kalman filter to prediction of finite population quantities. We conclude in Section 3.5 by presenting the theory of empirical Bayes predictors.

3.1. The Multivariate Normal Model

Consider the superpopulation regression model $\boldsymbol{\psi} = (\boldsymbol{\beta}, \mathbf{V})$ with normal distribution for the errors. Accordingly, the joint distribution of $\mathbf{y}$ given $\boldsymbol{\psi}$

is the N–variate normal,

$$\mathbf{y}|\psi, \mathbf{X} \sim N(\mathbf{X}\boldsymbol{\beta}; \mathbf{V}). \tag{3.1.1}$$

We assume here that $\mathbf{V}$ is known and $\boldsymbol{\beta}$ is unknown. The Bayes model assumes that $\boldsymbol{\beta}$ is a normal random vector, with mean vector $\mathbf{b}$ and covariance matrix $\mathbf{B}$; that is,

$$\boldsymbol{\beta} \sim N(\mathbf{b}; \mathbf{B}). \tag{3.1.2}$$

The model defined by Eqs. (3.1.1) and (3.1.2) is designated in the sequel as the ψ_B model. The next theorem specifies the Bayes predictive distribution of $\mathbf{y}_r$ given $\mathbf{y}_s$, for the case where the covariance matrix $\mathbf{V}$ is known (Bolfarine et al., 1987).

Theorem 3.1.1. *Under the Bayesian model ψ_B, the Bayes predictive distribution of $\mathbf{y}_r$ given $\mathbf{y}_s$ is multivariate normal, with mean vector*

$$E_{\psi_B}[\mathbf{y}_r|\mathbf{y}_s] = \mathbf{X}_r\hat{\boldsymbol{\beta}}_B + \mathbf{V}_{rs}\mathbf{V}_s^{-1}(\mathbf{y}_s - \mathbf{X}_s\hat{\boldsymbol{\beta}}_B), \tag{3.1.3}$$

and covariance matrix

$$\begin{aligned} Var_{\psi_B}[\mathbf{y}_r|\mathbf{y}_s] = \mathbf{V}_r - \mathbf{V}_{rs}\mathbf{V}_s^{-1}\mathbf{V}_{sr} + (\mathbf{X}_r - \mathbf{V}_{rs}\mathbf{V}_s^{-1}\mathbf{X}_s) \\ (\mathbf{X}_s'\mathbf{V}_s^{-1}\mathbf{X}_s + \mathbf{B}^{-1})^{-1}(\mathbf{X}_r - \mathbf{V}_{rs}\mathbf{V}_s^{-1}\mathbf{X}_s)', \end{aligned} \tag{3.1.4}$$

where

$$\hat{\boldsymbol{\beta}}_B = (\mathbf{X}_s'\mathbf{V}_s^{-1}\mathbf{X}_s + \mathbf{B}^{-1})^{-1}(\mathbf{X}_s'\mathbf{V}_s^{-1}\mathbf{y}_s + \mathbf{B}^{-1}\mathbf{b}). \tag{3.1.5}$$

Proof. Since $\mathbf{V}$ is known, a complete and sufficient statistics for the model (3.1.1) is the WLSE $\hat{\boldsymbol{\beta}}_s$ given in Eq. (1.3.14). Thus, $\hat{\boldsymbol{\beta}}_s$ is also Bayes sufficient for $\boldsymbol{\beta}$ (Zacks, 1971, p. 90), and the posterior distribution of $\boldsymbol{\beta}$ given $\mathbf{y}_s$ is the same as that of $\boldsymbol{\beta}$ given $\hat{\boldsymbol{\beta}}_s$. Furthermore, since,

$$\hat{\boldsymbol{\beta}}_s|\boldsymbol{\beta} \sim N[\boldsymbol{\beta}, (\mathbf{X}_s'\mathbf{V}_s^{-1}\mathbf{X}_s)^{-1}],$$

the joint distribution of $(\hat{\boldsymbol{\beta}}_s; \boldsymbol{\beta})$ is, as implied from Eqs. (3.1.2) and (3.1.3),

$$\begin{pmatrix} \hat{\boldsymbol{\beta}}_s \\ \boldsymbol{\beta} \end{pmatrix} \sim N\left[\begin{pmatrix} \mathbf{b} \\ \mathbf{b} \end{pmatrix}, \begin{pmatrix} (\mathbf{X}_s'\mathbf{V}_s^{-1}\mathbf{X}_s)^{-1} + \mathbf{B} & \mathbf{B} \\ \mathbf{B} & \mathbf{B} \end{pmatrix}\right].$$

Hence, the posterior distribution of $\boldsymbol{\beta}$, given $\hat{\boldsymbol{\beta}}_s$, is normal with mean

$$\hat{\boldsymbol{\beta}}_B = E_{\psi_B}[\boldsymbol{\beta}|\hat{\boldsymbol{\beta}}_s] = \mathbf{b} + \mathbf{B}[(\mathbf{X}_s'\mathbf{V}_s^{-1}\mathbf{X}_s)^{-1} + \mathbf{B}]^{-1}(\hat{\boldsymbol{\beta}}_s - \mathbf{b}),$$

and covariance matrix

$$\mathrm{Var}_{\psi_B}[\boldsymbol{\beta}|\hat{\boldsymbol{\beta}}_s] = \mathbf{B} - \mathbf{B}[(\mathbf{X}_s'\mathbf{V}_s^{-1}\mathbf{X}_s)^{-1} + \mathbf{B}]^{-1}\mathbf{B}. \tag{3.1.6}$$

Let $\hat{\boldsymbol{\beta}}_B = E_{\psi_B}[\boldsymbol{\beta}|\hat{\boldsymbol{\beta}}_s]$. Notice that $\hat{\boldsymbol{\beta}}_B$ is the Bayes estimator of $\boldsymbol{\beta}$ corresponding to the loss function $||\hat{\boldsymbol{\beta}}-\boldsymbol{\beta}||^2 = (\hat{\boldsymbol{\beta}}-\boldsymbol{\beta})'(\hat{\boldsymbol{\beta}}-\boldsymbol{\beta})$. Several algebraic manipulations (see Exercise 3.1) yield the formula (3.1.5). Furthermore, formula (3.1.6) is reduced to

$$\text{Var}_{\psi_B}[\boldsymbol{\beta}|\hat{\boldsymbol{\beta}}] = (\mathbf{X}_s'\mathbf{V}_s^{-1}\mathbf{X}_s + \mathbf{B}^{-1})^{-1} \tag{3.1.7}$$

(see Exercise 3.1). From the general theory of multivariate normal distribution, one can derive that the conditional distribution of $\mathbf{y}_r$ given $(\mathbf{y}_s, \boldsymbol{\beta})$ is multivariate normal with mean

$$E_{\psi}[\mathbf{y}_r|\mathbf{y}_s, \boldsymbol{\beta}] = \mathbf{X}_r\boldsymbol{\beta} + \mathbf{V}_{rs}\mathbf{V}_s^{-1}(\mathbf{y}_s - \mathbf{X}_s\boldsymbol{\beta}),$$

and covariance matrix

$$\text{Var}_{\psi}[\mathbf{y}_r|\mathbf{y}_s, \boldsymbol{\beta}] = \mathbf{V}_r - \mathbf{V}_{rs}\mathbf{V}_s^{-1}\mathbf{V}_{sr}.$$

Finally, combining all these results, the conditional distribution of $\mathbf{y}_r$ given $\hat{\boldsymbol{\beta}}_s$ is the multivariate normal distribution with mean

$$\begin{aligned} E_{\psi_B}[\mathbf{y}_r|\hat{\boldsymbol{\beta}}_s] =& E_{\psi_B}\{E_{\psi}[\mathbf{y}_r|\mathbf{y}_s, \boldsymbol{\beta}]|\hat{\boldsymbol{\beta}}_s\} \\ =& \mathbf{X}_r E_{\psi_B}[\boldsymbol{\beta}|\hat{\boldsymbol{\beta}}_s] + \mathbf{V}_{rs}\mathbf{V}_s^{-1}(\mathbf{y}_s - \mathbf{X}_s E_{\psi_B}[\boldsymbol{\beta}|\hat{\boldsymbol{\beta}}_s]). \end{aligned} \tag{3.1.8}$$

Notice that Eq. (3.1.8) is equivalent to Eq. (3.1.3). Notice also that $E_{\psi_B}[\boldsymbol{\beta}|\mathbf{y}_s] = E_{\psi_B}[\boldsymbol{\beta}|\hat{\boldsymbol{\beta}}_s]$, and therefore

$$E_{\psi_B}[\mathbf{y}_r|\mathbf{y}_s] = E_{\psi_B}[\mathbf{y}_r|\hat{\boldsymbol{\beta}}_s].$$

Similarly, since the distribution of $\mathbf{y}_r$ given $\mathbf{y}_s$ is the same as the distribution of $\mathbf{y}_r$ given $\hat{\boldsymbol{\beta}}_s$, it follows that

(3.1.9)

$$\begin{aligned} \text{Var}_{\psi_B}[\mathbf{y}_r|\mathbf{y}_s] &= \text{Var}_{\psi_B}[\mathbf{y}_r|\hat{\boldsymbol{\beta}}_s] \\ &= E_{\psi_B}\{\text{Var}_{\psi}[\mathbf{y}_r|\mathbf{y}_s, \boldsymbol{\beta}]|\hat{\boldsymbol{\beta}}_s\} + \text{Var}_{\psi_B}\{E_{\psi}[\mathbf{y}_r|\mathbf{y}_s, \boldsymbol{\beta}]|\hat{\boldsymbol{\beta}}_s\} \\ &= (\mathbf{V}_r - \mathbf{V}_{rs}\mathbf{V}_s^{-1}\mathbf{V}_{sr}) + \text{Var}_{\psi_B}[(\mathbf{X}_r - \mathbf{V}_{rs}\mathbf{V}_{rs}\mathbf{V}_s^{-1}\mathbf{X}_s)\boldsymbol{\beta}|\hat{\boldsymbol{\beta}}_s] \\ &= (\mathbf{V}_r - \mathbf{V}_{rs}\mathbf{V}_s^{-1}\mathbf{V}_{sr}) + (\mathbf{X}_r - \mathbf{V}_{rs}\mathbf{V}_s^{-1}\mathbf{X}_s) \\ &\quad (\mathbf{X}_s'\mathbf{V}_s^{-1}\mathbf{X}_s + \mathbf{B}^{-1})^{-1}(\mathbf{X}_r - \mathbf{V}_{rs}\mathbf{V}_s^{-1}\mathbf{X}_s)'. \end{aligned}$$

(Q.E.D.)

A result of Royall and Pfeffermann (1982) is a special case of the above theorem, for a noninformative prior, obtained as the limit of $N(\mathbf{b}, \mathbf{B})$ when $\mathbf{B}^{-1} \to \mathbf{0}$.

Theorem 3.1.2. *Consider the normal model ψ_B with $\mathbf{V} = \sigma^2\mathbf{W}$, where $\mathbf{W}$ is known and $\mathbf{W}_{rs} = \mathbf{0}$, but σ^2 is unknown. We also consider noninformative prior distribution on $(\boldsymbol{\beta}; \sigma^2)$, according to which,*

$$\zeta(\boldsymbol{\beta}; \sigma^2) \propto \frac{1}{\sigma}. \tag{3.1.10}$$

The posterior distribution of $\mathbf{y}_r$ given $\mathbf{y}_s$ is such that

$$E_{\psi_B}[\mathbf{y}_r|\mathbf{y}_s] = \mathbf{X}_r\hat{\boldsymbol{\beta}}_s \tag{3.1.11}$$

and

$$Var_{\psi_B}[\mathbf{y}_r|\mathbf{y}_s] = \frac{\nu}{\nu-2}\hat{\sigma}_s^2[\mathbf{W}_r + \mathbf{X}_r(\mathbf{X}_s'\mathbf{W}_s^{-1}\mathbf{X}_s)^{-1}\mathbf{X}_r'],$$

where

$$\hat{\sigma}_s^2 = (\mathbf{y}_s - \mathbf{X}_s\hat{\boldsymbol{\beta}}_s)'\mathbf{W}_s^{-1}(\mathbf{y}_s - \mathbf{X}_s\hat{\boldsymbol{\beta}}_s)/\nu, \tag{3.1.12}$$

with $\nu = n - p$.

Proof. From the previous theorem, with $\mathbf{B}^{-1}$ replaced by $\mathbf{0}$, for noninformative prior for $\boldsymbol{\beta}$, with a given σ, we obtain Eq. (3.1.11) and

$$\text{Var}_{\psi_B}[\mathbf{y}_r|\mathbf{y}_s, \sigma] = \sigma^2[\mathbf{W}_r + \mathbf{X}_r(\mathbf{X}_s'\mathbf{W}_s^{-1}\mathbf{X}_s)^{-1}\mathbf{X}_r'].$$

Furthermore, since $E_\psi[\mathbf{y}_r \mid \mathbf{y}_s, \sigma]$ is independent of σ,

$$\begin{aligned}\text{Var}_{\psi_B}[\mathbf{y}_r|\mathbf{y}_s] &= E_{\psi_B}[\text{Var}_\psi[\mathbf{y}_r|\mathbf{y}_s,\sigma]|\mathbf{y}_s] + \text{Var}_{\psi_B}[E_\psi[\mathbf{y}_r|\mathbf{y}_s,\sigma]|\mathbf{y}_s] \\ &= E_{\psi_B}[\sigma^2|\mathbf{y}_s][\mathbf{W}_r + \mathbf{X}_r(\mathbf{X}_s'\mathbf{W}_s^{-1}\mathbf{X}_s)^{-1}\mathbf{X}_r'].\end{aligned}$$

Finally, as required to prove in Exercise 3.2,

$$E_{\psi_B}[\sigma^2|\mathbf{y}_s] = \frac{\nu}{\nu-2}\hat{\sigma}^2.$$

(Q.E.D.)

In analogy to the notion of prediction MSE, we define, in the Bayesian framework, the notion of Bayes prediction risk. Let $\theta(\mathbf{y})$ be a population quantity of interest. Let $L(\hat{\theta}(\mathbf{y}_s), \theta(\mathbf{y}))$ be a loss function for predicting $\theta(\mathbf{y})$ by $\hat{\theta}(\mathbf{y}_s)$. For a Bayes model ψ_B, the *Bayes prediction* risk of $\hat{\theta}(\mathbf{y}_s)$ is defined as

$$E_{\psi_B}[L(\hat{\theta}(\mathbf{y}_s), \theta(\mathbf{y}))].$$

Notice that the expectation operator in the above expression is performed with respect to the joint distribution of $\mathbf{y}$ and ψ. In particular, for the squared-error loss function (the Bayesian analogue of the prediction MSE), the Bayes predictor of $\theta(\mathbf{y})$ is

$$\hat{\theta}_B(\mathbf{y}_s) = E_{\psi_B}[\theta(\mathbf{y})|\mathbf{y}_s], \tag{3.1.13}$$

and the Bayes prediction risk is

$$E_{\psi_B}[\hat{\theta}_B(\mathbf{y}_s) - \theta(\mathbf{y})]^2 = E_{\psi_B}[\text{Var}_{\psi_B}[\theta(\mathbf{y})|\mathbf{y}_s]]. \tag{3.1.14}$$

3.1.1. Bayes Predictors of T

The next theorem follows immediately from Eqs. (3.1.13) and (3.1.14).

Theorem 3.1.3. *For any linear quantity $\theta_L = \mathbf{l}'\mathbf{y}$, the Bayes predictor under the squared-error loss and any model ψ_B for which $\mathrm{Var}_{\psi_B}[\mathbf{y}_r|\mathbf{y}_s]$ exists, is*

$$\hat{\theta}_{BL}(\mathbf{y}_s) = \mathbf{l}_s'\mathbf{y}_s + \mathbf{l}_r' E_{\psi_B}[\mathbf{y}_r|\mathbf{y}_s].$$

The Bayes prediction risk of this predictor is

$$E_{\psi_B}[\hat{\theta}_{BL}(\mathbf{y}_s) - \theta_L(\mathbf{y})]^2 = \mathbf{l}_r' E_{\psi_B}[\mathrm{Var}_{\psi_B}[\mathbf{y}_r|\mathbf{y}_s]]\mathbf{l}_r.$$

Corollary 3.1.1. *The Bayes predictor of the population total $T(\mathbf{y})$ under the normal regression model (3.1.1)–(3.1.2) and squared-error loss function is*

$$\hat{T}_B(\mathbf{y}_s) = \mathbf{1}_s'\mathbf{y}_s + \mathbf{1}_r'[\mathbf{X}_r\hat{\boldsymbol{\beta}}_B + \mathbf{V}_{rs}\mathbf{V}_s^{-1}(\mathbf{y}_s - \mathbf{X}_s\hat{\boldsymbol{\beta}}_B)]. \tag{3.1.15}$$

The Bayes prediction risk of $\hat{T}_B$ is

$$\begin{aligned} E_{\psi_B}[\hat{T}_B(\mathbf{y}_s) - T(\mathbf{y})]^2 &= \mathbf{1}_r'(\mathbf{V}_r - \mathbf{V}_{rs}\mathbf{V}_s^{-1}\mathbf{V}_{rs})\mathbf{1}_r \\ &+\mathbf{1}_r'(\mathbf{X}_r - \mathbf{V}_{rs}\mathbf{V}_s^{-1}\mathbf{X}_s)(\mathbf{X}_s'\mathbf{V}_s^{-1}\mathbf{X}_s + \mathbf{B}^{-1})^{-1} \\ &(\mathbf{X}_r - \mathbf{V}_{rs}\mathbf{V}_s^{-1}\mathbf{X}_s)'\mathbf{1}_r. \end{aligned} \tag{3.1.16}$$

Corollary 3.1.2. *The Bayes predictor of T under the Bayes model of Theorem 3.1.2 is given by*

$$\hat{T}_B = \mathbf{1}_s'\mathbf{y}_s + \mathbf{1}_r'\mathbf{X}_r\hat{\boldsymbol{\beta}}_s.$$

The Bayes prediction risk is

$$E_{\psi_B}[\hat{T}_B - T]^2 = \frac{\nu}{\nu - 2}\hat{\sigma}^2\mathbf{1}_r'\left[\mathbf{W}_r + \mathbf{X}_r(\mathbf{X}_s'\mathbf{W}_s^{-1}\mathbf{X}_s)^{-1}\mathbf{X}_r'\right]\mathbf{1}_r.$$

Example 3.1.1. Consider model SM2 with the additional assumption of normal distribution for errors and known σ^2. In this case, β is a scalar having a prior $N(b, B)$ distribution. The Bayes estimator of β is

$$\hat{\beta}_B = \left(\frac{\sum_{i\in s} x_i}{\sigma^2} + \frac{1}{B}\right)^{-1}\left(\frac{\sum_{i\in s} y_i}{\sigma^2} + \frac{b}{B}\right).$$

The Bayes predictor of T, according to Eq. (3.1.15) is

$$\hat{T}_B = \sum_{i\in s} y_i + \hat{\beta}_B \sum_{i\in s} x_i.$$

The Bayes prediction risk of this predictor follows from Eq. (3.1.16) and is given by

$$E_{\psi_B}[\hat{T}_B - T]^2 = \sigma^2(N-n)\bar{x}_r \frac{N\bar{x} + \sigma^2/B}{n\bar{x}_s + \sigma^2/B}.$$

The corresponding formulae for the noninformative case are obtained as limits of the above, as $B \to \infty$. In this case, $\hat{T}_B \to \hat{T}_R$, the ratio estimator. Notice that if the Bayes prediction risk is regarded as a good measure of uncertainty, a Bayesian should maximize $\bar{x}_s$, by properly choosing the sample.

Furthermore, if σ^2 is unknown and noninformative prior (3.1.10) is considered for $(\boldsymbol{\beta}, \sigma^2)$, then, it follows from Corollary 3.1.2 that $\hat{T}_R$ is the Bayes predictor of T. The Bayes prediction risk is given by

$$E_{\psi_B}[\hat{T}_R - T]^2 = \frac{n-1}{n-3}\hat{\sigma}_s^2 \frac{\sum_{i=1}^N x_i \sum_{i \in \mathbf{r}} x_i}{\sum_{i \in \mathbf{s}} x_i},$$

where $\hat{\sigma}_s^2$ is as in Eq. (2.2.11).

Example 3.1.2. Consider model SM6 with the additional assumption of normality. Accordingly,

$$\mathbf{y} \sim N(\mu \mathbf{1}_N, \mathbf{V}).$$

In the Bayesian framework, let $\mu \sim N(\mu_0, \tau^2)$. We assume here that the parameters σ_h^2 and σ_v^2 are known. After the cluster sample $\mathbf{s}$ and $\{s_j,\ j = 1, \ldots, k\}$ have been selected, we rearrange $\mathbf{V}$ and make the partition

$$\mathbf{V} = \begin{pmatrix} \mathbf{V}_s & \mathbf{V}_{sr} \\ \mathbf{V}_{rs} & \mathbf{V}_r \end{pmatrix},$$

where $\mathbf{V}_s$, $\mathbf{V}_{sr}$, $\mathbf{V}_{rs}$, and $\mathbf{V}_r$ are specified in Example 2.1.2. Applying the derivations in Example 2.1.2 we obtain

$$\mathbf{X}_s'\mathbf{V}_s^{-1}\mathbf{X}_s = \frac{1}{\sigma_v^2}\sum_{j=1}^k w_j$$

and

$$\mathbf{X}_s'\mathbf{V}_s^{-1}\mathbf{y}_s = \frac{1}{\sigma_v^2}\sum_{j=1}^k w_j \bar{y}_{n_j},$$

where w_j is as given in Eq. (2.1.8). Thus, according to Eq. (3.1.5), the Bayes estimator of μ, given $\mathbf{y}_s$, is

$$\hat{\mu}_B = \frac{\tau^2 \sum_{j=1}^k w_j \bar{y}_{n_j} + \sigma_v^2 \mu_0}{\tau^2 \sum_{j=1}^k w_j + \sigma_v^2}. \tag{3.1.17}$$

In particular, if we let $\tau^2 \to \infty$ then $\hat{\mu}_B \to \hat{\mu}_s$, which is the LSE of μ given by Eq. (2.1.8).

The Bayes predictor of the population total T, $\hat{T}_B$, can be obtained from Eq. (2.1.9) by replacing $\hat{\mu}_s$ by $\hat{\mu}_B$. In the limiting case, when $\tau^2 \to \infty$, $\hat{T}_B \to \hat{T}_{\text{BLU}}$.

The prediction variance can be obtained as in Example 2.1.2. Notice that Theorem 3.1.2 cannot be applied as is, since V_{rs} is not a matrix of zeros.

If σ_h^2 and σ_v^2 are unknown, one could construct consistent estimators of these parameters, as in Ghosh and Meeden (1986). Model SM6 can be considered as a random effect model in which, for each h, $h = 1, \dots, K$, $y_{h_j} \mid \mu_h \sim N(\mu_h, \sigma_h^2)$ and $\mu_1, \dots, \mu_K$ are i.i.d. random variables having a common $N(\mu, \sigma_v^2)$ distribution. The marginal distribution of $\mathbf{y}$ is $N(\mu \mathbf{1}_N, \mathbf{V})$. There is vast literature on the estimation of the variance components σ_h^2 and σ_v^2. The reader is referred to Searle (1971).

3.1.2. Bayes Predictors of $\boldsymbol{\beta}_N$

As seen in the previous chapters, if $\mathbf{V}_{sr} = \mathbf{0}$, one can write

$$\boldsymbol{\beta}_N = \mathbf{E}_s \hat{\boldsymbol{\beta}}_s + \mathbf{E}_r \boldsymbol{\beta}_r.$$

We now extend formula (3.1.14). In analogy to the definition of the ψ–generalized prediction risk (or MSE), we define the ψ_B–generalized prediction risk of a predictor $\hat{\boldsymbol{\beta}}_N$ as

$$E_{\psi_B}[\mathbf{1}_p'(\hat{\boldsymbol{\beta}}_N - \boldsymbol{\beta}_N)(\hat{\boldsymbol{\beta}}_N - \boldsymbol{\beta}_N)'\mathbf{1}_p]. \tag{3.1.18}$$

This ψ_B–generalized prediction risk is minimized by

$$\hat{\boldsymbol{\beta}}_{BN} = E_{\psi_B}[\boldsymbol{\beta}_N | \mathbf{y}_s]. \tag{3.1.19}$$

Predictor $\hat{\boldsymbol{\beta}}_{BN}$ in Eq. (3.1.19) is the Bayes predictor of $\boldsymbol{\beta}_N$, with respect to expression (3.1.18). Its ψ_B–generalized prediction risk is given by

$$E_{\psi_B}[\mathbf{1}_p'(\hat{\boldsymbol{\beta}}_{BN} - \boldsymbol{\beta}_N)(\hat{\boldsymbol{\beta}}_{BN} - \boldsymbol{\beta}_N)'\mathbf{1}_p] = E_{\psi_B}[\mathbf{1}_p' \text{Var}_{\psi_B}[\boldsymbol{\beta}_N | \mathbf{y}_s] \mathbf{1}_p]. \tag{3.1.20}$$

In the next theorem, we consider Bayesian prediction of $\boldsymbol{\beta}_N$. The proof follows from Eqs. (3.1.19) and (3.1.20) and Theorem 3.1.1 (see Exercise 3.5).

Theorem 3.1.3. *Consider the Bayes normal model ψ_B of Theorem 3.1.1 with $\mathbf{V}_{sr} = \mathbf{0}$. The Bayes predictor of $\boldsymbol{\beta}_N$ with respect to the risk (3.1.18) is given by*

$$\hat{\boldsymbol{\beta}}_{BN} = E_{\psi_B}[\boldsymbol{\beta}_N | \mathbf{y}_s] = \mathbf{E}_s \hat{\boldsymbol{\beta}}_s + \mathbf{E}_r \hat{\boldsymbol{\beta}}_B. \tag{3.1.21}$$

The associated ψ_B–generalized prediction risk is

$$(3.1.22)\qquad \begin{aligned} &E_{\psi_B}[\mathbf{1}'_p(\hat{\boldsymbol{\beta}}_{BN} - \boldsymbol{\beta}_N)(\hat{\boldsymbol{\beta}}_{BN} - \boldsymbol{\beta}_N)'\mathbf{1}_p] \\ &= \mathbf{1}'_p(\mathbf{X}'\mathbf{V}^{-1}\mathbf{X})^{-1}\mathbf{X}'_r\mathbf{V}_r^{-1}\boldsymbol{\Sigma}_r\mathbf{V}_r^{-1}\mathbf{X}_r(\mathbf{X}'\mathbf{V}^{-1}\mathbf{X})^{-1}\mathbf{1}_p, \end{aligned}$$

where $\hat{\boldsymbol{\beta}}_B$ and $\boldsymbol{\Sigma}_r = Var_{\psi_B}[\mathbf{y}_r|\mathbf{y}_s]$ are given in (3.1.5) and (3.1.4), respectively.

Notice that $\hat{\boldsymbol{\beta}}_s$ is obtained as a limiting case of $\hat{\boldsymbol{\beta}}_{BN}$, when $\mathbf{B}^{-1} \to \mathbf{0}$ (noninformative prior on $\boldsymbol{\beta}$).

Example 3.1.3. Consider the ψ_B model of Example 3.1.1. It follows from Theorem 3.1.3 that the Bayes predictor of $\beta_N = \bar{y}/\bar{x}$ is

$$\hat{\beta}_{BN} = \frac{1}{\sum_{i=1}^N x_i}\left[\sum_{i\in\mathbf{s}} y_i + (\frac{\sum_{i\in\mathbf{s}} x_i}{\sigma^2} + \frac{1}{B})^{-1}(\frac{\sum_{i\in\mathbf{s}} y_i}{\sigma^2} + \frac{b}{B})\sum_{i\in\mathbf{r}} x_i\right].$$

The Bayes prediction risk is

$$E_{\psi_B}[\hat{\beta}_{BN} - \beta_N]^2 = \frac{\sum_{i\in\mathbf{r}} x_i}{(\sum_{i=1}^N x_i)^2}\left(1 + \frac{\sum_{i\in\mathbf{r}} x_i}{\sum x_i/\sigma^2 + 1/B}\right).$$

3.1.3. Bayes Predictors of S_y^2

We derive in the present section the general formula of the Bayes predictor of S_y^2, under the squared-error loss, for the normal regression model (3.1.1)–(3.1.2). Again, starting with the expression

$$S_y^2 = \frac{n}{N}s_y^2 + (1 - \frac{n}{N})[S_{ry}^2 + \frac{n}{N}(\bar{y}_s - \bar{y}_r)^2],$$

we derive first the Bayes predictor of $S_{ry}^2 + \frac{n}{N}(\bar{y}_s - \bar{y}_r)^2$, which is then substituted above. In the previous section, we showed that the Bayes predictive distribution of $\mathbf{y}_r$ given $\mathbf{y}_s$ for this model is normal with mean vector given by Eq. (3.1.3) and covariance matrix given by Eq. (3.1.4). Let $\boldsymbol{\eta}_r(\mathbf{y}_s) = E[\mathbf{y}_r|\mathbf{y}_s]$ as in Eq. (3.1.3) and $\boldsymbol{\Sigma}_r$ be the covariance matrix (3.1.4). The Bayes predictive distribution of $\bar{y}_r$ given $\mathbf{y}_s$ is normal with mean

$$h(\mathbf{y}_s) = \frac{1}{N-n}\mathbf{1}'_r\boldsymbol{\eta}_r(\mathbf{y}_s)$$

and variance

$$\mathbf{D}_r^2 = \frac{1}{(N-n)^2}\mathbf{1}'_r\boldsymbol{\Sigma}_r\mathbf{1}_r.$$

It follows that the Bayes predictive distribution of $\frac{n}{N}(\bar{y}_r - \bar{y}_s)^2$, given $\mathbf{y}_s$, is like that of

$$\frac{n}{N}D_r^2\chi^2[1;\lambda],$$

where

$$\lambda = \frac{[h(\mathbf{y}_s) - \bar{y}_s]^2}{2D_r^2}.$$

All these imply that,

$$E_{\psi_B}[\frac{n}{N}(\bar{y}_r - \bar{y}_s)^2|\mathbf{y}_s] = \frac{n}{N}\{D_r^2 + [h(\mathbf{y}_s) - \bar{y}_s]^2\}. \tag{3.1.23}$$

Using the theorem about the expected value of a symmetric quadratic form (see Lemma 1.3.2), we obtain

$$E_{\psi_B}[S_{ry}^2|\mathbf{y}_s] = tr\{\mathbf{A}_r\boldsymbol{\Sigma}_r\} + \boldsymbol{\eta}_r(\mathbf{y}_s)'\mathbf{A}_r\boldsymbol{\eta}_r(\mathbf{y}_s), \tag{3.1.24}$$

where

$$\mathbf{A}_r = \frac{1}{N-n}(\mathbf{I}_r - \frac{1}{N-n}\mathbf{J}_r),$$

$\mathbf{I}_r$ is the identity matrix of dimension $N-n$ and $\mathbf{J}_r = \mathbf{1}_r\mathbf{1}_r'$. Substituting into Eqs. (3.1.23) and (3.1.24) above, we obtain the following theorem.

Theorem 3.1.4. *The Bayes predictor of S_y^2 under the normal model ψ_B (3.1.1)–(3.1.2) and squared-error loss is*

$$\begin{aligned}\hat{S}_B^2 = E_{\psi_B}[S_y^2|\mathbf{y}_s] = \frac{n}{N}s_y^2 + (1-\frac{n}{N})\Big(& tr\{\mathbf{A}_r\boldsymbol{\Sigma}_r\} \\ & + \boldsymbol{\eta}_r(\mathbf{y}_s)'\mathbf{A}_r\boldsymbol{\eta}_r(\mathbf{y}_s) + \frac{n}{N}\{D_r^2 + [h(\mathbf{y}_s) - \bar{y}_s]^2\}\Big).\end{aligned} \tag{3.1.25}$$

Derivation of the corresponding Bayes risk is similar to that of Example 2.2.2. Theorems 3.1.2 and 3.1.4. can be employed to derive the Bayes predictor for the case of $\mathbf{V} = \sigma^2\mathbf{W}$, with σ^2 unknown, for the noninformative prior of $(\boldsymbol{\beta}, \sigma^2)$ and squared-error loss. This we obtain by replacing $\mathbf{V}$ by $\sigma^2\mathbf{W}$ in Eq. (3.1.25); taking the limit as $\mathbf{B}^{-1} \to \mathbf{0}$ and replacing σ^2 by $E_{\psi_B}[\sigma^2|\mathbf{y}_s] = \nu\hat{\sigma}_s^2/(\nu-2)$, with $\hat{\sigma}_s^2$ given in (3.1.12). Accordingly,

Theorem 3.1.5. *Under the normal model (3.1.2), with $\mathbf{V} = \sigma^2\mathbf{W}$ and noninformative prior on $(\boldsymbol{\beta}, \sigma^2)$, the Bayes predictor of S_y^2 is given by*

$$\hat{S}_B^2 = E_{\psi_B}\left[E_{\psi}[S_y^2|\sigma^2, \mathbf{y}_s]|\mathbf{y}_s\right],$$

where $E_{\psi}[S_y^2|\sigma^2, \mathbf{y}_s]$ follows from Eq. (3.1.25) by letting $\mathbf{V} = \sigma^2\mathbf{W}$ and $\mathbf{B}^{-1} \to \mathbf{0}$.

Example 3.1.3. Consider the location model SM1 with normality. In this case, $\mathbf{V} = \sigma^2\mathbf{I}_N$, where σ^2 is known, $\mathbf{X} = \mathbf{1}_N$, and β is a scalar with prior distribution $N(b, B)$. The Bayes estimator of β, $\hat{\beta}_B$ is given by

$$\hat{\beta}_B = \frac{\sum_{i\in\mathbf{s}} y_i/\sigma^2 + b/B}{n/\sigma^2 + 1/B}.$$

Thus, in this special case,

$$\boldsymbol{\eta}_r(\mathbf{y}_s) = \hat{\beta}_B \mathbf{1}_r$$

and

$$\boldsymbol{\Sigma}_r = \sigma^2(\mathbf{I}_r + \frac{1}{n+\sigma^2/B}\mathbf{J}_r).$$

Moreover, $h(\mathbf{y}_s) = \hat{\beta}_B$,

$$D_r^2 = \frac{\sigma^2}{N-n}(1+\frac{N-n}{n+\sigma^2/B}),\ tr\{\mathbf{A}_r\boldsymbol{\Sigma}_r\} = \sigma^2(1-\frac{1}{N-n})$$

and

$$\boldsymbol{\eta}_r(\mathbf{y}_s)'\mathbf{A}_r\boldsymbol{\eta}_r(\mathbf{y}_s) = 0.$$

Substituting all these terms into Eq. (3.1.25), we obtain the Bayes predictor of S_y^2,

$$\hat{S}_B^2 = \frac{n}{N}s_y^2 + (1-\frac{n}{N})\sigma^2\Big[1 - \frac{\sigma^2}{(N-n)B}(\frac{1}{n}-\frac{1}{N}) (1+\frac{\sigma^2}{nB})^{-1} + \frac{n}{N}(\frac{\hat{\beta}_B - \bar{y}_s}{\sigma})^2\Big]. \tag{3.1.26}$$

In the case of the noninformative prior, we obtain as a limiting case (when $B \to \infty$),

$$\hat{S}_B^2 = \frac{n}{N}s_y^2 + (1-\frac{n}{N})\sigma^2. \tag{3.1.27}$$

In addition, if σ^2 is also unknown, the noninformative prior (3.1.10) and Theorem 3.1.5 yield the Bayes predictor

$$\hat{S}_B^2 = \frac{N-3}{N}\frac{n}{n-3}s_y^2, \tag{3.1.28}$$

as derived by Ericson (1969) and Zacks and Solomon (1981).

Example 3.1.4. Consider the superpopulation model SM2, where σ^2 is known. It can be shown, after some algebraic manipulations, that the Bayes predictor of S_y^2 with respect to noninformative prior distribution on β is given by

$$\hat{S}_B^2 = \frac{n}{N}s_y^2 + (1-\frac{n}{N})\{(1-\frac{1}{N})\sigma^2\bar{x}_r + S_{rx}^2[\hat{\beta}_s^2 + V(\hat{\beta}_s)] + \frac{n}{N}[(\bar{y}_s - \hat{\beta}_s\bar{x}_r)^2 + \bar{x}_r^2 V(\hat{\beta}_s)]\}, \tag{3.1.29}$$

where

$$\hat{\beta}_s = \frac{\bar{y}_s}{\bar{x}_s} \text{ and } V(\hat{\beta}_s) = \frac{\sigma^2}{\sum_{i\in s} x_i}.$$

If σ^2 is unknown, Theorem 3.1.5 may be used to derive the Bayes predictor of S_y^2. It is given by Eq. (3.1.29) with σ^2 replaced by $(n-1)\hat{\sigma}_s^2/(n-3)$, where $\hat{\sigma}_s^2$ is as specified in Eq. (2.2.11).

Example 3.1.5. Consider the superpopulation model SM3, with $\beta_o = 0$, $g = 0$, and σ^2 known. Consider a noninformative prior on β. After some algebraic manipulations, it follows that the Bayes predictor of S_y^2 is given by

$$\begin{aligned} \hat{S}_B^2 =& \frac{n}{N}s_y^2 + (1-\frac{n}{N})\{(1-\frac{1}{N})\sigma^2 \\ &+S_{rx}^2[\hat{\beta}_s^2 + V(\hat{\beta}_s)] + \frac{n}{N}[(\bar{y}_s - \hat{\beta}_s\bar{x}_r)^2 + \bar{x}_r^2 V(\hat{\beta}_s)]\}, \end{aligned} \tag{3.1.30}$$

with

$$\hat{\beta}_s = \frac{\sum_{i\in s} x_i y_i}{\sum_{i\in s} x_i^2} \text{ and } V(\hat{\beta}_s) = \frac{\sigma^2}{\sum_{i\in s} x_i^2}. \tag{3.1.31}$$

If σ^2 is unknown, it follows from Theorem 3.1.5 that the Bayes predictor of S_y^2 is given by formula (3.1.30), with $\hat{\beta}_s$ and $V(\hat{\beta}_s)$ given in Eq. (3.1.31), but with σ^2 replaced by

$$\frac{n-1}{n-3}\hat{\sigma}_s^2, \text{ where } \hat{\sigma}_s^2 = \frac{1}{n-1}\sum_{i\in s}(y_i - \hat{\beta}_s x_i)^2.$$

3.2. Bayes Linear Predictors

In the previous section we studied the Bayes predictors of the population total, T, under the normal model, that is, where $\mathbf{y} \sim N(\mathbf{X}\boldsymbol{\beta}, \mathbf{V})$ and $\boldsymbol{\beta} \sim N(\mathbf{b}, \mathbf{B})$. In particular, for squared–error loss, we have shown that the Bayes predictor is a linear one, that is,

$$\hat{T}_B = \mathbf{1}_s'\mathbf{y}_s + \mathbf{1}_r'[\mathbf{X}_r\hat{\boldsymbol{\beta}}_B + \mathbf{V}_{rs}\mathbf{V}_s^{-1}(\mathbf{y}_s - \mathbf{X}_s\hat{\boldsymbol{\beta}}_B)].$$

In this section, we drop the assumption of normality and assume only that

$$E[\mathbf{y}|\boldsymbol{\beta}] = \mathbf{X}\boldsymbol{\beta}, \text{ Var}[\mathbf{y}|\boldsymbol{\beta}] = \mathbf{V}$$

and that

$$E[\boldsymbol{\beta}] = \mathbf{b}, \text{ Var}[\boldsymbol{\beta}] = \mathbf{B}.$$

The covariance matrix $\mathbf{V}$ is assumed to be known. This general model will be designated by ψ_G. We introduce now the notions of ψ_G-unbiasedness and ψ_G-MSE of predictors of T, as considered in Rodrigues (1988).

Definition 3.2.1. *A predictor $\hat{T}$ of T is called ψ_G-unbiased if*

$$E_{\psi_G}[\hat{T} - T] = 0 \quad \text{for all} \quad \psi_G.$$

Definition 3.2.2. *The ψ_G-MSE of a predictor $\hat{T}$ of T is*

$$MSE_{\psi_G}[\hat{T}] = E_{\psi_G}[\hat{T} - T]^2.$$

Example 3.2.1. Let $\mathbf{X} = (x_1, \ldots, x_N)'$ and $\mathbf{V} = \text{diag}\{x_1, \ldots, x_N\}$ as in SM2. In this model, β is a scalar with prior mean b and prior variance $B = \tau^2$. Under this model the ratio predictor is ψ_G-unbiased. Indeed, since

$$\hat{T}_R = n\bar{y}_s + \frac{\mathbf{1}_s'\mathbf{y}_s}{\mathbf{1}_s'\mathbf{X}_s}\mathbf{1}_r'\mathbf{X}_r,$$

it follows that

$$E_{\psi_G}[\hat{T}_R] = b\mathbf{1}_s'\mathbf{X}_s + b\mathbf{1}_r'\mathbf{X}_r = E_{\psi_G}[T].$$

Moreover, the ψ_G-MSE of the ratio predictor is

$$\begin{aligned}
\text{MSE}_{\psi_G}[\hat{T}_R] &= \text{Var}_{\psi_G}[\hat{T}_R - T] \\
&= E_{\psi_G}\left[\text{Var}_{\psi}[\mathbf{1}_r'\mathbf{X}_r\frac{\mathbf{1}_s'\mathbf{y}_s}{\mathbf{1}_s'\mathbf{X}_s} - \mathbf{1}_r'\mathbf{y}_r|\beta]\right] \\
&\quad + \text{Var}_{\psi_G}\left[E_{\psi}[\mathbf{1}_r'\mathbf{X}_r\frac{\mathbf{1}_s'\mathbf{y}_s}{\mathbf{1}_s'\mathbf{X}_s} - \mathbf{1}_r'\mathbf{y}_r|\beta]\right] \\
&= \sigma^2\frac{(\mathbf{1}_r'\mathbf{X}_r)^2}{\mathbf{1}_s'\mathbf{X}_s} + \sigma^2\mathbf{1}_r'\mathbf{X}_r \\
&= \sigma^2\frac{\mathbf{1}_r'\mathbf{X}_r\mathbf{1}_N'\mathbf{X}}{\mathbf{1}_s'\mathbf{X}_s}.
\end{aligned}$$

Definition 3.2.3. *A predictor $\hat{T} = \mathbf{1}_s'\mathbf{y}_s + \hat{T}_r(\mathbf{y}_s)$ is called ψ_G-linear if*

$$\hat{T}_r(\mathbf{y}_s) = a + \mathbf{l}_{sr}'\mathbf{y}_s.$$

Definition 3.2.4. *A ψ_G-linear predictor is called ψ_G-best if it minimizes the ψ_G-MSE.*

Example 3.2.2. Consider again the ψ_G-model of Example 3.1.1. Let $\hat{T}_{LG} = n\bar{y}_s + a + \mathbf{l}_{sr}'\mathbf{y}_s$ be any ψ_G-linear predictor. Since

$$E_{\psi_G}[\hat{T}_{LG} - T] = a + \mathbf{l}_{sr}'\mathbf{X}_s b - \mathbf{1}_r'\mathbf{X}_r b,$$

it follows that $\hat{T}_{LG}$ is ψ_G-unbiased if, and only if,

$$a + \mathbf{l}'_{sr}\mathbf{X}_s b - \mathbf{1}'_r\mathbf{X}_r b = 0.$$

Moreover, the ψ_G-MSE of any ψ_G-unbiased predictor $\hat{T}_{LG}$ may be written as

$$(3.2.1) \qquad E_{\psi_G}[\hat{T}_{LG} - T]^2 = (\mathbf{l}'_{sr}\mathbf{X}_s - \mathbf{1}'_r\mathbf{X}_r)^2\tau^2 + \mathbf{l}'_{sr}\mathbf{V}_s\mathbf{l}_{sr} + \sigma^2\sum_{i\in r} x_i,$$

where $\mathbf{V}_s = \sigma^2\text{diag}\{x_1, \ldots, x_N\}$.

Notice that if the normality assumption is added to the ψ_G-model (ψ_B-model of Example 3.1.1), then, any ψ_G linear-unbiased predictor is also ψ_B-unbiased. Moreover, it is not difficult to see that

$$E_{\psi_B}[\hat{T}_{LG} - T]^2 = E_{\psi_G}[\hat{T}_{LG} - T]^2,$$

which is given in Eq. (3.2.1). Let $\hat{T}_B$ be defined as in Example 3.1.1. Thus,

$$E_{\psi_G}[\hat{T}_B - T]^2 = E_{\psi_B}[\hat{T}_B - T]^2 \leq E_{\psi_B}[\hat{T}_{LG} - T]^2 = E_{\psi_G}[\hat{T}_{LG} - T]^2,$$

where the inequality is a consequence of the fact that $\hat{T}_B$ is the Bayes predictor of T under the ψ_B model. Therefore, $\hat{T}_B$ is the ψ_G-best predictor of T. A generalization of this result is provided in the next theorem.

Theorem 3.2.1. *The ψ_G-best linear predictor (BLP) of T is $\hat{T}_B$ given in Eq. (3.1.12). The ψ_G-MSE of $\hat{T}_B$ is as given in Eq. (3.1.13).*

Proof. The proof is a direct consequence of the fact that $\hat{T}_B$ minimizes [under the model ψ_B defined by Eqs. (3.1.1) and (3.1.2)], the Bayes risk for the squared-error loss, which is equivalent to the ψ_G-MSE. Since $\hat{T}_B$ is ψ_G-linear, it must be the ψ_G-best linear predictor.

(Q.E.D.)

In order to relate the ψ_G-best linear predictor (ψ_G-BLP) to other classes of Bayesian least-squares predictors, let us consider the following particular cases of model ψ_G:

$$\psi_{GV} = \psi_G(\mathbf{V}; \mathbf{B} = \mathbf{0}), \ \psi_{GB} = \psi_G(\mathbf{V} = \mathbf{0}; \mathbf{B}),$$

$$\bar{\psi}_{GV} = \psi_G(\mathbf{V}; \mathbf{B}^{-1} = \mathbf{0}), \ \bar{\psi}_{GB} = (\mathbf{V}^{-1} = \mathbf{0}; \mathbf{B}).$$

Let

$$\boldsymbol{\Omega} = \mathbf{X}'\mathbf{B}\mathbf{X}, \ \boldsymbol{\Omega}_s = \mathbf{X}'_s\mathbf{B}\mathbf{X}_s, \ \boldsymbol{\Omega}_{sr} = \mathbf{X}'_s\mathbf{B}\mathbf{X}_r,$$

$$\mathbf{V}^* = \mathbf{V} + \boldsymbol{\Omega}, \ \mathbf{V}^*_s = \mathbf{V}_s + \boldsymbol{\Omega}_s, \ \mathbf{V}^*_{sr} = \mathbf{V}_{sr} + \boldsymbol{\Omega}_{sr},$$

and so on. Consider the following models examined by Smouse (1984):

$$\psi_{GS}(\mathbf{V};\mathbf{B}) = \begin{cases} E[\mathbf{y}|\mathbf{b}] = \mathbf{X}\mathbf{b} \\ \mathrm{Var}[\mathbf{y}|\mathbf{b}] = \mathbf{V}^*, \end{cases}$$

$$\psi_{GSV} = \psi(\mathbf{V};\mathbf{B}=\mathbf{0}) \text{ and } \psi_{GSB} = \psi(\mathbf{V}=\mathbf{0};\mathbf{B}).$$

Note that $\psi_{GSV} = \psi_{GV}$, $\psi_{GSB} = \psi_{GB}$. If the distribution of $\underline{\mathbf{y}}$ given $\boldsymbol{\beta}$ is independent of $\mathbf{b}$, then ψ_G implies ψ_{GS}. Models $\bar{\psi}_{GB}$ and $\bar{\psi}_{GV}$ are noninformative with respect to $\boldsymbol{\beta}$ and $\mathbf{b}$, respectively. The model ψ_{GB} with $\mathbf{X} = \mathbf{I}$, where $\mathbf{I}$ is the identity matrix is considered by O'Haggan (1984). Smouse (1984) and O'Haggan (1984) considered $\mathbf{b}$ to be known and arrived, respectively, at the following predictors

$$\hat{T}_{GS} = \mathbf{1}_s'\mathbf{y}_s + \mathbf{1}_r'[\mathbf{X}_r\mathbf{b} + \mathbf{V}_{rs}^*\mathbf{V}_s^{*-1}(\mathbf{y}_s - \mathbf{X}_s\mathbf{b})], \tag{3.2.2}$$

and

$$\hat{T}_{GO} = \mathbf{1}_s'\mathbf{y}_s + \mathbf{1}_r'[\mathbf{b} + \boldsymbol{\Omega}_{rs}\boldsymbol{\Omega}_s{}^{-1}(\mathbf{y}_s - \mathbf{b})]. \tag{3.2.3}$$

Notice that after some algebraic manipulations, $\hat{T}_{GS}$ reduces to the ψ_G-BLP. Moreover, $\hat{T}_G$ and $\hat{T}_G$ are ψ_G-BLP under their respective models. Table 3.1 (from Rodrigues, 1988) below outlines some predictors derived according to the approach proposed above.

Table 3.1. Some Linear Bayes Predictors.

Model	ψ_G BLP
ψ_G	$\mathbf{1}_s'\mathbf{y}_s + \mathbf{1}_r'\{\mathbf{X}_r\hat{\boldsymbol{\beta}}_B + \mathbf{V}_{rs}\mathbf{V}_s^{-1}(\mathbf{y}_s - \mathbf{X}_s\hat{\boldsymbol{\beta}}_B)\}$
ψ_{GV}	$\mathbf{1}_s'\mathbf{y}_s + \mathbf{1}_r'\{\mathbf{X}_r\mathbf{b} + \mathbf{V}_{rs}\mathbf{V}_s^{-1}(\mathbf{y}_s - \mathbf{X}_s\mathbf{b})\}$
ψ_{GB}	$\mathbf{1}_s'\mathbf{y}_s + \mathbf{1}_r'\{\mathbf{b} + \boldsymbol{\Omega}_{rs}\boldsymbol{\Omega}_s^{-1}(\mathbf{y}_s - \mathbf{b})\}$
ψ_{GV}	$\mathbf{1}_s'\mathbf{y}_s + \mathbf{1}_r'\{\mathbf{X}_r\hat{\boldsymbol{\beta}}_s + \mathbf{V}_{rs}\mathbf{V}_s^{-1}(\mathbf{y}_s - \mathbf{X}_s\hat{\boldsymbol{\beta}}_s)\}$
ψ_{GB}	$\mathbf{1}_s'\mathbf{y}_s + \mathbf{1}_r'\mathbf{X}_r\mathbf{b}$

Some others Bayes linear predictors, which are found in Ericson (1988), can also be derived using the approach developed in this section. Furthermore, it can be shown, by using the approach considered in Theorem 3.2.1, that $\hat{\boldsymbol{\beta}}_{BN}$ is a Bayes linear predictor of $\boldsymbol{\beta}_N$, when $\mathbf{V}_{rs} = \mathbf{0}$ with respect to the ψ_B-generalized risk.

3.3. Minimax and Admissible Predictors

Some results in the theory of minimax and admissible estimators (see, for example, Zacks, 1971, Ch. 6–8; Lehmann, 1983, Ch. 4) can be directly translated to prediction theory (Bolfarine and Zacks, 1990). Accordingly, let $\theta(\mathbf{y})$ be a quantity of interest and $\hat{\theta}(\mathbf{y}_s)$ a corresponding predictor. Let $L(\hat{\theta}(\mathbf{y}_s), \theta(\mathbf{y}))$ be a predictor loss function, and let

$$R_\psi[\hat{\theta};\theta] = E_\psi[L(\hat{\theta}(\mathbf{y}_s), \theta(\mathbf{y}))],$$

be the corresponding prediction risk function. We assume that the parameter ψ of the superpopulation model belongs to a parameter space $\mathbf{\Psi}$ and that the predictor $\hat{\theta}(\mathbf{y}_s)$ belongs to a class $\mathcal{H}$ of suitable predictors. We assume that the parameter space $\mathbf{\Psi}$ is such that, for each $\hat{\theta}$ of $\mathcal{H}$,

$$\rho^*(\hat{\theta}) = \sup_{\psi\in\mathbf{\Psi}} R_\psi[\hat{\theta};\theta] < \infty.$$

A predictor $\hat{\theta}_*$ is called minimax in $\mathcal{H}$, if

$$\rho^*(\hat{\theta}_*) = \inf_{\hat{\theta}\in\mathcal{H}} \rho^*(\hat{\theta}),$$

and if

$$\inf_{\theta\in\mathcal{H}} \sup_{\psi\in\mathbf{\Psi}} R_\psi[\hat{\theta};\theta] = \sup_{\psi\in\mathbf{\Psi}} \inf_{\hat{\theta}\in\mathcal{H}} R_\psi[\hat{\theta};\theta].$$

We bring here, without proof, several key results in minimax prediction in finite populations, for which there are equivalent results in minimax estimation theory. Let $\zeta(\psi)$ be a prior (density or probability function) over $\mathbf{\Psi}$, and let $\hat{\theta}_B(\mathbf{y}_s;\zeta)$ be the corresponding Bayes predictor, having a Bayes risk $\rho(\hat{\theta}_B;\zeta)$.

Result 3.3.1. *If $\hat{\theta}_B(\mathbf{y}_s;\zeta)$ is a Bayes predictor and if $R_\psi[\hat{\theta}_B(\cdot;\zeta);\theta]$ is independent of ψ, then, $\hat{\theta}_B(\mathbf{y}_s;\zeta)$ is minimax.*

Result 3.3.2. *Let $\{\zeta_k; k = 1, 2, \dots\}$ be a sequence of prior distributions over $\mathbf{\Psi}$, and let $\{\hat{\theta}_B(.;\zeta_k), k = 1, 2, \dots\}$ and $\{\rho(\hat{\theta}_B;\zeta_k), k = 1, 2, \dots\}$ be the corresponding Bayes predictors and Bayes risks. If $\hat{\theta}(\mathbf{y}_s)$ is a predictor such that*

$$\sup_{\psi\in\mathbf{\Psi}} R_\psi[\hat{\theta};\theta] \le \limsup_{k\to\infty} \rho(\hat{\theta}_B;\zeta_k),$$

then $\hat{\theta}$ is a minimax predictor.

Result 3.3.3. *If the prediction risk of $\hat{\theta}$ is constant over $\mathbf{\Psi}$, and if there exists a sequence $\{\zeta_k; k = 1, 2, \dots\}$ of prior distributions over $\mathbf{\Psi}$, such that*

$$\lim_{k\to\infty} \rho(\hat{\theta}_B;\zeta_k) = R_\psi[\hat{\theta};\theta] = \rho^*(\hat{\theta}),$$

then $\hat{\theta}$ is a minimax predictor.

Example 3.3.1. Consider the Binomial superpopulation model (Bolfarine, 1987). According to this model, $y_1, \dots, y_N$ are independent and identically distributed, having a common $Binomial(\psi, 1)$ distribution, that is, $\psi = P[y_j = 1]$ and $1 - \psi = P[y_j = 0]$, $j = 1, \ldots, N$ and $0 < \psi < 1$. We wish to predict the population total T. Let ψ have a prior $Beta(a, b)$ distribution. The Bayes predictor of T under the squared-error loss is then

$$\hat{T}_B(a,b) = n\bar{y}_s + (N-n)\frac{a + n\bar{y}_s}{n+a+b}. \tag{3.3.1}$$

The prediction risk of Eq. (3.3.1) is

$$\begin{aligned} R_\psi[\hat{T}_B(a,b);T] = (N-n)^2\Bigg\{ & \frac{n\psi(1-\psi)}{(n+a+b)^2} \\ & + \frac{\psi(1-\psi)}{N-n} + \frac{1}{(n+a+b)^2}[a - \psi(a+b)]^2\Bigg\}. \end{aligned} \tag{3.3.2}$$

Prediction risk (3.3.2) is independent of ψ if

$$a = b = \sqrt{n}\frac{\sqrt{n} + \sqrt{n + N(N-n-1)}}{2(N-n-1)}. \tag{3.3.3}$$

Thus, the Bayes predictor (3.3.1), with $a = b$ as in Eq. (3.3.3) is minimax. This minimax predictor is denoted by $\hat{T}_M$. Using the results of Exercise 3.10, Table 3.2 (from Bolfarine, 1987) illustrates the performance of predictors $\hat{T}_M$ and $\hat{T}_E$ for several values of n and $N = 200$.

Table 3.2. Efficiency of $\hat{T}_M$ with respect to $\hat{T}_E$.

n	$a = b$	R^{*a}	$R^*(E)^b$	$\mathbf{I}_n$
9	1.56	149.5	1061.1	(0.01,0.99)
16	2.14	94.2	575.0	(0.04,0.96)
32	3.44	43.3	227.8	(0.05,0.95)

[a] $R^* = E_\psi[\hat{T}_M - T]^2$.
[b] $R^*(E) = \sup_\psi E_\psi[\hat{T}_E - T]^2$.

$\mathbf{I}_n$ is the interval of values of ψ where $\hat{T}_M$ is more efficient than $\hat{T}_E$. It follows that, if the fraction n/N is small, then $\hat{T}_M$ is considerably better. Of course, if n/N is close to one, $\hat{T}_E$ is better.

Example 3.3.2. Consider the normal superpopulation Bayes model (3.1.1)–(3.1.2) with diagonal covariance matrix $\mathbf{V}$. In this case, the Bayes predictor of the population total, T, is given in Eq. (3.1.15), with a Bayes risk given in Eq. (3.1.16). Consider a sequence of prior distributions $N(\mathbf{b}, \mathbf{B}_k)$ such that $||\mathbf{B}_k|| = k$, where the norm of the covariance matrix $\mathbf{B}$ is $||\mathbf{B}|| = \sum_{i=1}^p B_{ii}$. The corresponding Bayes predictors $\hat{T}_{B_k}$ converge as $k \to \infty$ to the BUP

$$\hat{T}_{BU} = \mathbf{1}_s'\mathbf{y}_s + \mathbf{1}_r'\mathbf{X}_r\hat{\boldsymbol{\beta}}_s.$$

Moreover, the Bayes risks $\rho(\hat{T}_{B_k}; \mathbf{B}_k)$ converge as $k \to \infty$ to the prediction risk of $\hat{T}_{BU}$, namely

$$R_\psi[\hat{T}_{BU}; T] = \mathbf{1}_r'\mathbf{V}_r\mathbf{1}_r + \mathbf{1}_r'\mathbf{X}_r(\mathbf{X}_s'\mathbf{V}_s^{-1}\mathbf{X}_s)^{-1}\mathbf{X}_r'\mathbf{1}_r.$$

Finally, since this prediction risk is independent of $\boldsymbol{\beta}$, it follows from Result 3.3.2 that $\hat{T}_{BU}$ is minimax.

Special cases of Example 3.3.2, are the expansion, ratio, and regression predictors.

Example 3.3.3. As in Example 3.1.3, consider the location model SM1 with normality. We show here that predictor (3.1.27) of S_y^2 is minimax, for the squared-error loss. Under this model, the unknown parameter is β; σ^2 is known.

The Bayes predictor, $\hat{S}_B^2$ of S_y^2 for the squared–error loss, when $\beta \sim N(b, B)$ is given in Eq. (3.1.26). We derive now the Bayes risk $\rho(\hat{S}_B^2; b, B)$. In Example 3.1.3, we have shown that the Bayes predictive distribution of $\mathbf{y}_r$ given $\mathbf{y}_s$ is $N(\hat{\beta}_B\mathbf{1}_r, \boldsymbol{\Sigma}_r)$, where

$$\boldsymbol{\Sigma}_r = \sigma^2[\mathbf{I}_r + \mathbf{J}_r/(n + \sigma^2/B)].$$

The Bayes risk of $\hat{S}_B^2$ is the expected value of the posterior variance of S_y^2, that is,

$$\rho(\hat{S}_B^2; b, B) = E_{\psi_B}\left\{\mathrm{Var}_{\psi_B}\left[(1 - \frac{n}{N})\left(S_{ry}^2 + \frac{n}{N}(\bar{y}_r - \bar{y}_s)^2\right)|\mathbf{y}_s\right]\right\}.$$

Let $\mathbf{y}'\mathbf{A}\mathbf{y}$ be a symmetric quadratic form, $\mathbf{l}'\mathbf{y}$ be a linear form, $\boldsymbol{\mu} = E[\mathbf{y}]$ and $\boldsymbol{\Sigma} = \mathrm{Var}[\mathbf{y}]$. We have

(i) $\mathbf{y}'\mathbf{A}\mathbf{y} \sim \chi^2[p; \lambda]$ if, and only if, $\mathbf{A}\boldsymbol{\Sigma}$ is idempotent of rank p. Moreover, $\lambda = \frac{1}{2}\boldsymbol{\mu}'\mathbf{A}\boldsymbol{\mu}$.

(ii) If $\mathbf{A}\boldsymbol{\Sigma}\mathbf{l} = \mathbf{0}$, then $\mathbf{y}'\mathbf{A}\mathbf{y}$ and $\mathbf{l}'\mathbf{y}$ are independent.

Accordingly, since $S_{ry}^2 = \mathbf{y}_r'\mathbf{A}_r\mathbf{y}_r$, with $\mathbf{A}_r = \dfrac{1}{N-n}[\mathbf{I}_r - \mathbf{J}_r/(N-n)]$, and, since

$$\frac{1}{\sigma^2}\mathbf{A}_r\boldsymbol{\Sigma}_r = \frac{1}{N-n}(\mathbf{I}_r - \frac{1}{N-n}\mathbf{J}_r)(\mathbf{I}_r + \frac{1}{n+\sigma^2/B}\mathbf{J}_r) = \mathbf{A}_r,$$

which is idempotent of rank $N-n$, the Bayes predictive distribution of S_{ry}^2 given $\mathbf{y}_s$ is like that of $[\sigma^2/(N-n)]\chi^2[N-n-1]$. Indeed,

$$\lambda = \frac{1}{2\sigma^2}\hat{\beta}_B^2\mathbf{1}_r'\mathbf{A}_r\mathbf{1}_r = 0.$$

Hence,

$$\mathrm{Var}_{\psi_B}[S_{ry}^2|\mathbf{y}_s] = \frac{2\sigma^4}{(N-n)^2}(N-n-1). \tag{3.3.4}$$

Moreover, since $\bar{y}_r = \mathbf{1}_r'\mathbf{y}_r/(N-n)$ and $\mathbf{A}_r\mathbf{\Sigma}_r\mathbf{1}_r = \mathbf{0}$, S_{ry}^2 and $\bar{y}_r$ are conditionally independent given $\mathbf{y}_s$. Thus,

$$\mathrm{Cov}_{\psi_B}[S_{ry}^2, (\bar{y}_s - \bar{y}_r)^2|\mathbf{y}_s] = 0. \tag{3.3.5}$$

It remains to compute $\mathrm{Var}_{\psi_B}[(\bar{y}_s - \bar{y}_r)^2|\mathbf{y}_s]$. The Bayes predictive distribution of $\bar{y}_r - \bar{y}_s$ given $\mathbf{y}_s$ is normal with mean and variance given, respectively, by

$$\hat{\beta}_B - \bar{y}_s \quad \text{and} \quad \frac{\sigma^2}{N-n} + \frac{\sigma^2}{n+\sigma^2/B}.$$

Therefore,

$$\begin{aligned}(\bar{y}_r - \bar{y}_s)^2|\mathbf{y}_s \sim \sigma^2\left(\frac{1}{N-n} + \frac{1}{n+\sigma^2/B}\right) \\ \chi^2\left[1, \frac{(\hat{\beta}_B - \bar{y}_s)^2}{2\sigma^2(\frac{1}{N-n} + \frac{1}{n+\sigma^2/B})}\right].\end{aligned} \tag{3.3.6}$$

It follows that

$$\begin{aligned}\mathrm{Var}_{\psi_B}[(\bar{y}_r - \bar{y}_s)^2|\mathbf{y}_s] = 2\sigma^4\left(\frac{1}{N-n} + \frac{1}{n+\sigma^2/B}\right)^2 \\ \left[\frac{1+2(\hat{\beta}_B - \bar{y}_s)^2}{\sigma^2(\frac{1}{N-n} + \frac{1}{n+\sigma^2/B})}\right].\end{aligned} \tag{3.3.7}$$

By taking the expected value of Eq. (3.3.7) with respect to the marginal distribution of $\mathbf{y}_s$, it follows from Eqs. (3.3.4)–(3.3.6) that the Bayes risk of $\hat{S}_B^2$ is

$$\begin{aligned}\rho(\hat{S}_B^2; b, B) = (1-\frac{n}{N})^2\frac{2\sigma^4}{(N-n)^2}\Bigg\{N-n-1+(\frac{n}{N})^2\left(1+\frac{N-n}{n+\sigma^2/B}\right)^2 \\ \left[1+\frac{2(N-n)}{\sigma^2(1+\frac{N-n}{n+\sigma^2/B})(\frac{\sigma^2}{n}+B)}\right]\Bigg\}.\end{aligned}$$

Hence,

$$\lim_{B\to\infty}\rho(\hat{S}_B^2; b, B) = \frac{2\sigma^4(N-n)}{N^2}. \tag{3.3.8}$$

But, the right-hand side of Eq. (3.3.8) is the risk function of Eq. (3.1.27) (see Exercise 2.15). Moreover, the right-hand side of Eq. (3.3.8) is independent of β. Hence, according to Result 3.3.2,

$$\hat{S}^2_M = \frac{n}{N}s^2_y + (1 - \frac{n}{N})\sigma^2 \tag{3.3.9}$$

is a minimax predictor of S^2_y.

The following result from the theory of minimaxity allows us to show that the above minimax predictors of T are minimax also for distribution–free superpopulation models, with bounded variances (could be unknown).

Result 3.3.4. *Let $y_1, \ldots, y_N$ be jointly distributed according to the distribution F, belonging to a family of distributions $\mathcal{F}_1$. Suppose that $\hat{T}_M$ is a minimax predictor of T when $F \in \mathcal{F}_o \subset \mathcal{F}_1$. If*

$$\sup_{F\in\mathcal{F}_o} E_F[\hat{T}_M - T]^2 = \sup_{F\in\mathcal{F}_1} E_F[\hat{T}_M - T]^2,$$

$\hat{T}_M$ is minimax for $\mathcal{F}_1$.

Example 3.3.4. Consider the normal model of Example 3.3.2, where the matrix $\mathbf{V} = \sigma^2\mathbf{W}$, $0 < \sigma^2 < M < \infty$ (which could be unknown), and $\mathbf{W}$ is known and diagonal. Then, by using Result 3.3.4, it follows that $\hat{T}_{BU} = \mathbf{1}_s'\mathbf{y}_s + \mathbf{1}_r'\mathbf{X}_r\hat{\boldsymbol{\beta}}_s$, is a minimax predictor under these assumptions. The minimaxity property of $\hat{T}_{BU}$ continue to hold if the normality assumption is dropped, but $\mathbf{V}$ is as described above (Exercise 3.25).

Consider the superpopulation model with a parameter space $\boldsymbol{\Psi}$, and a class of predictors $\mathcal{H}$ of a population quantity vector $\theta(\mathbf{y})$. The GMSE prediction risk is as defined in Eq. (1.3.18). We now consider minimax prediction of $\boldsymbol{\beta}_N$ under the $\boldsymbol{\psi}$ model (3.1.1), with $\mathbf{V}_{rs} = \mathbf{0}$. First we extend the Result 3.3.2 to the present case.

Result 3.3.5. *Let $\{\zeta_k, k \geq 1\}$ be a sequence of prior distributions over $\boldsymbol{\Psi}$ and let $\{\hat{\boldsymbol{\beta}}_{B_kN}, k \geq 1\}$ and $\{\rho_G(\hat{\boldsymbol{\beta}}_{B_kN}; \zeta_k), k \geq 1\}$ be the corresponding Bayes predictors of $\boldsymbol{\beta}_N$ and their corresponding prediction risks. If there is a predictor $\hat{\boldsymbol{\beta}}_{MN}$ such that*

$$\sup_{\boldsymbol{\psi}\in\boldsymbol{\Psi}} R_{G\boldsymbol{\psi}}[\hat{\boldsymbol{\beta}}_{MN}; \boldsymbol{\beta}_N] \leq \limsup_{k\to\infty} \rho_G(\hat{\boldsymbol{\beta}}_{B_k}; \zeta_k),$$

then $\hat{\boldsymbol{\beta}}_{MN}$ is a minimax predictor of $\boldsymbol{\beta}_N$.

We are now in position to prove that $\hat{\boldsymbol{\beta}}_s$ is a minimax predictor of $\boldsymbol{\beta}_N$ under the generalized prediction risk (2.2.19).

Example 3.3.5. As in Example 3.3.2, we consider the sequence of prior distributions $N(\mathbf{b}, \mathbf{B}_k)$ such that $||\mathbf{B}|| = k$. It follows from Eq. (3.1.21) that the corresponding Bayes predictors $\hat{\boldsymbol{\beta}}_{B_kN}$ converge as $k \to \infty$ to $\hat{\boldsymbol{\beta}}_s$. Moreover, it follows from Eq. (3.1.22) that $\rho_G(\hat{\boldsymbol{\beta}}_{B_kN}; \mathbf{b}, \mathbf{B}_k)$ converge as $k \to \infty$ to the generalized prediction risk (2.2.21), which is independent of $\boldsymbol{\beta}$. Result 3.3.5 implies that $\hat{\boldsymbol{\beta}}_s$ is a minimax predictor of $\boldsymbol{\beta}_N$.

Result 3.3.4 can be extended to the case of vector valued quantities by using the generalized prediction risk (2.2.19). By using these extensions, it can be shown that $\hat{\boldsymbol{\beta}}_s$ is also minimax when $\mathbf{V} = \sigma^2\mathbf{W}$, $\mathbf{W}$ is known and diagonal, and σ^2 (which could be unknown), belongs to a closed interval $[0, M]$, where, $0 < M < \infty$ (Exercise 3.26).

As before, let $R_\psi[\hat{\theta}(\mathbf{y}_s); \theta]$ denote the risk function of predictor $\hat{\theta}(\mathbf{y}_s)$.

Definition 3.3.1. *A predictor $\hat{\theta}(\mathbf{y}_s)$ is called admissible with respect to $(\mathcal{H}, \Psi, R)$ if there exists no predictor $\hat{\hat{\theta}}(\mathbf{y}_s)$ in $\mathcal{H}$, such that*

$$R_\psi[\hat{\hat{\theta}}(\mathbf{y}_s); \theta] \leq R_\psi[\hat{\theta}(\mathbf{y}_s); \theta], \text{ for all } \psi \in \Psi,$$

with strict inequality for some values of ψ.

Example 3.3.6. Consider the superpopulation model SM2. We wish to predict the population total T. The prediction risk for the squared–error loss of the expansion predictor $\hat{T}_E$ is

$$R_\psi[\hat{T}_E; T] = (N-n)^2 \left[\sigma^2(\frac{\bar{x}_s}{n} + \frac{\bar{x}_r}{N-n}) + \beta^2(\bar{x}_r - \bar{x}_s)^2\right], \tag{3.3.10}$$

where $\psi = (\beta, \sigma^2)$. On the other hand, the prediction risk of the ratio predictor $\hat{T}_R$ may be written as

$$R_\psi[\hat{T}_R; T] = (N-n)^2\sigma^2 \left[(\frac{\bar{x}_r}{\bar{x}_s})^2\frac{\bar{x}_s}{n} + \frac{\bar{x}_r}{N-n}\right]. \tag{3.3.11}$$

Thus, if $\frac{\bar{x}_r}{\bar{x}_s} < 1$, then, according to Eqs. (3.3.10) and (3.3.11),

$$R_\psi[\hat{T}_R; T] < R_\psi[\hat{T}_E; T],$$

for all ψ. Thus, if $\bar{x}_s > \bar{x}_r$, then $\hat{T}_E$ is inadmissible.

A celebrated result of Blyth (1951) can be applied to the prediction theory, to yield the following result.

Result 3.3.6. *If the prediction risk function $R_\psi[\hat{\theta}(\mathbf{y}_s);\theta]$ is continuous in ψ for each $\hat{\theta} \in \mathcal{H}$ and if the support of the prior distribution ζ is Ψ, then the Bayes predictor with respect to ζ is admissible.*

The Bayes predictors derived in the previous examples with respect to informative priors are all admissible.

Example 3.3.7. It is now shown that $\hat{T}_R$ is admissible under model SM2 and squared-error loss by using the limiting Bayes risk method as considered in Bolfarine (1987). By negation, suppose that $\hat{T}_R$ is not admissible and without loss of generality that $\sigma^2 = 1$. Then, there is a predictor $\hat{T}^*$ such that

$$E_\beta[\hat{T}^* - T]^2 \le \frac{N(N-n)}{n}\frac{\bar{x}\bar{x}_r}{\bar{x}_s},$$

for all β, with strict inequality for at least one β. Since, for any predictor $\hat{T}$, $E_\psi[\hat{T}-T]^2$ is a continuous function of β (Ferguson, 1966), there is $\epsilon > 0$ and $\beta_o < \beta_1$, such that

$$E_\psi[\hat{T}^* - T]^2 \le \frac{N(N-n)}{n}\frac{\bar{x}\bar{x}_r}{\bar{x}_s} - \epsilon,$$

for all $\beta_o < \beta < \beta_1$. Let r_ζ^* be the Bayes risk of $\hat{T}^*$ with respect to the prior $\beta \sim N(0, B)$ (prior ζ) and let r_ζ be the Bayes risk of the Bayes predictor with respect to the prior ζ. Hence, from Example 3.1.1 with $\sigma = 1$, we have

$$\frac{\dfrac{N(N-n)}{n}\dfrac{\bar{x}\bar{x}_r}{\bar{x}_s} - r_\zeta^*}{\dfrac{N(N-n)}{n}\dfrac{\bar{x}\bar{x}_r}{\bar{x}_s} - r_\zeta} = \frac{\dfrac{1}{\sqrt{2\pi\beta}}\displaystyle\int\left\{\frac{N(N-n)}{n}\frac{\bar{x}\bar{x}_r}{\bar{x}_s} - E_\psi[\hat{T}^* - T]^2\right\}e^{-\beta^2/2B}d\beta}{\dfrac{(\sum_{i\in r} x_i)^2}{\sum_{i\in s} x_i} - \dfrac{(\sum_{i\in r} x_i)^2}{(\sum_{i\in s} x_i) + 1/B}}$$

$$\ge \frac{(\sum_{i\in s} x_i)(B\sum_{i\in s} x_i + 1)}{(\sum_{i\in r} x_i)^2\sqrt{2\pi B}}\epsilon\int_{\beta_0}^{\beta_1} e^{-\beta^2/2B}d\beta \to \infty$$

as $B \to \infty$. Hence, for sufficiently large B,

$$\frac{N(N-n)}{n}\frac{\bar{x}\bar{x}_r}{\bar{x}_s} - r_\zeta^* > \frac{N(N-n)}{n}\frac{\bar{x}\bar{x}_r}{\bar{x}_s} - r_\zeta,$$

which implies that $r_\zeta^* < r_\zeta$. This is a contradiction since r_ζ is the risk of the Bayes predictor with respect to ζ. Therefore, $\hat{T}_R$ must be admissible.

3.4. Dynamic Bayesian Prediction

In the present section, we consider Bayes prediction of the population total and population variance by using data from repeated sampling of the finite population over time. Dynamic Bayes prediction of the finite population regression coefficient $\boldsymbol{\beta}_N$ is considered in Exercise 3.27. It is assumed that the unknown parameters of the superpopulation model at each time period follow a stochastic model.

3.4.1. The Multinormal Dynamic Model

Consider a sequence of repeated samples carried out at time points $t = 1, 2, \ldots,$ on a population $\mathcal{P}_t$ of N_t units. At time point t, we associate with the jth unit of $\mathcal{P}_t$ the unknown quantity y_{tj} and the known vector $\mathbf{X}_{tj} = (x_{tj1}, \ldots, x_{tjp})$. As before, we assume that the vector $\mathbf{y}_t' = (y_{t1}, \ldots, y_{tN_t})$ is a realization of a random vector and is related to the known matrix $\mathbf{X}_t' = (\mathbf{X}_{t1}, \ldots, \mathbf{X}_{tN_t})$ through the model

$$\mathbf{y}_t = \mathbf{X}_t\boldsymbol{\beta}_t + \mathbf{e}_t, \tag{3.4.1}$$

where $\boldsymbol{\beta}_t$ is a $p \times 1$ vector of unknown parameters and $\mathbf{e}_t \sim N(\mathbf{0}_t, \mathbf{V}_t)$, $t = 1, 2, \ldots.$

In order to obtain information about the population total and the population variance at time t, $T_t = \sum_{j=1}^{N_t} y_{tj}$ and $S_{ty}^2 = \sum_{j=1}^{N_t}(y_{tj} - \bar{y}_t)^2/N_t$, where $\bar{y}_t = T_t/N_t$, a sample of n_t units is selected from $\mathcal{P}_t$ according to some specified sampling design. As before, the observed and unobserved parts of $\mathcal{P}_t$ are denoted by $\mathbf{s}_t$ and $\mathbf{r}_t$, which are shortened to $\mathbf{s}$ and $\mathbf{r}$. Therefore, after $\mathbf{s}_t$ has been selected, we have the partition

$$\mathbf{y}_t = \begin{pmatrix} \mathbf{y}_{ts} \\ \mathbf{y}_{tr} \end{pmatrix}, \quad \mathbf{X}_t = \begin{pmatrix} \mathbf{X}_{ts} \\ \mathbf{X}_{tr} \end{pmatrix}, \quad \text{and} \quad \mathbf{V}_t = \begin{pmatrix} \mathbf{V}_{ts} & \mathbf{V}_{tsr} \\ \mathbf{V}_{trs} & \mathbf{V}_{tr} \end{pmatrix}.$$

Let $\mathcal{D}_t = \{\mathcal{D}_{t-1}, \mathbf{y}_{ts}\}$ denote all the information available on $\mathbf{y}$ up to time t. $\mathcal{D}_o$ denote the prior information at time t_o.

The changes in the parameters $\boldsymbol{\beta}_t$ follow the linear equation

$$\boldsymbol{\beta}_t = \boldsymbol{\Lambda}_t\boldsymbol{\beta}_{t-1} + \mathbf{u}_t, \quad t = 1, 2, \ldots, \tag{3.4.2}$$

where $\mathbf{u}_t \sim N(\mathbf{0}_t, \boldsymbol{\Delta}_t)$, $\mathrm{Cov}(\mathbf{e}_t, \mathbf{u}_t) = \mathbf{0}_t$, and $\mathbf{V}_t$, $\boldsymbol{\Lambda}_t$ and $\boldsymbol{\Delta}_t$ are assumed to be known, for all $t = 1, 2, \ldots.$ Equation (3.4.2) is usually known as the *system equation*, and it describes the way the process $\boldsymbol{\beta}_t$ evolves through time. The superpopulation model specified by Eqs. (3.4.1) and (3.4.2) is named model ψ_{DB}.

3.4.1.1. Dynamic Prediction of T_t

Suppose that the state of the system at time $t-1$ is such that

$$\boldsymbol{\beta}_{t-1}|\mathcal{D}_{t-1} \sim N(\mathbf{m}_{t-1}, \mathbf{C}_{t-1}). \tag{3.4.3}$$

Accordingly, before observing $\mathbf{y}_{ts}$, the prior information about $\boldsymbol{\beta}_t$ follows directly from Eqs. (3.4.2) and (3.4.3), namely,

$$\boldsymbol{\beta}_t|\mathcal{D}_{t-1} \sim N(\boldsymbol{\Lambda}_t\mathbf{m}_{t-1}, \mathbf{R}_t), \tag{3.4.4}$$

where $\mathbf{R}_t = \boldsymbol{\Lambda}_t\mathbf{C}_{t-1}\boldsymbol{\Lambda}_t' + \boldsymbol{\Delta}_t$. After $\mathbf{y}_t$ is observed, we convert the prior distribution (3.4.4) to the posterior distribution given $\mathcal{D}_t$; that is,

$$\boldsymbol{\beta}_t|\mathcal{D}_t \sim N(\mathbf{m}_t, \mathbf{C}_t), \tag{3.4.5}$$

where

$$\begin{aligned}\mathbf{m}_t = E_{\psi_{DB}}[\boldsymbol{\beta}_t|\mathcal{D}_t] &= (\mathbf{X}_{ts}'\mathbf{V}_{ts}^{-1}\mathbf{X}_{ts} + \mathbf{R}_t^{-1})^{-1} \\ &(\mathbf{X}_{st}\mathbf{V}_{st}^{-1}\mathbf{y}_{ts} + \mathbf{R}_t^{-1}\boldsymbol{\Lambda}_t\mathbf{m}_{t-1})\end{aligned} \tag{3.4.6}$$

and

$$\mathbf{C}_t = \mathrm{Var}_{\psi_{DB}}[\boldsymbol{\beta}_t|\mathcal{D}_t] = (\mathbf{X}_{ts}'\mathbf{V}_{ts}^{-1}\mathbf{X}_{ts} + \mathbf{R}_t^{-1})^{-1}. \tag{3.4.7}$$

Notice that formulas (3.4.6) and (3.4.7) are analogous to formulas (3.1.5)–(3.1.7). Moreover, the dynamic Bayes predictor of the population total T_t and its Bayes risk at time t are given, respectively, by

$$\hat{T}_{DBt} = n_t\bar{y}_{ts} + \mathbf{1}_{tr}'[\mathbf{X}_{tr}\mathbf{m}_t + \mathbf{V}_{trs}\mathbf{V}_{ts}^{-1}(\mathbf{y}_s - \mathbf{X}_{ts}\mathbf{m}_t)], \tag{3.4.8}$$

and

$$\mathrm{Var}_{\psi_{DB}}[T_t|\mathcal{D}_t] = \mathbf{1}_{tr}'(\mathbf{V}_{tr} - \mathbf{V}_{trs}\mathbf{V}_{ts}^{-1}\mathbf{V}_{tsr})\mathbf{1}_{tr} \tag{3.4.9}$$

$$+\mathbf{1}_{tr}'(\mathbf{X}_{tr} - \mathbf{V}_{trs}\mathbf{V}_{ts}^{-1}\mathbf{X}_{ts})(\mathbf{X}_{ts}'\mathbf{V}_{ts}^{-1}\mathbf{X}_{ts} + \mathbf{R}_t^{-1})^{-1}(\mathbf{X}_{tr} - \mathbf{V}_{trs}\mathbf{V}_{ts}^{-1}\mathbf{X}_{ts})'\mathbf{1}_{tr},$$

where $\bar{y}_{ts}$ denotes the sample mean at time t.

If the covariance matrix $\mathbf{V}_t = \sigma_t^2\mathbf{W}_t$ where $\mathbf{W}_t$ is diagonal and σ_t^2 is unknown, one can use the results in Bolfarine (1989a). Tam (1987b) obtained similar results by using a non-Bayesian approach.

Example 3.4.1. Suppose that at time t,

$$y_{ti} = \mu_t + e_{ti}$$

and

$$\mu_t = \lambda\mu_{t-1} + u_t,$$

e_{ti} and u_t are all independent; $e_{ti} \sim N(0, \sigma_t^2)$; $u_t \sim N(0, \Delta_t)$; λ, σ_t^2, and Δ_t are all known; $i = 1, \ldots, N_t$; and $t = 1, \ldots$. The observation equation (3.4.1) may be written as

$$\bar{y}_{ts} = \mu_t + \bar{e}_{ts},$$

where $\bar{y}_{ts}$ denotes the sample mean at time t,

$$E[\bar{e}_{ts}] = 0 \text{ and } \mathrm{Var}[\bar{e}_{ts}] = \sigma_t^2/n_t.$$

Since in this particular case $\mathbf{X}_t = \mathbf{1}_{N_t}$ and $\mathbf{V}_t = \mathbf{I}_{N_t}\sigma_t^2$, where $\mathbf{I}_{N_t}$ denotes the identity matrix of dimension N_t, it follows from Eq. (3.4.5) that the posterior distribution of μ_t given $\mathbf{y}_{st}$ is normal with mean and variance given, respectively, by

$$m_t = \lambda m_{t-1} - \frac{n_t R_t}{\sigma_t^2 + n_t R_t}(\bar{y}_{ts} - \lambda m_{t-1}) \tag{3.4.10}$$

and

$$C_t = \frac{\sigma_t^2 R_t}{\sigma_t^2 + n_t R_t}. \tag{3.4.11}$$

Therefore, the Bayes dynamic predictor of T_t and its Bayes risk are given by

$$\hat{T}_{DBt} = n_t \bar{y}_{ts} + (N_t - n_t) m_t \tag{3.4.12}$$

and

$$\mathrm{Var}_{\psi_{DB}}[T_t|\mathcal{D}_t] = (N_t - n_t)^2 C_t + (N_t - n_t)\sigma_t^2.$$

The prediction variance of the usual expansion predictor at time t, $\hat{T}_{Et} = N_t \bar{y}_t$, is

$$\begin{aligned}\mathrm{Var}_{\psi}[\hat{T}_{Et} - T_t] =& (N_t - n_t)^2 \frac{\sigma_t^2}{n_t} + (N_t - n_t)\sigma_t^2 \\ \geq& \mathrm{Var}_{\psi_{DB}}[T_t|\mathcal{D}_t].\end{aligned}$$

Thus, since $\mathrm{Var}_{\psi_{DB}}[T_t|\mathcal{D}_t] < \mathrm{Var}_{\psi}[\hat{T}_{Et} - T]$, for all $0 < \sigma_t^2 < \infty$, the expansion estimator $\hat{T}_{Et}$ is inadmissible in the repeated sampling case.

3.4.1.2. Dynamic Prediction of S_{ty}^2

Using the notation of the previous section, the Bayes predictive distribution of $\mathbf{y}_{tr}$, given $\mathcal{D}_t$, is normal with mean

$$E_{\psi_{DB}}[\mathbf{y}_{tr}|\mathcal{D}_t] = \mathbf{X}_{tr}\mathbf{m}_t + \mathbf{V}_{trs}\mathbf{V}_{ts}^{-1}(\mathbf{y}_{ts} - \mathbf{X}_{ts}\mathbf{m}_t)$$

and covariance matrix

$$\begin{aligned}\mathrm{Var}_{\psi_{DB}}[\mathbf{y}_{tr}|\mathcal{D}_t] =& (\mathbf{V}_{tr} - \mathbf{V}_{trs}\mathbf{V}_{ts}^{-1}\mathbf{V}_{tsr}) \\ &+ (\mathbf{X}_{tr} - \mathbf{V}_{trs}\mathbf{V}_{ts}^{-1}\mathbf{V}_{tsr})\mathbf{C}_t(\mathbf{X}_{tr} - \mathbf{V}_{trs}\mathbf{V}_{ts}^{-1}\mathbf{V}_{tsr})'.\end{aligned}$$

Let

$$s_{ty}^2 = \sum_{i \in s_t} (y_{ti} - \bar{y}_{ts})^2 / n_t \tag{3.4.13}$$

be the tth sample variance. Therefore, the Bayes predictor of S_{ty}^2, for the squared-error loss, is in analogy to Eq. (3.1.25),

$$\begin{aligned} \hat{S}_{DBt}^2 = & \frac{n_t}{N_t} s_{ty}^2 + (1 - \frac{n_t}{N_t}) \Big\{ tr\{\mathbf{A}_{tr} \boldsymbol{\Sigma}_{trs}\} \\ & + \boldsymbol{\eta}_{trs}' \mathbf{A}_{tr} \boldsymbol{\eta}_{trs} + \frac{n_t}{N_t} (D_{tr}^2 + [h_{tr}(\mathbf{y}_{ts}) - \bar{y}_{ts}]^2) \Big\}, \end{aligned} \tag{3.4.14}$$

where

$$\mathbf{A}_{tr} = \frac{1}{N_t - n_t} (\mathbf{I}_{r_t} - \frac{1}{N_t - n_t} \mathbf{J}_{r_t}), \ \boldsymbol{\Sigma}_{trs} = \text{Var}_{\psi_{DB}} [\mathbf{y}_{tr} | \mathcal{D}_t],$$

$$D_{tr}^2 = \frac{1}{(N_t - n_t)^2} \mathbf{1}_{tr}' \boldsymbol{\Sigma}_{tr} \mathbf{1}_{tr}, \ h_{tr}(\mathbf{y}_{ts}) = \frac{1}{N_t - n_t} \mathbf{1}_{tr}' \boldsymbol{\eta}_{trs}$$

and $\boldsymbol{\eta}_{trs} = E[\mathbf{y}_{tr} | \mathcal{D}_t]$.

In the simpler case, when there is no system evolution, that is, $\boldsymbol{\beta}_t = \boldsymbol{\beta}$ for all $t = 1, 2, \ldots$, the formulae of the dynamic Bayes predictor of T_{ty} and S_{ty}^2 reduce to Eqs. (3.1.15) and (3.1.25) in which the prior parameter at the tth time period, $\mathbf{b}_t$ and $\mathbf{B}_t$, are the posterior mean and posterior variance of $\boldsymbol{\beta}$ after the $(t-1)$st sampling epoch.

Example 3.4.2. Consider the model ψ_{DB} of Example 3.4.1. It follows that

$$\boldsymbol{\eta}_{trs} = E[\mathbf{y}_r | \mathcal{D}_t] = \mathbf{1}_r m_t,$$

$$\mathbf{D}_{tr}^2 = \frac{1}{(N - t - n_t)^2} \mathbf{1}_{tr}' \mathbf{1}_{tr},$$

and

$$\boldsymbol{\Sigma}_{trs} = \mathbf{I}_{tr} \sigma_t^2 + C_t \mathbf{1}_{tr} \mathbf{1}_{tr}',$$

where m_t and C_t are given in Eqs. (3.4.10) and (3.4.11). Collecting all these results and using Eq. (3.4.14), it may be shown that

$$\begin{aligned} \hat{S}_{DBt}^2 = E[S_{ty}^2 | \mathcal{D}_t] = & \frac{n_t}{N_t} s_{ty}^2 \\ & + (1 - \frac{n_t}{N_t}) \{ \frac{1}{N_t} [C_t^* + n_t (\bar{y}_{ts} - m_t)^2] + (1 - \frac{1}{N_t}) \sigma_t^2 \}, \end{aligned} \tag{3.4.15}$$

where $C_t^* = n_t C_t$ and s_{ty}^2 is given in Eq. (3.4.1).

Dynamic Bayes predictors of the population total under generalized dynamic exponential families superpopulation models is considered in Bolfarine (1988a). Applications are considered to dynamic binomial and Poisson superpopulation models. Rodrigues and Bolfarine (1987) considered a Kalman filter approach for the problem of predicting T_t, where it is considered that the population total itself follows a stochastic model. A time series approach is considered in Scott and Smith (1974) and Blight and Scott (1973). A least-squares approach, which is based on some results in Rao (1973, p. 234), is considered in Jones (1980). Bolfarine (1990) extended the approach of the present section to the case of a linear dynamic superpopulation model, where the covariance matrix $\mathbf{V}_t$ may depend on unknown parameters.

3.5. Empirical Bayes Predictors

As shown in the previous sections, Bayes predictors of population quantities depend on the prior distribution of the model parameters. The Bayes prediction risk is often sensitive to the correct choice of the prior distribution. In the following example, we illustrate this sensitivity in a simple case.

Example 3.5.1. Consider model SM1 under normality, that is, $\mathbf{y} \mid \mu \sim N(\mu\mathbf{1}, \sigma^2\mathbf{I})$, and assume further that the prior distribution of μ is $N(\mu_0, \tau^2)$ and σ is known.

The Bayes predictor of the population total, for a squared-error loss is

$$\hat{T}_B = n\bar{y}_s + (N-n)[w\bar{y}_s + (1-w)\mu_0],$$

where $\bar{y}_s$ is the mean of a sample of size n; N is the population size and $w = n\lambda/(1+n\lambda)$, $\lambda = \tau^2/\sigma^2$. The corresponding Bayes prediction risk is

$$(3.5.1) \qquad E_{\psi_B}[\hat{T}_B - T]^2 = \sigma^2 N \frac{1-f}{f}[w + f(1-w)],$$

where $f = n/N$.

If one does not know the values of μ_0 and w, and uses instead the values μ_0' and w', the corresponding predictor is

$$\hat{T}_{B'} = n\bar{y}_s + (N-n)[w'\bar{y}_s + (1-w')\mu_0'].$$

The Bayes prediction risk of $\hat{T}_{B'}$ can be written as

$$(3.5.2) \qquad E_{\psi_B}[\hat{T}_{B'} - T]^2 = E_{\psi_B}[\hat{T}_B - T]^2 + E_{\psi_B}[\hat{T}_{B'} - \hat{T}_B]^2.$$

Indeed,

$$\begin{aligned} E_{\psi_B}[(\hat{T}_B - T)(\hat{T}_{B'} - \hat{T}_B)] &= E_{\psi_B}[(\hat{T}_{B'} - \hat{T}_B)E_{\psi_B}\{\hat{T}_B - T \mid \mathbf{y}_s\}] \\ &= 0. \end{aligned}$$

Thus, the relative prediction risk (RPR) of $\hat{T}_{B'}$ versus $\hat{T}_B$ is

$$\text{(3.5.3)} \qquad \frac{E_{\psi_B}[\hat{T}_{B'} - T]^2}{E_{\psi_B}[\hat{T}_B - T]^2} = 1 + \frac{E_{\psi_B}[\hat{T}_{B'} - \hat{T}_B]^2}{E_{\psi_B}[\hat{T}_B - T]^2}.$$

Let $\delta = (\mu_0' - \mu_0)/\sigma$ then

$$\hat{T}_{\mathrm{B'}} - \hat{T}_{\mathrm{B}} = (N-n)[(w'-w)(\bar{y}_s - \mu_0) + \delta\sigma(1-w')].$$

Notice that under ψ_B the predictive distribution of $\bar{y}_s$ is $N\big(\mu_0, \sigma^2/n(1-w)\big)$. Hence,
(3.5.4)

$$\begin{aligned} E_{\psi_B}[\hat{T}_{B'} - \hat{T}_B]^2 &= (N-n)^2[(w'-w)^2\frac{\sigma^2}{n(1-w)} + \delta^2\sigma^2(1-w')^2] \\ &= N^2(1-f^2)\sigma^2[\frac{(w'-w)^2}{n(1-w)} + \delta^2(1-w')^2] \end{aligned}$$

Thus, from Eqs. (3.5.1), (3.5.3), and (3.5.4), the RPR of $\hat{T}_{B'}$ with respect to $\hat{T}_B$ is

$$\text{(3.5.5)} \quad \mathrm{RPR}_{\psi_B}(\hat{T}_{B'}; \hat{T}_B) = 1 + \frac{1-f}{1-w} \cdot \frac{(w'-w)^2 + \delta^2 n(1-w)(1-w')^2}{w + f(1-w)}.$$

In Table 3.5.1, we present the RPR values for the case of $f = 0.1$, $w = 0.1(0.1)0.4$, $n\delta^2 = 0, 1$.

In the first part of the table, we see the effect of a wrong choice of w' when μ_0 is chosen correctly ($\delta = 0$). The second part ($n\delta^2 = 1$) shows the combined effects of wrong choice of w' and of μ_0'. We see that if the choice of the prior parameters is very erroneous, the Bayes prediction risk might be more than five times larger than the Bayes prediction risk of the correct Bayes predictor.

If the prediction problem is repetitive or if the population is partitioned to similar strata or clusters, the empirical Bayes (EB) procedure may yield consistent estimators of certain prior parameters, and correspondingly predictors whose Bayes risk approaches that of the correct Bayes predictor. The method of empirical Bayes was introduced by Robbins (1955). There is much literature on the subject of EB estimation. Ghosh and Meeden (1986) and Ghosh and Lahiri (1987) developed EB predictors of finite population means. We outline here the procedure in terms of predicting population quantities and provide a few examples.

Table 3.5.1. Relative Prediction Risk (RPR) for $f = 0.1$, $n\delta^2 = 0, 1$.

	$n\delta^2 = 0$			
$w' \backslash w$	0.1	0.2	0.3	0.4
0.1	1	1.040	1.139	1.293
0.2	1.053	1	1.035	1.130
0.3	1.211	1.040	1	1.033
0.4	1.474	1.161	1.035	1
0.5	1.842	1.362	1.139	1.033
0.6	2.316	1.643	1.313	1.130
0.7	2.895	2.004	1.556	1.293
0.8	3.579	2.446	1.869	1.522
0.9	4.368	2.969	2.251	1.815
1	5.263	3.571	2.703	2.174
	$n\delta^2 = 1$			
$w' \backslash w$	0.1	0.2	0.3	0.4
0.1	4.837	3.644	3.109	2.878
0.2	4.084	3.057	2.591	2.383
0.3	3.532	2.615	2.192	1.991
0.4	3.179	2.318	1.910	1.704
0.5	3.026	2.165	1.747	1.522
0.6	3.074	2.157	1.702	1.443
0.7	3.321	2.294	1.775	1.470
0.8	3.768	2.575	1.966	1.600
0.9	4.416	3.001	2.275	1.835
1	5.263	3.571	2.703	2.174

Consider a sequence of repetitive predictions of certain population quantities. More specifically, let $\mathcal{P}_t$, $t = 1, 2, \ldots$ be a sequence of finite populations. Let θ_t be a specified quantity of $\mathcal{P}_t$. Let $\mathbf{s}_t$ be a sample from $\mathcal{P}_t$; $\mathbf{y}_{s_t}$ and ψ_t the observed sample vector and the parameters of the superpopulation model, corresponding to $\mathcal{P}_t$, $t = 1, 2, \ldots$. The empirical Bayes (EB) model assumes that $\mathbf{y}_{s_1}, \mathbf{y}_{s_2}, \ldots$ are *conditionally independent*, given $\psi_1, \psi_2, \ldots$ and that $\psi_1, \psi_2, \ldots$ is a sequence of independent random vectors having a common distribution $\zeta(\psi)$, which is the prior distribution of ψ. This distribution is completely or partially unknown.

Let $\hat{\theta}_{B,t}(\mathbf{y}_{s_t}; \zeta)$ denote the Bayes predictor of θ_t, with respect to ζ and a specified loss function. Let $\rho(\hat{\theta}_{B,t}; \zeta)$ denote the corresponding Bayes prediction risk.

The empirical distribution of $\psi_1, \psi_2, \ldots, \psi_m$ is not directly observable. However, in certain cases, some functionals of this empirical distribution can be consistently estimated. In such cases, suppose that one can construct a sequence of predictors $\{\hat{\theta}_{\text{EB}}(\mathbf{y}_{s_{m+1}}; \mathbf{y}_{s_1}, \ldots, \mathbf{y}_{s_m}),\ m \geq 1\}$ having Bayes prediction risks $\rho[\hat{\theta}_{\text{EB}}(\mathbf{y}_{s_{m+1}}; \mathbf{y}_{s_1}, \ldots, \mathbf{y}_{s_m}); \zeta]$.

Definition 3.5.1. *A sequence of predictors* $\{\hat{\theta}_{\text{EB}}(\mathbf{y}_{s_{m+1}}; \mathbf{y}_{s_1}, \ldots, \mathbf{y}_{s_m});\ m \geq 1\}$ *is called asymptotically optimal if* $\lim_{m\to\infty} \{\rho[\hat{\theta}_{\text{EB}}(\mathbf{y}_{s_{m+1}}; \cdot), \zeta] - \rho(\hat{\theta}_{B,m+1}; \zeta)\} = 0$.

The predictors $\hat{\theta}_{\text{EB}}(\mathbf{y}_{s_{m+1}}; \mathbf{y}_{s_1}, \ldots, \mathbf{y}_{s_m})$ are called *empirical Bayes predictors* (EBP). In the following example we illustrate a sequence of EBPs, which is asymptotically optimal.

Example 3.5.2. Consider a population $\mathcal{P}$ of size N. Let $\mathbf{y}_t$ denote the vector of values of its elements at time t, where $t = 1, 2, \ldots$. We assume that $\mathbf{y}_t$ changes its values at random at different time epochs, according to the following model;

$$\mathbf{y}_t \mid \mu_t \sim N(\mu_t \mathbf{1}_N, \sigma^2 \mathbf{I}_N), \qquad t = 1, 2, \ldots,$$

where $\mu_1, \mu_2, \ldots$ is a sequence of i.i.d. random variables having a common normal (prior) distribution $N(\mu_0, \tau^2)$. At every time epoch, a sample $\mathbf{s}$ is drawn from $\mathcal{P}$. All samples are of equal size n.

Let $\lambda = \tau^2/\sigma^2$. As in Example 3.5.1, the Bayes predictor of T_t is

$$\hat{T}_{B,t} = n\bar{y}_{s_t} + (N - n)[w\bar{y}_{s_t} + (1 - w)\mu_0].$$

If the prior parameters (w, μ_0) are unknown, we could estimate them from the sequence of sample vectors $\{\mathbf{y}_{s_t};\ t \geq 1\}$ and obtain EBPs. Indeed, given $\{\mathbf{y}_{s_1}, \ldots, \mathbf{y}_{s_m}\}$ define

$$\text{(3.5.6)} \qquad \hat{\sigma}_m^2 = \frac{1}{(n-1)m} \sum_{t=1}^{m} \sum_{i \in s_t} (y_{i,t} - \bar{y}_{s_t})^2,$$

and the mean squared error between epochs (MSB) statistic

$$\text{MSB}_m = \frac{n}{m-1} \sum_{t=1}^{m} (\bar{y}_{s_t} - \hat{\mu}_{0,m})^2,$$

where

$$\hat{\mu}_{0,m} = \frac{1}{m} \sum_{t=1}^{m} \bar{y}_{s,t}.$$

According to the empirical Bayes model, ψ_{EB},

$$\hat{\sigma}_m^2 \sim \frac{\sigma^2}{(n-1)m} \chi^2[(n-1)m]$$

and

$$\text{MSB}_m \sim (\sigma^2 + n\tau^2) \frac{1}{m-1} \chi^2[m-1],$$

Moreover, $\hat{\sigma}_m^2$ and MSB_m are independent. Thus, by the strong law of large numbers, $\hat{\mu}_{0,m} \underset{m\to\infty}{\longrightarrow} \mu_0$ almost surely (a.s.) $[\psi_{\text{EB}}]$, and

$$\frac{\hat{\sigma}_m^2}{\text{MSB}_m} \underset{m\to\infty}{\longrightarrow} \frac{1}{1+n\lambda} = 1-w \quad \text{a.s. } [\psi_{\text{EB}}].$$

We thus define the estimator $\hat{w}_m = 1 - \hat{\sigma}_m^2/\text{MSB}_m$. This estimator converges a.s. $[\psi_{\text{EB}}]$ to w. Thus, the sequence of predictors

$$\text{(3.5.7)} \qquad \hat{T}_{\text{EB},m+1} = n\bar{y}_{s_{m+1}} + (N-n)[\hat{w}_m \bar{y}_{s_{m+1}} + (1-\hat{w}_m)\hat{\mu}_{0,m}]$$

converges a.s. $[\psi_{\text{EB}}]$, as $m \to \infty$, to the Bayes predictor. Furthermore,

$$E_{\psi_{\text{EB}}}[\hat{T}_{\text{EB},m+1} - \hat{T}_{B,m+1}]^2 \to 0 \quad \text{as} \quad m \to \infty.$$

Hence, $\{\hat{T}_{\text{EB},m+1}, m \geq 1\}$ is asymptotically optimal and $\hat{T}_{\text{EB},m+1}$ is an EBP.

3.6. Exercises

[3.1] Consider the Bayesian model ψ_B given in Eqs. (3.1.1) and (3.1.2). Show that
(i) $(\mathbf{X}_s'\mathbf{V}_s^{-1}\mathbf{X}_s + \mathbf{B}^{-1})^{-1}(\mathbf{X}_s'\mathbf{V}_s^{-1}\mathbf{y}_s + \mathbf{B}^{-1}\mathbf{b}) = \mathbf{b} + \mathbf{B}[(\mathbf{X}_s'\mathbf{V}_s^{-1}\mathbf{X}_s)^{-1} + \mathbf{B}^{-1}]^{-1}(\hat{\boldsymbol{\beta}}_s - \mathbf{b})$ and
(ii) $(\mathbf{X}_s'\mathbf{V}_s^{-1}\mathbf{X}_s + \mathbf{B}^{-1})^{-1} = \mathbf{B} - \mathbf{B}((\mathbf{X}_s'\mathbf{V}_s^{-1}\mathbf{X}_s)^{-1} + \mathbf{B}]^{-1}\mathbf{B}$.

[3.2] Consider the Bayes model ψ_B of Theorem 3.1.2. Show that

$$E_{\psi_B}[\sigma^2|\mathbf{y}_s] = \frac{\nu}{\nu-2}\hat{\sigma}_s^2,$$

where $\hat{\sigma}_s^2$ is as defined in Eq. (3.1.12).

[3.3] (Scott and Smith, 1969) Consider model SM6. Let $\mathbf{y}_h$ be the vector of quantities associated with the elements of the hth primary unit $(h = 1, \ldots, k)$. Furthermore assume that, given μ_h and σ_h,

$$\mathbf{y}_h \mid \mu_h, \sigma_h \sim N(\mu_h 1_{N_h}, \sigma_h^2 I_{N_h}).$$

The means $\mu_1, \ldots, \mu_k$ are considered i.i.d. random variables, distributed like $N(\mu, \sigma_v^2)$.
(i) Show that the distribution of $\boldsymbol{\beta} = (\mu_1, \ldots, \mu_k)'$ given $\mathbf{y}_s$ is multivariate normal with mean $\hat{\boldsymbol{\beta}} = (\hat{\mu}_1, \ldots, \hat{\mu}_k)$, where

$$\hat{\mu}_h = \begin{cases} w_h \bar{y}_{s_h} + (1-w_h)\bar{y}_s, & \text{if } h \in \mathbf{s}, \\ \\ \bar{y}_s, & \text{if } h \notin \mathbf{s}, \end{cases}$$

where w_h is given in Eq. (2.1.8) and $\bar{y}_s = \sum_{h\in s} w_h \bar{y}_{s_h} / \sum_{h\in s} w_h$.

(ii) Show that the covariance matrix of $\boldsymbol{\beta}$ has elements

$$c_{hq} = \begin{cases} (1-w_h)^2\tau^2 + (1-w_h)\sigma_v^2, & \text{if } h = q, \\ (1-w_h)(1-w_q)\sigma_v^2, & \text{if } h \neq q, \end{cases}$$

where

$$\tau^2 = \left[\sum_{h\in s} \frac{m_h}{\sigma_h^2 + m_h\sigma_v^2}\right]^{-1}.$$

[3.4] Consider the superpopulation model SM3, where $g = 0$ and σ^2 is known. Consider noninformative prior on $\boldsymbol{\beta} = (\beta_o, \beta_1)'$. Find the Bayes predictor of T and its posterior variance.

[3.5] Consider the Bayes model (3.1.1) and (3.1.2) with $\mathbf{V}_{sr} = \mathbf{0}$. Show that the posterior distribution of $\boldsymbol{\beta}_N$ given $\mathbf{y}_s$ is multivariate normal with mean vector

$$\hat{\boldsymbol{\beta}}_{BN} = (\mathbf{X}'\mathbf{V}^{-1}\mathbf{X})^{-1}(\mathbf{X}_s'\mathbf{V}_s^{-1}\mathbf{y}_s + \mathbf{X}_r'\mathbf{V}_r^{-1}\mathbf{X}_r\hat{\boldsymbol{\beta}}_B)$$

and covariance matrix given by

$$\text{Var}_{\psi_B}[\boldsymbol{\beta}_N|\mathbf{y}_s] = (\mathbf{X}'\mathbf{V}^{-1}\mathbf{X})^{-1}\mathbf{X}_r'\mathbf{V}_r^{-1}\boldsymbol{\Sigma}_r\mathbf{V}_r^{-1}\mathbf{X}_r(\mathbf{X}'\mathbf{V}^{-1}\mathbf{X})^{-1},$$

where $\boldsymbol{\Sigma}_r$ is given in Eq. (3.1.4).

[3.6] Consider model SM3 with errors normally distributed, $g = 0$, $\beta_o = 0$, and $\beta_1 \sim N(b, B)$. Find the Bayes predictor of β_N and its prediction variance.

[3.7] Find the Bayes prediction risks of predictors (3.1.27) and (3.1.28).

[3.8] Consider the superpopulation model SM2 with errors normally distributed and $\beta \sim N(b, B)$. Find the Bayes predictor of S_y^2 and its Bayes prediction risk.

[3.9] Consider the superpopulation model where

$$E[\mathbf{y}|\boldsymbol{\beta}] = \mathbf{X}\boldsymbol{\beta} \text{ and } \text{Var}[\mathbf{y}|\boldsymbol{\beta}] = \mathbf{V}(\boldsymbol{\beta}).$$

Suppose also that the prior information on $\boldsymbol{\beta}$ is specified by

$$E[\boldsymbol{\beta}] = \boldsymbol{\mu} \text{ and } \text{Var}[\boldsymbol{\beta}] = \boldsymbol{\Sigma},$$

with both $\boldsymbol{\mu}$ and $\boldsymbol{\Sigma}$ known. Let $\bar{\mathbf{V}} = E[\mathbf{V}(\boldsymbol{\beta})]$. Find the Bayes linear predictor of T and its posterior variance (see Smouse, 1984).

[3.10] Derive the predictors listed in Table 3.1.

[3.11] (Rodrigues, 1988) Consider model ψ_G defined in Section 3.2 and let $\hat{T}_{LG}$ be any ψ_G-linear predictor and $\mathbf{K} = (\mathbf{V}_s, \mathbf{V}_{sr})$.
(i) Show that $\hat{T}_{LG}$ is ψ_G-unbiased if, and only if,

$$E_\psi[a + \mathbf{F}'\mathbf{y}_s - \mathbf{P}'\boldsymbol{\beta}] = 0,$$

where

$$\mathbf{F} = \mathbf{h}_s - \mathbf{1}_N'\mathbf{V}_s^{-1}\mathbf{K}\mathbf{1}_N \text{ and } \mathbf{P} = \mathbf{1}_N'(\mathbf{X} - \mathbf{X}_s\mathbf{V}_s^{-1}\mathbf{K})\mathbf{1}_N.$$

(ii) For any ψ_G-linear predictor $\hat{T}_{LG}$, show that

$$\begin{aligned} E_\psi[\hat{T}_{LG} - T]^2 &= \text{Var}_\psi[a + \mathbf{F}'\mathbf{y}_s - \mathbf{P}'\boldsymbol{\beta}] \\ &+ \mathbf{1}_N'\mathbf{V}'\mathbf{1}_N - \mathbf{1}_N'\mathbf{K}'\mathbf{V}_s^{-1}\mathbf{K}\mathbf{1}_N + [a + (\mathbf{h}_s'\mathbf{X}_s - \mathbf{1}_N'\mathbf{X})\boldsymbol{\mu}]^2, \end{aligned}$$

where $\mathbf{F}$ and $\mathbf{P}$ are as in (i).

[3.12] (Rodrigues, 1988) Use Exercise 3.11 and result 4a.11.4 in Rao (1973) to show, alternatively, that the ψ_G-best predictor and its ψ_G-MSE are as given in Theorem 3.2.1. If $\mathbf{b}$ is unknown and fixed, show that the ψ_G-best predictor of T is the $\hat{T}_{BLU}$.

[3.13] Consider the binomial superpopulation model of Example 3.3.1. Assuming that ψ has a prior beta distribution with known parameters a and b, show that the Bayes predictor of T is given by Eq. (3.3.1). Find the posterior variance of T.

[3.14] (Bolfarine, 1987) Consider the binomial superpopulation model of Example 3.3.1. Find the minimax predictor of the finite population distribution function defined in Exercise 1.14.

[3.15] (Bolfarine, 1987) Consider the Binomial model of Example 3.3.1. Let $\hat{T}_M$ be the minimax predictor of T.
(i) Show that the ψ-BUP of T is $\hat{T}_E$.
(ii) Show that

$$E_\psi[\hat{T}_E - T]^2 = \frac{N(N-n)}{n}\psi(1-\psi).$$

(iii) Show that

$$E_\psi[\hat{T}_M - T]^2 \leq E_\psi[\hat{T}_E - T]^2$$

in the interval $I_n = [\frac{1}{2} \mp \sqrt{\frac{1}{4} + b_n}]$, where

$$b_n = \frac{(N-n)[\sqrt{n} + \sqrt{n + N(N-n-1)}]}{N[\sqrt{N}(N-n) + \sqrt{n + N(N-n-1)}]}.$$

[3.16] (Bolfarine, 1989c) Consider the superpopulation model ψ in Eq. (3.1.1) and the linex loss function (Zellner, 1986)

$$L(\Delta) = b[e^{\Delta} - a\Delta - 1], \; a \neq 0, \; b > 0. \tag{3.6.1}$$

(i) Suppose that $\boldsymbol{\beta} \sim N(\mathbf{b}, \mathbf{B})$. Find the Bayes predictor of T and its posterior variance with respect to the loss function (3.5.1).
(ii) Suppose that the superpopulation model in (i) is such that $\mathbf{X}' = (x_1, \ldots, x_N)$ and $\mathbf{V} = \sigma^2 \text{diag}\{x_1, \ldots, x_N\}$. Use (i) to derive the Bayes predictor of T and its posterior variance.
(iii) Find a minimax predictor of T, for the superpopulation model described in (ii), under the loss function (3.6.1).
(iv) Using the limiting Bayesian risk approach (Example 3.3.5), prove that the minimax predictor derived in (iii) is admissible with respect to the loss (3.6.1).

[3.17] (Bolfarine, 1989c) Consider the binomial superpopulation model of Exercise 3.7. Find the Bayes predictor of T with respect to the loss function (3.6.1).

[3.18] (Bolfarine, 1989c) Consider the two-stage sampling superpopulation model SM6. Find the Bayes predictor of T and its posterior variance with respect to the loss function (3.6.1).

[3.19] Consider model SM2 with $\sigma^2 \leq M < \infty$ and no other assumption on the distribution of $\mathbf{y}$. Show that $\hat{T}_R$ is a minimax predictor of T.

[3.20] Let $y_1, \ldots, y_N$ be independent and $y_i \sim N(\beta x_i, \sigma^2)$, with σ^2 known. Find the minimax predictors of T with respect to the squared error loss and with respect to the loss (3.6.1).

[3.21] Show that the minimax predictors in Exercise 3.16(iii) and Exercise 3.20 are still minimax predictors if σ^2 is unknown, but $\sigma^2 \leq M < \infty$.

[3.22] Derive the dynamic Bayes predictors considered in Examples 3.4.1 and 3.4.2.

[3.23] Consider the stratified superpopulation model defined in Exercise 1.13. Find a minimax predictor of S_y^2.

[3.24] Prove Results 3.3.1, 3.3.2, 3.3.3, 3.3.4, and 3.3.5.

[3.25] Consider the regression model $\boldsymbol{\psi} = (\boldsymbol{\beta}, \mathbf{V})$, where $\mathbf{V} = \sigma^2 \mathbf{W}$ and $\mathbf{W}$ is known and diagonal. Suppose that $0 < \sigma^2 < M < \infty$. Show that under the above assumptions $\hat{T}_{BU} = \mathbf{1}_s' \mathbf{y}_s + \mathbf{1}_r' \mathbf{X}_r \hat{\boldsymbol{\beta}}_s$ is a minimax predictor of T with respect to the squared-error loss.

[3.26] Extend and prove Result 3.3.4 to the case of vector-valued quantities by using the GMSE given in Eq. (1.3.18). Show that $\hat{\boldsymbol{\beta}}_s$ is a minimax predictor of $\boldsymbol{\beta}_N$ under the regression model $\boldsymbol{\psi} = (\boldsymbol{\beta}, \mathbf{V})$, where $\mathbf{V} = \sigma^2 \mathbf{W}$, $\mathbf{W}$ is known and diagonal and σ^2 (which could be unknown) belong to a closed interval $[0, M]$, $0 < M < \infty$.

[3.27] Consider the normal dynamic regression model given in Eqs. (3.4.1) and (3.4.2). Using Theorem 3.1.3 and Eqs. (3.4.6) and (3.4.7) derive the dynamic Bayes predictor of $\boldsymbol{\beta}_N$.

4
Maximum–Likelihood Predictors

In this chapter, we study the structure of maximum–likelihood predictors (MLP). There are several notions in the literature of predictive likelihoods. The reader is referred to the illuminating review paper of Bjørnstad (1990) on the subject. We focus attention on three types of predictive likelihoods and the ensuing predictors.

Let $\mathbf{y}_r$ be the quantity vector corresponding to the elements that are not in the sample. As defined in this book, the predictive likelihood of $\mathbf{y}_r$ given a superpopulation model under ψ is $\text{Lik}(\mathbf{y}_r \mid \psi) \propto f(\mathbf{y}_r \mid \psi)$, where $f(\mathbf{y}_r \mid \psi)$ is the joint probability density function (p.d.f.) of $\mathbf{y}_r$ under ψ. The parameter ψ is generally unknown and the predictor of $\theta(\mathbf{y})$ should be a function only of the observed sample values $\mathbf{y}_s$. We have therefore to eliminate the nuisance parameter ψ. In the previous chapter, we saw the Bayes predictive distribution of $\mathbf{y}_r$ given $\mathbf{y}_s$, for the Bayes normal model. Generally, if $\xi(\psi)$ is the prior p.d.f. of ψ and $\xi(\psi \mid \mathbf{y}_s)$ is the posterior p.d.f. of ψ given $\mathbf{y}_s$, the Bayes predictive likelihood of $\mathbf{y}_r$ given $\mathbf{y}_s$ under $\xi(\cdot)$ is $\text{Lik}_B(\mathbf{y}_r \mid \mathbf{y}_s) \propto \int f(\mathbf{y}_r \mid \psi)\xi(\psi \mid \mathbf{y}_s)d\psi$.

In this chapter, we develop other types of predictive likelihood functions and the resulting predictors. In particular, we consider *estimative predictive likelihoods*, $\text{Lik}_E(\mathbf{y}_r \mid \hat{\psi}_s)$, in which $\hat{\psi}_s$ is the maximum-likelihood estimator (MLE) of ψ; profile predictive likelihoods, $\text{Lik}_p(\mathbf{y}_r \mid \mathbf{y}_s)$; *Lauritzen–Hinkley predictive likelihoods* (see Hinkley, 1979); and *Royall joint predictive likelihoods* (Royall, 1976 b).

4.1. Predictive Likelihoods

4.1.1. Estimative Predictive Likelihoods

Consider a superpopulation model in which the joint p.d.f. of $\mathbf{y} = (\mathbf{y}'_s, \mathbf{y}'_r)$, under ψ, is $f(\mathbf{y}_s, \mathbf{y}_r \mid \psi)$, $\psi \in \Psi$. Let $f_s(\mathbf{y}_s \mid \psi)$ be the marginal p.d.f. of

$\mathbf{y}_s$ under ψ. The likelihood function of ψ given $\mathbf{y}_s$ is

$$\text{Lik}(\psi; \mathbf{y}_s) = f_s(\mathbf{y}_s \mid \psi), \quad \psi \in \Psi. \tag{4.1.1}$$

An MLE of ψ given $\mathbf{y}_s$, if it exists, is a point in Ψ, $\hat{\psi}_s$, for which

$$\text{Lik}(\hat{\psi}_s; \mathbf{y}_s) = \sup_{\psi \in \Psi} \text{Lik}(\psi; \mathbf{y}_s).$$

Definition 4.1.1. *Consider a superpopulation model $\{f(\mathbf{y}_s, \mathbf{y}_r \mid \psi); \psi \in \Psi\}$. Let $\hat{\psi}_s$ be an MLE of ψ. An estimative predictive likelihood of $\mathbf{y}_r$ given $\mathbf{y}_s$ is*

$$Lik_E(\mathbf{y}_r \mid \mathbf{y}_s) = f_r(\mathbf{y}_r; \hat{\psi}_s), \quad \mathbf{y}_r \in \mathcal{Y}^{(r)}, \tag{4.1.2}$$

where $f_r(\mathbf{y}_r; \psi)$ is the marginal p.d.f. of $\mathbf{y}_r$ under ψ, and $\mathcal{Y}^{(r)}$ is the space of $\mathbf{y}_r$.

Example 4.1.1. (i) Consider the general normal regression model, that is, $\mathbf{y} \sim N(\mathbf{X}\boldsymbol{\beta}, \sigma^2\mathbf{W})$, where $\psi = (\boldsymbol{\beta}, \sigma)$. The MLE of $(\boldsymbol{\beta}, \sigma)$ is $(\hat{\boldsymbol{\beta}}_s, \hat{\sigma}_s)$ (see Srivastava and Khatri, 1979). Hence

$$\begin{aligned} \text{Lik}_E(\mathbf{y}_r \mid \mathbf{y}_s) = {} & \frac{1}{(2\pi)^{\frac{N-n}{2}} \hat{\sigma}_s^{(N-n)} |\mathbf{W}_r|^{1/2}} \\ & \exp\left\{ -\frac{1}{2\hat{\sigma}_s^2} (\mathbf{y}_r - \mathbf{X}_r\hat{\boldsymbol{\beta}}_s)' \mathbf{W}_r^{-1} (\mathbf{y}_r - \mathbf{X}_r\hat{\boldsymbol{\beta}}_s) \right\}. \end{aligned} \tag{4.1.3}$$

(ii) Consider a location–scale parameter exponential distribution, according to which, $y_1, \cdots, y_N$ are *independent*, and the p.d.f. of y_i is

$$f(y_i; \mu, \beta, x_i) = \frac{1}{\beta x_i} \exp\left\{ -\frac{(y_i - \mu)}{\beta x_i} \right\} I_{(\mu,\infty)}(y_i),$$

$i = 1, \cdots, N$; where $x_1, \cdots, x_N$ are known positive quantities and β and μ are unknown parameters, $-\infty < \mu < \infty$, $0 < \beta < \infty$. $I_A(y)$ is the indicator function for the set A.

The MLE of $\psi = (\mu, \beta)$ are

$$\hat{\beta}_s = \frac{1}{n} \sum_{i \in s} \frac{y_i - \hat{\mu}_s}{x_i}, \quad \text{and} \quad \hat{\mu}_s = y_{s(1)}.$$

Hence, the estimative predictive likelihood of $\mathbf{y}_r$, given $\mathbf{y}_s$, is

$$\begin{aligned} \text{Lik}_E(\mathbf{y}_r \mid \mathbf{y}_s) = {} & \frac{1}{(\hat{\beta}_s)^{N-n} \prod_{i \in r} x_i} \exp\left\{ -\frac{1}{\hat{\beta}_s} \sum_{i \in r} \frac{y_i - \hat{\mu}_s}{x_i} \right\} \times \\ & I_{(\hat{\mu}_s,\infty)}(y_{r(1)}). \end{aligned} \tag{4.1.4}$$

where $y_{r(1)} = \min\{y_i, i \in \mathbf{r}\}$.

Let $\theta(\mathbf{y})$ be a quantity of interest. Let $\mathcal{T}$ be the space of $\theta(\mathbf{y})$.

Definition 4.1.2. *An estimative maximum–likelihood predictor (EMLP) of $\theta(\mathbf{y})$, if it exists, is a point in $\mathcal{T}$, $\hat{\theta}_{\mathrm{EML}} = \theta(\mathbf{y}_s, \hat{\mathbf{y}}_{\mathrm{EML}})$, where*

$$Lik_E(\hat{\mathbf{y}}_{\mathrm{EML}} \mid \mathbf{y}_s) = \sup_{\mathbf{y}_r \in \mathcal{Y}^{(r)}} Lik_E(\mathbf{y}_r \mid \mathbf{y}_s). \tag{4.1.5}$$

Example 4.1.2. In continuation of Example 4.1.1 (ii), let $\beta_N = 1/N\sum_{i=1}^{N} y_i/x_i$ be the population quantity of interest. Notice that $T_r = \sum_{i\in r}(y_i/x_i) - \mu\sum_{i\in r}(1/x_i)$ has the gamma distribution with parameters $(\beta, N-n)$. Let $U_r = \sum_{i\in r} y_i/x_i$. The p.d.f. of U_r, given (μ,β), is

$$f(u_r \mid \beta, \mu) = \frac{1}{\beta^{N-n}\Gamma(N-n)}\left(u_r - \mu\sum_{i\in r}\frac{1}{x_i}\right)^{N-n-1}$$
$$\exp\left[-\frac{1}{\beta}\left(U_r - \mu\sum_{i\in r}(1/x_i)\right)\right] I_{(\mu\sum_{i\in r}\frac{1}{x_i},\infty)}(u_r).$$

Accordingly, the estimative likelihood of U_r given $(\hat{\beta}_s, \hat{\mu}_s)$ is

$$\mathrm{Lik}_E(u_r \mid \hat{\beta}_s, \hat{\mu}_s) \propto \left(u_r - \hat{\mu}_s\sum_{i\in r}\frac{1}{x_i}\right)^{N-n-1}$$
$$\exp\left[-\frac{1}{\hat{\beta}_s}\left(U_r - \hat{\mu}_s\sum_{i\in r}(1/x_i)\right)\right] I_{(\hat{\mu}_s\sum_{i\in r}\frac{1}{x_i},\infty)}(u_r).$$

Accordingly, the EMLP of $\hat{T}_r = U_r - \hat{\mu}_s\sum_{i\in r}\frac{1}{x_i}$ is $(N-n-1)\hat{\beta}_s$, and the EMLP of $\hat{\beta}_N$ is

$$\hat{\beta}_{\mathrm{EML}} = (1 - \frac{1}{N})\hat{\beta}_s + \hat{\mu}_s\frac{1}{N-n}\sum_{i\in r}\frac{1}{x_i}. \tag{4.1.6}$$

It is not clear what is the EMLP of the population total T in the above example. We therefore introduce an additional definition of an estimative predictor.

Definition 4.1.3. *Let $L[\hat{\theta}(\mathbf{y}_s), \theta(\mathbf{y})]$ be a loss function for predictors of $\theta(\mathbf{y})$. Let $h(\theta \mid \mathbf{y}_s, \psi)$ be the conditional p.d.f. of $\theta(\mathbf{y})$, given $\mathbf{y}_s$ under ψ. Let $\hat{\theta}_\psi(\mathbf{y}_s)$ be a predictor minimizing*

$$R(\hat{\theta}, \theta; \psi) = \int L[\hat{\theta}(\mathbf{y}_s), \theta] h(\theta \mid \mathbf{y}_s, \psi) d\theta.$$

Let $\hat{\psi}_s$ be the MLE of ψ, then $\hat{\theta}_{\hat{\psi}_s}(\mathbf{y}_s)$ is called best estimative predictor (BEP) of θ.

Example 4.1.3. In continuation of Example 4.1.2, the BEP of T, for the squared–error loss, is

$$\hat{T}_{\text{BE}} = n\bar{y}_s + (N-n)(\hat{\mu}_s + \hat{\beta}_s \bar{x}_r). \tag{4.1.7}$$

4.1.2 Profile Predictive Likelihoods

Another type of predictive likelihood is that of a *profile predictive likelihood.* Let $\theta(\mathbf{y}) = \theta_s + \theta_{sr}$ be a population quantity. Let $f(\mathbf{y}_s \mid \psi)$ be the p.d.f. of $\mathbf{y}_s$ under ψ and $h(\theta_{sr} \mid \mathbf{y}_s, \psi)$ be the p.d.f. of θ_{sr} under ψ, given $\mathbf{y}_s$.

Definition 4.1.4. *Under the above assumptions, the profile predictive likelihood function of θ_{sr} is given by*

$$\text{Lik}_P(\theta_{sr} \mid \mathbf{y}_s) \propto \sup_{\psi \in \Psi} f(\mathbf{y}_s \mid \psi) h(\theta_{sr} \mid \mathbf{y}_s, \psi). \tag{4.1.8}$$

Definition 4.1.5. *The maximum profile likelihood predictor (MPLP) of $\theta(\mathbf{y})$ is $\hat{\theta}_P(\mathbf{y}_s) = \theta_s(\mathbf{y}_s) + \hat{\theta}_{P,sr}(\mathbf{y}_s)$, where $\hat{\theta}_{P,sr}(\mathbf{y}_s)$ maximizes $\text{Lik}_P(\theta_{sr} \mid \mathbf{y}_s)$.*

We illustrate this approach with an example.

Example 4.1.4. Consider the normal regression model with $\mathbf{X} = \mathbf{1}_N$, an N–dimensional vector of ones, and $\mathbf{V} = \sigma^2 \mathbf{I}_N$, and suppose that σ^2 is known. We wish to predict the population total T. The distribution of $\mathbf{y}_s$ under $\psi = \beta$ is $N(\beta \mathbf{1}_N, \sigma^2 \mathbf{I}_n)$. Furthermore, the distribution of $\theta_{sr} = T_r = \sum_{i \in r} y_i$ under ψ is $N[(N-n)\beta, (N-n)\sigma^2]$. Thus, the profile likelihood predictor of T_r is given by

$$\begin{aligned} \text{Lik}_P(T_r \mid \mathbf{y}_s) &\propto \sup_{-\infty<\beta<\infty} \exp\left\{ -\frac{1}{2\sigma^2} \left(n(\bar{\mathbf{y}}_s - \beta)^2 \right.\right. \\ &\quad \left.\left. + \frac{1}{N-n} [T_r - (N-n)\beta]^2 \right) \right\} \\ &= \exp\left\{ -\frac{1}{2\sigma^2} \left[n(\bar{y}_s - \bar{y})^2 + (N-n) \left(\frac{T_r}{N-n} - \bar{y} \right)^2 \right] \right\}, \end{aligned} \tag{4.1.9}$$

where $\bar{y}_s$ is the sample mean and $\bar{y}$ is the population mean. Formula (4.1.9) can be written more explicitly as

$$\begin{aligned}\mathrm{Lik}_P(T_r \mid \mathbf{y}_s) &\propto \exp\left\{-\frac{1}{\sigma^2}\left[n\left(\bar{y}_s - \frac{n\bar{y}_s + T_r}{N}\right)^2 \right.\right.\\ &\qquad\left.\left. + (N-n)\left(\frac{T_r}{N-n} - \frac{n\bar{y}_s + T_r}{N}\right)^2\right]\right\} \\ &= \exp\left\{-\frac{1}{2\sigma^2}\frac{n}{N}\frac{2N-n}{(N-n)N}[T_r - (N-n)\bar{y}_s]^2\right\}.\end{aligned} \tag{4.1.10}$$

It follows that the MPLP of T_r is $(N-n)\bar{y}_s$, and the MPLP of T is $N\bar{y}_s$, which is the well-known expansion predictor.

4.1.3. The Lauritzen–Hinkley Predictive Likelihoods

Suppose that the superpopulation model is represented by a family of distributions of $\mathbf{y}$, $\mathcal{F} = \{F(\cdot, \psi), \psi \in \Psi\}$. Suppose that $\mathcal{F}_s$ and $\mathcal{F}_r$ are the corresponding families of marginal distributions of $\mathbf{y}_s$ and $\mathbf{y}_r$, respectively. In this section, we consider the case where $\mathbf{y}_s$ and $\mathbf{y}_r$ are *independent*.

Suppose that $\mathcal{F}_s$, $\mathcal{F}_r$, and $\mathcal{F}$ admit nontrivial sufficient statistics and let U_s, U_r, and U be the corresponding minimal sufficient statistics. It is further assumed that U_r can be uniquely determined by U and U_s. Notice that since U is sufficient for $\mathcal{F}$, the conditional distribution of U_s given U is independent of ψ. Similarly, the conditional distribution of $\mathbf{y}_r$ given U_r is independent of ψ. Let $h(u_s \mid u)$ denote the conditional p.d.f. of U_s given $U = u$. Let $f(\mathbf{y}_r \mid u_r)$ denote the conditional p.d.f. of $\mathbf{y}_r$ given $U_r = u_r$.

Definition 4.1.6. *The Lauritzen–Hinkley predictive likelihood of U_r given U_s is*

$$Lik_{\mathrm{LH}}(u_r \mid u_s) \propto h(u_s \mid u), \tag{4.1.11}$$

and the Lauritzen–Hinkley predictive likelihood of $\mathbf{y}_r$ given U_s is

$$Lik_{\mathrm{LH}}(\mathbf{y}_r \mid u_s) \propto f(\mathbf{y}_r \mid u_r)h(u_s \mid u). \tag{4.1.12}$$

Example 4.1.5. We derive here the Lauritzen–Hinkley (LH) predictive likelihood for the exponential model of Example 4.1.1(ii) with $\mu = 0$. In this case, $\mathcal{F}_s$ is the family of exponential distributions of $\mathbf{y}_s$, as $0 < \beta < \infty$, and $\mathcal{F}_r$ is the corresponding family of exponential distributions for $\mathbf{y}_r$. The minimal sufficient statistics are, respectively,

$$\begin{aligned}&U_s \equiv \hat{\beta}_s = \frac{1}{n}\sum_{i\in s}\frac{y_i}{x_i}, \quad U_r \equiv \beta_r = \frac{1}{N-n}\sum_{i\in r}\frac{y_i}{x_i}, \quad \text{and}\\ &U \equiv \beta_N = \frac{n}{N}\hat{\beta}_s + (1 - \frac{n}{N})\beta_r.\end{aligned}$$

Thus $\beta_r = (\beta_N - \frac{n}{N}\hat{\beta}_s)/(1-\frac{n}{N})$. Furthermore, $\hat{\beta}_s \sim \frac{\beta}{n}G(1,n)$, $\beta_r \sim \frac{\beta}{N-n}G(1,N-n)$, where $G(1,\nu)$ designates a r.v. having a gamma distribution with scale parameter $\sigma = 1$ and shape parameter ν. Therefore, since $\hat{\beta}_s$ and β_r are independent,

$$\frac{n\hat{\beta}_s}{n\hat{\beta}_s + (N-n)\beta_r} = \frac{n\hat{\beta}_s}{N\beta_N} \sim \text{Beta}(n, N-n)$$

or $\hat{\beta}_s/\beta_N \sim \frac{N}{n}\text{Beta}(n, N-n)$ where Beta(p,q) designates a random variable having a beta distribution with parameters (p,q). Thus, the LH predictive likelihood of β_r is

$$\text{Lik}_{\text{LH}}(\beta_r \mid \hat{\beta}_s) \propto \frac{1}{n\hat{\beta}_s + (N-n)\beta_r}\left(\frac{n\hat{\beta}_s}{n\hat{\beta}_s + (N-n)\beta_r}\right)^{n-1} \left(1 - \frac{n\hat{\beta}_s}{n\hat{\beta}_s + (N-n)\beta_r}\right)^{N-n-1} \tag{4.1.13}$$

Definition 4.1.7. *Let U_r be a minimal sufficient statistic for $\mathcal{F}_r$, and let $\theta(\mathbf{y}) = \theta_s(\mathbf{y}_s) + \theta_{sr}(\mathbf{y}_s, U_r)$ be a population quantity that depends on $\mathbf{y}_r$ only through U_r. Let U_r^0 be a value of U_r maximizing Eq. (4.1.11). Then the maximum LH predictor (MLHP) of θ is $\hat{\theta}_{\text{MLH}} = \theta_s(\mathbf{y}_s) + \theta_{sr}(\mathbf{y}_s, U_r^0)$.*

Example 4.1.6. In continuation of Example 4.1.5, consider the population quantity $\beta_N = \frac{n}{N}\hat{\beta}_s + (1-\frac{n}{N})\beta_r$. From the maximization of $\text{Lik}_{\text{LH}}(\beta_r \mid \hat{\beta}_s)$, the MLHP of $(N-n)\beta_r$ is $(N-n-1)\hat{\beta}_s$, and, consequently, the MLHP of β_N is $(1-\frac{1}{N})\hat{\beta}_s$. This predictor coincides with the EMLP when $\mu = 0$.

4.1.4. The Royall Predictive Likelihoods

Royall (1976b) introduced the following joint likelihood function for (θ, ψ), where θ is a population quantity. Let $f(\mathbf{y};\psi)$ be the joint p.d.f. of $\mathbf{y}$ under ψ, and let $\theta = \theta(\mathbf{y})$ be a quantity of interest. Let $h(\mathbf{y}_s \mid \theta, \psi)$ denote the conditional p.d.f. of $\mathbf{y}_s$ given θ under ψ.

Definition 4.1.8. *Let $\mathcal{F} = \{f(\mathbf{y};\psi); \psi \in \Psi\}$ be a superpopulation model. Let $\theta(\mathbf{y})$ be a population quantity. The Royall joint predictive likelihood (RJPL) of (θ, ψ) is*

$$Lik_R(\theta, \psi; \mathbf{y}_s) \propto h(\mathbf{y}_s \mid \theta(\mathbf{y}), \psi), \tag{4.1.14}$$

for $(\theta, \psi) \in \Theta \times \Psi$.

Definition 4.1.7. *Let $Lik_R(\theta, \psi; \mathbf{y}_s)$, $(\theta, \psi) \in \Theta \times \Psi$ be an RJPL. If (θ^0, ψ^0) is a point in $\Theta \times \Psi$ maximizing $Lik_R(\theta, \psi; \mathbf{y}_s)$, then $\hat{\theta}_{\text{RML}}(\mathbf{y}_s) = \theta^0$ is called the Royall maximum-likelihood predictor (RMLP) of θ.*

Example 4.1.7. Consider the model of Example 4.1.1(ii) with $\mu = 0$. Let $\theta = \beta_N = \frac{1}{N}\sum_{i=1}^{n} \frac{y_i}{x_i}$. Let $\hat{\beta}_s = \frac{1}{n}\sum_{i \in s} \frac{y_i}{x_i}$. Since $\hat{\beta}_s$ is sufficient for $\mathcal{F}_s$, and β_N is sufficient for $\mathcal{F}$, the conditional p.d.f. of $\mathbf{y}_s$, given β_N is independent of β, and

$$h(\mathbf{y}_s \mid \beta_N) = g(\mathbf{y}_s \mid \hat{\beta}_s) k(\hat{\beta}_s \mid \beta_N),$$

where $g(\mathbf{y}_s \mid \hat{\beta}_s)$ is the conditional p.d.f. of $\mathbf{y}_s$ given $\hat{\beta}_s$. The conditional p.d.f. of $\hat{\beta}_s$, given β_N, $k(\hat{\beta}_s \mid \beta_N)$ is proportional to Eq. (4.1.13). Accordingly,

$$\text{(4.1.15)} \qquad \text{Lik}_R(\beta_N; \mathbf{y}_s) \propto g(\mathbf{y}_s \mid \hat{\beta}_s)\text{Lik}_{\text{LH}}(\hat{\beta}_r \mid \hat{\beta}_s).$$

Hence the RMLP of β_N coincides with the MLHP, which is $\hat{\beta}_N = (1 - \frac{1}{N})\hat{\beta}_s$.

In Examples 4.1.2, 4.1.6, and 4.1.7 we have demonstrated cases where different types of maximum-likelihood predictors coincide. In the next section, we investigate the structure of these maximum-likelihood predictors of the population total, T, under the normal regression model. Section 4.3 is devoted to the maximum-likelihood predictors of the population variance S_y^2, under the normal regression model.

4.2. Maximum–Likelihood Predictors of T Under the Normal Superpopulation Model

We present the results in this section for cases where $\mathbf{y}_s$ and $\mathbf{y}_r$ are independent; that is, the superpopulation model is:

$$\text{(4.2.1)} \qquad \mathbf{y} \sim N(\mathbf{X}\boldsymbol{\beta}, \sigma^2 \text{diag}\{\mathbf{W}_s, \mathbf{W}_r\}),$$

where $\psi = (\boldsymbol{\beta}, \sigma) \in \mathbb{R}^{(p)} \times (0, \infty)$. The population quantity of interest is $T = \mathbf{1}'_N \mathbf{y}$.

4.2.1. Estimative Likelihood Predictors

The MLE of $\boldsymbol{\beta}$ and σ^2 are, as mentioned earlier, the WLSE, $\hat{\boldsymbol{\beta}}_s$, and $\hat{\sigma}_s^2$ given by Eqs. (1.3.14) and (2.2.5). Hence the estimative prediction likelihood is given by Eq. (4.1.2). An EMLP of $\mathbf{y}_r$ is $\mathbf{X}_r\hat{\boldsymbol{\beta}}_s$ and an EMLP of T is

$$\text{(4.2.2)} \qquad \hat{T}_{\text{EMLP}} = n\bar{y}_s + \mathbf{1}'_r \mathbf{X}_r \hat{\boldsymbol{\beta}}_s.$$

This predictor coincides with the BUP of T.

4.2.2. Profile Likelihood Predictors

We assume that $\sigma^2 = 1$ (known). The profile predictive likelihood of T_r is

$$\text{Lik}_p(T_r \mid \mathbf{y}_s) \propto \exp\Bigg\{ -\frac{1}{2}\Bigg[\frac{T_r^2}{D_r^2} + \mathbf{y}_s'\mathbf{W}_s^{-1}\mathbf{y}_s - \left(\mathbf{X}_s'\mathbf{W}_s^{-1}\mathbf{y}_s + \frac{T_r}{D_r^2}\mathbf{X}_r'\mathbf{1}_r\right)' \left(\mathbf{X}_s'\mathbf{W}_s^{-1}\mathbf{X}_s + \frac{1}{D_r^2}\mathbf{X}_r'\mathbf{J}_r\mathbf{X}_r\right)^{-1} \left(\mathbf{X}_s'\mathbf{W}_s^{-1}\mathbf{y}_s + \frac{T_r}{D_r^2}\mathbf{X}_r'\mathbf{1}_r\right)\Bigg]\Bigg\}, \tag{4.2.3}$$

$-\infty < T_r < \infty$, where $D_r^2 = \mathbf{1}_r'\mathbf{W}_r\mathbf{1}_r$. This predictive likelihood yields the MPLP of T_r,

$$\hat{T}_{r,\text{MPL}} = \frac{\mathbf{1}_r'\mathbf{X}_r(\mathbf{X}_s'\mathbf{W}^{-1}\mathbf{X}_s + \frac{1}{D_r^2}\mathbf{X}_r'\mathbf{J}_r\mathbf{X}_r)^{-1}\mathbf{X}_s'\mathbf{W}_s^{-1}\mathbf{y}_s}{1 - \frac{1}{D_r^2}\mathbf{1}_r'\mathbf{X}_r(\mathbf{X}_s'\mathbf{W}_s^{-1}\mathbf{X}_s + \frac{1}{D_r^2}\mathbf{X}_r'\mathbf{J}_r\mathbf{X}_r)^{-1}\mathbf{X}_r'\mathbf{1}_r} \tag{4.2.4}$$

Theorem 4.2.1. *The MPLP of T under the assumed normal regression model coincides with the BUP of T; that is,*

$$\hat{T}_{\text{MPL}} = n\bar{y}_s + \mathbf{1}_r'\mathbf{X}_r\hat{\boldsymbol{\beta}}_s. \tag{4.2.5}$$

Proof. Let $\mathbf{M} = [\mathbf{X}_s'\mathbf{W}_s^{-1}\mathbf{X}_s + \frac{1}{D_r^2}\mathbf{X}_r'\mathbf{J}_r\mathbf{X}_r]$ and $q = \frac{1}{D_r^2}\mathbf{1}_r'\mathbf{X}_r\mathbf{M}^{-1}\mathbf{X}_r'\mathbf{1}_r$. Furthermore, let $\sigma_{BU}^2 = \mathbf{1}_r'\mathbf{X}_r(\mathbf{X}_s'\mathbf{W}_s^{-1}\mathbf{X}_s)^{-1}\mathbf{X}_r\mathbf{1}_r$. Then, $\mathbf{M}^{-1} = (\mathbf{I}_p + \frac{1}{D_r^2}\mathbf{B})^{-1}(\mathbf{X}_s'\mathbf{W}_s^{-1}\mathbf{X}_s)^{-1}$, where $\mathbf{B} = (\mathbf{X}_s'\mathbf{W}_s^{-1}\mathbf{X}_s)^{-1}\mathbf{X}_r'\mathbf{J}_r\mathbf{X}_r$. Notice that $\mathbf{B}^2 = \sigma_{BU}^2\mathbf{B}$ and hence

$$\left(\mathbf{I}_p + \frac{1}{D_r^2}\mathbf{B}\right)^{-1} = \mathbf{I}_p - \frac{1}{D_r^2}\left(1 + \frac{\sigma_{BU}^2}{D_r^2}\right)^{-1}\mathbf{B}.$$

It is straightforward then to show that $q = (\sigma_{BU}^2/D_r^2)[1 + \sigma_{BU}^2/D_r^2]^{-1}$. It follows that

$$\begin{aligned}
\hat{T}_{r,\text{MPL}} &= \frac{1}{1-q}\mathbf{1}_r'\mathbf{X}_r\mathbf{M}^{-1}(\mathbf{X}_s'\mathbf{W}_s^{-1}\mathbf{X}_s)\hat{\boldsymbol{\beta}}_s \\
&= \left(1 + \frac{\sigma_{BU}^2}{D_r^2}\right)\mathbf{1}_r'\mathbf{X}_r\left[\mathbf{I}_p - \frac{1}{D_r^2}\left(1 + \frac{\sigma_{BU}^2}{D_r^2}\right)^{-1}\mathbf{B}\right]\hat{\boldsymbol{\beta}}_s \\
&= \left(1 + \frac{\sigma_{\text{BU}}^2}{D_r^2}\right)\mathbf{1}_r'\mathbf{X}_r\hat{\boldsymbol{\beta}}_s - \frac{1}{D_r^2}\mathbf{1}_r'\mathbf{X}_r\mathbf{B}\hat{\boldsymbol{\beta}}_s.
\end{aligned}$$

But,

$$\begin{aligned}
\mathbf{1}_r'\mathbf{X}_r\mathbf{B}\hat{\boldsymbol{\beta}}_s &= \mathbf{1}_r'\mathbf{X}_r(\mathbf{X}_s'\mathbf{W}_s^{-1}\mathbf{X}_s)\mathbf{X}_r\mathbf{1}_r\mathbf{1}_r'\mathbf{X}_r\hat{\boldsymbol{\beta}}_s \\
&= \sigma_{BU}^2\mathbf{1}_r'\mathbf{X}_r\hat{\boldsymbol{\beta}}_s.
\end{aligned}$$

Hence, $\hat{T}_{r,\text{MPL}} = \mathbf{1}'_r \mathbf{X}_r \hat{\boldsymbol{\beta}}_s$.

(Q.E.D.)

It is straightforward to derive the predictive mean–squared error of $\hat{T}_{BU}$, namely,

$$\text{PMSE}[\hat{T}_{BU}; \boldsymbol{\psi}] = \mathbf{1}'_r \mathbf{W}_r \mathbf{1}_r + \sigma^2_{\text{BU}}. \tag{4.2.6}$$

If $\sigma^2 \neq 1$ we have to multiply the right-hand side of Eq. (4.2.5) by σ^2.

4.2.3. The LH Likelihood Predictors

In this section, we derive the LH predictive likelihood for the special case of known σ^2. When σ^2 is unknown the computations become too cumbersome, and the usefulness of the method is doubtful. Thus, let $\mathcal{F}_s = \{\mathcal{N}(\mathbf{X}_s\boldsymbol{\beta}, \mathbf{W}_s);\ \boldsymbol{\beta} \in \mathbb{R}^{(p)}\}$, $\mathcal{F}_r = \{\mathcal{N}(\mathbf{X}_r\boldsymbol{\beta}, \mathbf{W}_r);\ \boldsymbol{\beta} \in \mathbb{R}^{(p)}\}$, and $\mathcal{F} = \{\mathcal{N}(\mathbf{X}\boldsymbol{\beta}, \text{diag}\{\mathbf{W}_s, \mathbf{W}_r\)\};\ \boldsymbol{\beta} \in \mathbb{R}^{(p)}\}$. The minimal sufficient statistics for $\mathcal{F}_s$, $\mathcal{F}_r$, and $\mathcal{F}$ are $\hat{\boldsymbol{\beta}}_s$, $\boldsymbol{\beta}_r$, and $\boldsymbol{\beta}_N$, respectively, where $\boldsymbol{\beta}_r = (\mathbf{X}'_r\mathbf{W}_r^{-1}\mathbf{X}_r)^{-1}$ $\mathbf{X}'_r\mathbf{W}_r^{-1}\mathbf{y}_s$. Notice that the conditional distribution of $\hat{\boldsymbol{\beta}}_s$ given $\boldsymbol{\beta}_N$ is independent of $\boldsymbol{\beta}$. We have to derive the p.d.f. of this conditional distribution in order to obtain $\text{Lik}_{\text{LH}}(\boldsymbol{\beta}_r; \hat{\boldsymbol{\beta}}_s)$.

According to Eq. (1.3.13), $\boldsymbol{\beta}_N = \mathbf{E}_s\hat{\boldsymbol{\beta}}_s + \mathbf{E}_r\boldsymbol{\beta}_r$, where $\mathbf{E}_s$ and $\mathbf{E}_r$ are given in Eqs. (1.3.11) and (1.3.12). Let $\mathbf{P}_{sr} = \mathbf{X}'_s\mathbf{W}_s^{-1}\mathbf{X}_s + \mathbf{X}'_r\mathbf{W}_r^{-1}\mathbf{X}_r$. Then, $\mathbf{E}_s = \mathbf{P}_{sr}^{-1}(\mathbf{X}'_s\mathbf{W}_s^{-1}\mathbf{X}_s)$ and $\mathbf{E}_r = \mathbf{P}_{sr}^{-1}(\mathbf{X}'_r\mathbf{W}_r^{-1}\mathbf{X}_r)$. Thus, the joint distribution of $\mathbf{E}_s\hat{\boldsymbol{\beta}}_s$ and $\mathbf{E}_r\boldsymbol{\beta}_r$ is normal with mean vector $\begin{pmatrix}\mathbf{E}_s\\ \mathbf{E}_r\end{pmatrix}\boldsymbol{\beta}$ and covariance matrix $\text{diag}\{\mathbf{P}_{sr}^{-1}\mathbf{E}'_s, \mathbf{P}_{sr}^{-1}\mathbf{E}'_r\}$. It follows that the covariance matrix of $\mathbf{E}_s\hat{\boldsymbol{\beta}}_s$ and $\boldsymbol{\beta}_N$ is

$$\begin{bmatrix} \mathbf{P}_{sr}^{-1}\mathbf{E}'_s & \mathbf{P}_{sr}^{-1}\mathbf{E}'_s \\ \mathbf{E}_s\mathbf{P}_{sr}^{-1} & \mathbf{P}_{sr}^{-1} \end{bmatrix}.$$

Routine calculations yield that the conditional distribution of $\mathbf{E}_s\hat{\boldsymbol{\beta}}_s$ given $\boldsymbol{\beta}_N$ is normal with mean $\mathbf{E}_s\boldsymbol{\beta}_N$ and covariance matrix $\mathbf{P}_{sr}^{-1}\mathbf{E}'_s(\mathbf{I}_p - \mathbf{E}'_s)$. Finally we obtain $\hat{\boldsymbol{\beta}}_s \mid \boldsymbol{\beta}_N \sim N(\boldsymbol{\beta}_N, \mathbf{I}_p - \mathbf{P}_{sr}^{-1})$. Accordingly,

(4.2.7)

$$\begin{aligned} \text{Lik}_{\text{LH}}(\boldsymbol{\beta}_r \mid \hat{\boldsymbol{\beta}}_s) &\propto \exp\{-\frac{1}{2}(\hat{\boldsymbol{\beta}}_s - \boldsymbol{\beta}_N)'(\mathbf{I} - \mathbf{P}_{sr}^{-1})^{-1}(\hat{\boldsymbol{\beta}}_s - \boldsymbol{\beta}_N)\} \\ &= \exp\{-\frac{1}{2}(\hat{\boldsymbol{\beta}}_s - \boldsymbol{\beta}_r)'\mathbf{E}'_r(\mathbf{I} - \mathbf{P}_{sr}^{-1})^{-1}\mathbf{E}_r(\hat{\boldsymbol{\beta}}_s - \boldsymbol{\beta}_r)\}. \end{aligned}$$

It follows that $\hat{\boldsymbol{\beta}}_s$ is an MLHP of $\boldsymbol{\beta}_r$, and an MLHP of the population total T is

$$\hat{T}_{\text{MLH}} = n\bar{y}_s + \mathbf{1}'_r\mathbf{X}_r\hat{\boldsymbol{\beta}}_s. \tag{4.2.8}$$

4.2.4. The Royall Maximum–Likelihood Predictors

Here too we consider the normal regression model, with a covariance matrix $\mathbf{W} = \text{diag}\{\mathbf{W}_s, \mathbf{W}_r\}$. The joint distribution of $\mathbf{y}_s$ and T under $\boldsymbol{\beta}$ is normal with mean

$$(4.2.9) \qquad E[\begin{matrix}\mathbf{y}_s \\ T\end{matrix} \mid \boldsymbol{\beta}] = \begin{bmatrix}\mathbf{X}_s \\ \mathbf{1}'_N\mathbf{X}\end{bmatrix}\boldsymbol{\beta}$$

and covariance matrix

$$(4.2.10) \qquad \text{Var}_\psi[\begin{matrix}\mathbf{y}_s \\ T\end{matrix} \mid \boldsymbol{\beta}] = \begin{bmatrix}\mathbf{W}_s & \mathbf{W}_s\mathbf{1}_s \\ \mathbf{1}'_s\mathbf{W}_s & \mathbf{1}'_s\mathbf{W}_s\mathbf{1}_s + \mathbf{1}'_r\mathbf{W}_r\mathbf{1}_r\end{bmatrix}.$$

Hence, the conditional distribution of $\mathbf{y}_s$ given $(T, \boldsymbol{\beta})$ is normal with mean

$$(4.2.11) \quad E_\psi[\mathbf{y}_s \mid T, \boldsymbol{\beta}] = \frac{T}{V}\mathbf{W}_s\mathbf{1}_s + \mathbf{X}_s\boldsymbol{\beta} - \frac{1}{V}(\mathbf{W}_s\mathbf{J}_s\mathbf{X}_s + \mathbf{W}_s\mathbf{J}_{sr}\mathbf{X}_r)\boldsymbol{\beta},$$

where $\mathbf{J}_{sr} = \mathbf{1}_s\mathbf{1}'_r$, and the conditional covariance matrix is

$$(4.2.12) \qquad \boldsymbol{\Sigma}_s = \mathbf{W}_s - \frac{1}{V}\mathbf{W}_s\mathbf{J}_s\mathbf{W}_s,$$

where

$$(4.2.13) \qquad V = \mathbf{1}'_s\mathbf{W}_s\mathbf{1}_s + \mathbf{1}'_r\mathbf{W}_r\mathbf{1}_r.$$

Let $\mathbf{g}_s = \frac{1}{V}\mathbf{W}_s\mathbf{1}_s$. The conditional mean vector can be written as

$$(4.2.14) \qquad E_\psi[\mathbf{y}_s \mid T, \boldsymbol{\beta}] = \mathbf{G}\begin{pmatrix}T \\ \boldsymbol{\beta}\end{pmatrix},$$

where

$$(4.2.15) \qquad \mathbf{G}_s = (\mathbf{g}_s, \mathbf{G}^*_s)$$

and

$$(4.2.16) \qquad \mathbf{G}^*_s = \mathbf{X}_s - \frac{1}{V}(\mathbf{W}_s\mathbf{J}_s\mathbf{X}_s + \mathbf{W}_s\mathbf{J}_{sr}\mathbf{X}_r).$$

The MLE theory for multivariate normal distributions implies that, if $\mathbf{G}$ is of full rank, then the MLE of $(T, \boldsymbol{\beta})$ is

$$(4.2.17) \qquad \begin{pmatrix}\hat{T} \\ \hat{\boldsymbol{\beta}}\end{pmatrix} = (\mathbf{G}'_s\boldsymbol{\Sigma}_s^{-1}\mathbf{G}_s)^{-1}\mathbf{G}'_s\boldsymbol{\Sigma}_s^{-1}\mathbf{y}_s.$$

Thus, if $\mathbf{G}$ is of full rank, the RMLP of T is $\hat{T}_{\text{RML}} = \hat{T}$.

There are several problems with Eq. (4.2.17). The first one is that for models that satisfy condition L, $\mathbf{G}_s$ is not of full rank, as can be easily demonstrated in the case of models SM1 and SM2. If, however, condition L is not satisfied then $\hat{T}_{\text{RML}}$ might be different from the BUP.

Royall (1976a) proved that, if the model satisfies condition L, then $\hat{T}_{\text{RML}} = \hat{T}_{\text{BU}}$. We prove here this result in the case of completely balanced samples; that is, $\bar{x}_s^{(i)} = \bar{x}_r^{(i)}$ $(i = 1, \cdots, p)$. We start with several lemmas.

Lemma 4.2.1.

$$\boldsymbol{\Sigma}_s^{-1} = \mathbf{W}_s^{-1} + \lambda \mathbf{W}_s^{-1}\mathbf{g}_s\mathbf{g}_s'\mathbf{W}_s^{-1}, \tag{4.2.18}$$

where

$$\lambda = V/(1 - V\mathbf{g}_s'\mathbf{W}_s^{-1}\mathbf{g}_s). \tag{4.2.19}$$

The proof is immediately obtained by multiplying $\boldsymbol{\Sigma}_s$ by $\boldsymbol{\Sigma}_s^{-1}$.

Lemma 4.2.2. *If the regression model satisfies condition L then $\mathbf{G}^*$ is not of full rank.*

Proof. Under condition L, there exists a p-dimensional vector $\boldsymbol{\lambda}$, $\boldsymbol{\lambda} \neq \mathbf{0}$, such that $\mathbf{W}\mathbf{1}_N = \mathbf{X}\boldsymbol{\lambda}$. Hence,

$$\mathbf{g} = \frac{1}{\mathbf{1}_N'\mathbf{X}\boldsymbol{\lambda}}\mathbf{X}_s\boldsymbol{\lambda},$$

and according to Eq. (4.2.16),

$$\mathbf{G}^* = \mathbf{X}_s - \frac{1}{\mathbf{1}_N'\mathbf{X}\boldsymbol{\lambda}}\mathbf{X}_s\boldsymbol{\lambda}\mathbf{1}_N'\mathbf{X}.$$

It follows immediately that $\mathbf{G}^*\boldsymbol{\lambda} = \mathbf{0}$. This proves that $rank(\mathbf{G}^*) < p$.

(Q.E.D.)

Corollary 4.2.1. *If the regression model satisfies condition L then $\mathbf{G}'\boldsymbol{\Sigma}_s^{-1}\mathbf{G}$ is singular.*

Lemma 4.2.3. *Assume that the regression model $\psi(\boldsymbol{\beta}, \mathbf{W})$ satisfies condition L, where $\mathbf{V}$ is diagonal. Then, the $(p+1)$th column of $\mathbf{G}$ is the vector $\mathbf{0}$.*

Proof. Condition L asserts that $\mathbf{W}\mathbf{1}_N = \mathbf{X}\boldsymbol{\lambda}$, for some vector $\boldsymbol{\lambda}$. Without loss of generality assume that $\mathbf{W} = \text{diag}\{\mathbf{X}^{(p)}\}$, where $\mathbf{X}^{(p)}$ is the pth column of $\mathbf{X}$. In this case, $\boldsymbol{\lambda} = (0, \ldots, 0, 1)'$. Let $\bar{x}^{(i)}$ be the mean of the ith column of $\mathbf{X}$, $i = 1, \ldots, p$. Let $\mathbf{m} = [\bar{x}^{(1)}, \bar{x}^{(2)}, \ldots, \bar{x}^{(p)}]'$. Then,

$$\mathbf{X}_s\boldsymbol{\lambda}\mathbf{1}_N'\mathbf{X} = N\mathbf{X}_s^{(p)}\mathbf{m}',$$

where $\mathbf{X}_s^{(p)}$ is the pth column of $\mathbf{X}_s$. Accordingly,

$$\mathbf{G}^* = \mathbf{X}_s - \frac{1}{\mathbf{1}_N'\mathbf{X}\boldsymbol{\lambda}}\mathbf{X}_s\boldsymbol{\lambda}\mathbf{1}_N'\mathbf{X} = [\mathbf{X}_s^{(1)} - \alpha_1\mathbf{X}_s^{(p)}, \ldots, \mathbf{X}_s^{(p)} - \alpha_p\mathbf{X}_s^{(p)}],$$

where $\alpha_i = \bar{x}^{(i)}/\bar{x}^{(p)}$, $i = 1, \ldots, p$. Thus, $\alpha_p = 1$ and the pth column of $\mathbf{G}^*$ is $\mathbf{0}$.

(Q.E.D.)

Corollary 4.2.2. *Under the conditions of Lemma 4.2.3, the joint likelihood* $Lik_R(T, \boldsymbol{\beta}; \mathbf{y}_s)$ *is independent of the pth component of* $\boldsymbol{\beta}$.

In the following example, we illustrate such a case.

Example 4.2.1. Consider model SM3 with normality and where the covariance matrix is $\mathbf{W} = \text{diag}\{x_1, \ldots, x_N\}$. Let $\mathbf{X}_{2s}$ and $\mathbf{X}_{2r}$ correspond to the observed and unobserved parts of $\mathbf{X}_2 = (x_1, \ldots, x_N)'$, respectively. Notice that $\mathbf{X} = (\mathbf{1}_N, \mathbf{X}_2)$. For this model, $\beta = (\beta_o, \beta_1)'$ is unknown and the ψ-BUP of T is

$$\hat{T}_{BU} = n\bar{y}_s + (N-n)\hat{\beta}_o + \hat{\beta}_1 \sum_{j \in \mathbf{r}} x_j,$$

where

$$\hat{\beta}_o = \frac{\sum_{j\in \mathbf{s}} x_j \sum_{j\in \mathbf{s}} y_j/x_j - n\sum_{j\in \mathbf{s}} y_j}{\sum_{j\in \mathbf{s}} 1/x_j \sum_{j\in \mathbf{s}} x_j - n^2}$$

and

$$\hat{\beta}_1 = \frac{\sum_{j\in \mathbf{s}} x_j^{-1} \sum_{j\in \mathbf{s}} y_j - n\sum_{j\in \mathbf{s}} y_j/x_j}{\sum_{j\in \mathbf{s}} 1/x_j \sum_{j\in \mathbf{s}} x_j - n^2}.$$

According to Lemma 4.2.3, the likelihood function $\text{Lik}_R(T, \beta \mid \mathbf{y}_s)$ is independent of β_1. We can therefore assume that $\beta_1 = 0$ and consider the reduced model

$$\mathbf{y}|\beta_o \sim N(\beta_o \mathbf{1}_N, \mathbf{W}).$$

By considering the reduced model, $\text{Lik}_R(T, \beta_o; \mathbf{y}_s)$ is proportional to the density function of the $N(\mathbf{H}_s\delta, \boldsymbol{\Sigma}_s)$, where $\delta = (T, \beta_o)'$,

$$\mathbf{H}_s = (\frac{\mathbf{X}_{2s}}{N\bar{x}_2}; \mathbf{1}_s - \frac{\mathbf{X}_{2s}}{\bar{x}_2}), \text{ and } \boldsymbol{\Sigma}_s = \text{diag}\{x_i, i \in \mathbf{s}\} - \frac{\mathbf{X}_{2s}\mathbf{X}_{2s}'}{N\bar{x}_2},$$

where $\bar{x}_2$ is the mean of $\mathbf{X}_2$. From Lemma 4.2.1, it follows that

$$\boldsymbol{\Sigma}_s^{-1} = (\text{diag}\{x_i, i \in \mathbf{s}\})^{-1} + \frac{\mathbf{J}_s}{(N-n)\bar{x}_{2r}},$$

where $\bar{x}_{2r}$ is the mean of $\mathbf{X}_{2r}$. As seen before, the RMLP of T, is the first element of $(\mathbf{H}_s'\boldsymbol{\Sigma}_s^{-1}\mathbf{H}_s)^{-1}\mathbf{H}_s'\boldsymbol{\Sigma}_s^{-1}\mathbf{y}_s$. Let

$$\mathbf{H}_s'\boldsymbol{\Sigma}_s^{-1}\mathbf{H}_s = \begin{pmatrix} \mathbf{A}_{11} & \mathbf{A}_{12} \\ \mathbf{A}_{21} & \mathbf{A}_{22} \end{pmatrix}.$$

From the above, it may be shown that

$$\mathbf{A}_{11} = \frac{n\bar{x}_{2s}}{N^2\bar{x}_2^2}(1 + \frac{n\bar{x}_{2s}}{(N-n)\bar{x}_{2r}})$$

$$\mathbf{A}_{12} = \mathbf{A}_{21} = -\frac{n(\bar{x}_2 - \bar{x}_{2s})}{N\bar{x}_2}(\frac{1}{\bar{x}_2} + \frac{n\bar{x}_{2s}}{(N-n)\bar{x}_{2r}\bar{x}_2})$$

and

$$\mathbf{A}_{22} = \sum_{i\in s}\frac{(x_i - \bar{x}_2)^2}{x_i\bar{x}_2^2} + \frac{n^2(\bar{x}_2 - \bar{x}_{2s})^2}{(N-n)\bar{x}_2^2\bar{x}_{2r}}.$$

Collecting all the above results, it follows using Eq. (4.2.18) that

$$\begin{aligned}\hat{T}_{\text{RML}} &= \frac{1}{D}\Bigg(\mathbf{A}_{22}\left[1 + \frac{n\bar{x}_{2s}}{(N-n)\bar{x}_{2r}}\right]\frac{n\bar{y}_s}{N\bar{x}_2} \\ &\quad - \mathbf{A}_{21}\left\{\sum_{j\in s} y_j/x_j - \left[1 + \frac{n(\bar{x}_2 - \bar{x}_{2s})}{(N-n)\bar{x}_{2r}}\right]\frac{n\bar{y}_s}{\bar{x}_2}\right\}\Bigg),\end{aligned} \quad (4.2.20)$$

where $D = \mathbf{A}_{11}\mathbf{A}_{22} - \mathbf{A}_{21}^2$. After some algebraic manipulations, it can be shown that $\hat{T}_{\text{BU}} = \hat{T}_{\text{ML}}$ (Exercise 4.8). Furthermore, if the sample is balanced, that is, $\bar{x}_2 = \bar{x}_{2s}$, then, $\hat{T}_{\text{ML}} = \hat{T}_{\text{BU}} = N\bar{y}_s$, which is the usual expansion predictor.

Let $\mathbf{X}^{(i)}$ denote the ith column of $\mathbf{X}$, $i = 1, \ldots, p$.

Theorem 4.2.2. *Under the general normal regression model, in which*

$$\mathbf{W} = diag\{x_{1p}, \ldots, x_{Np}\},$$

if the sample is completely balanced, that is, $\bar{x}^{(i)} = \bar{x}_s^{(i)} = \bar{x}_r^{(i)}$, *for all* $i = 1, \ldots, p$, *then,*

$$\hat{T}_{\text{RML}} = \hat{T}_{BU} = N\bar{y}_s.$$

Proof. Obviously, condition L is satisfied. Thus, by Theorem 2.1.2, we have to show that $\hat{T}_{\text{RML}} = \hat{T}_{SP} = N\bar{y}_s$.

Since $\mathbf{W} = \text{diag}\{x_{1p}, \ldots, x_{Np}\}$, the $(p+1)$ column of $\mathbf{G}$ is $\mathbf{0}$, and we consider the reduced model

$$\mathbf{y} = \mathbf{X}^*\boldsymbol{\beta}^* + \mathbf{e},$$

where $\mathbf{X}^*$ consists of the first $(p-1)$ columns of $\mathbf{X}$, and $\boldsymbol{\beta}^*$ consists of the corresponding $(p-1)$ components of $\boldsymbol{\beta}$. Let $\boldsymbol{\delta}' = (T, \boldsymbol{\beta}^{*\prime})$. The RMLP estimator of $\boldsymbol{\delta}$ is given by

$$\hat{\delta}_{ML} = (\mathbf{H}_s'\boldsymbol{\Sigma}_s^{-1}\mathbf{H}_s)^{-1}\mathbf{H}_s'\boldsymbol{\Sigma}_s^{-1}\mathbf{y}_s,$$

where

$$\mathbf{\Sigma}_s^{-1} = (\text{diag}\{x_{1p}, \ldots, x_{Np}\})^{-1} + \frac{1}{(N-n)\bar{x}^{(p)}}\mathbf{J}_s,$$

and $\mathbf{H}_s = (\mathbf{g}, \mathbf{G}^*)$. In this case of completely balanced sample,

$$\mathbf{g} = \frac{1}{N\bar{x}^{(p)}}\mathbf{X}_s^{(p)},$$

and

$$\mathbf{G}^* = \mathbf{X}_s^* - \frac{1}{\bar{x}^{(p)}}\mathbf{X}_s^{(p)}\mathbf{m}',$$

where $\mathbf{m}' = [\bar{x}^{(1)}, \ldots, \bar{x}^{(p-1)}]$.

Let

$$\mathbf{A} = \mathbf{H}_s'\mathbf{\Sigma}_s^{-1}\mathbf{H}_s = \begin{pmatrix} \mathbf{A}_{11} & \mathbf{A}_{12} \\ \mathbf{A}_{21} & \mathbf{A}_{22} \end{pmatrix}.$$

Some algebraic manipulations yield

$$\mathbf{A}_{11} = \frac{n}{N(N-n)\bar{x}^{(p)}},$$
$$\mathbf{A}_{12} = \mathbf{A}_{21} = \mathbf{0}$$

and

$$\mathbf{A}_{22} = \mathbf{X}_s^{*\prime}(\mathbf{W}_s^{-1} + \frac{1}{(N-n)\bar{x}^{(p)}}\mathbf{J}_s)\mathbf{X}_s^* - \frac{Nn}{(N-n)\bar{x}^{(p)}}\mathbf{m}\mathbf{m}',$$

where $\mathbf{W}_s = \sigma^2\text{diag}\{x_{1p}, \ldots, x_{Np}\}$. Hence,

$$\begin{aligned} \hat{T}_{\text{RML}} &= \mathbf{A}_{11}\mathbf{g}'\mathbf{\Sigma}_s^{-1}\mathbf{y}_s \\ &= \frac{\sigma^2(N-n)}{n}\mathbf{X}_s^{(p)\prime}\left[\mathbf{W}_s^{-1} + \frac{1}{(N-n)\bar{x}^{(p)}}\mathbf{J}_s\right]\mathbf{y}_s \\ &= \frac{(N-n)}{n}\left[1 + \frac{n}{(N-n)}\right]\mathbf{1}_s'\mathbf{y}_s \\ &= N\bar{y}_s. \end{aligned} \tag{4.2.21}$$

On the other hand, to show that $\hat{T}_{SP}$ is equal to $\hat{T}_{ML}$, write

$$\hat{T}_{\text{SP}} = \mathbf{1}_N'\mathbf{X}\hat{\boldsymbol{\beta}}_s = N\mathbf{m}'\hat{\boldsymbol{\beta}}_s.$$

Algebraic manipulations yield the formula

$$\hat{T}_{SP} = N[\mathbf{m}'\hat{\boldsymbol{\beta}}_s^* + \frac{n\mathbf{m}'(\mathbf{S}^*)^{-1}\mathbf{m}_s - \bar{x}^{(p)}}{\bar{x}_s^{(p)} - n\mathbf{m}_s'(\mathbf{S}^*)^{-1}\mathbf{m}_s}(\mathbf{m}_s'\hat{\boldsymbol{\beta}}_s^* - \bar{y}_s)], \tag{4.2.22}$$

where $\mathbf{m}'_s = [\bar{x}_s^{(1)}, \ldots, \bar{x}_s^{(p-1)}]$,

$$\mathbf{S}^* = \mathbf{X}_s^{*\prime}\mathbf{W}_s^{-1}\mathbf{X}_s^* \text{ and } \hat{\boldsymbol{\beta}}_s^* = (\mathbf{S}^*)^{-1}\mathbf{X}_s^{*\prime}\mathbf{W}_s^{-1}\mathbf{y}_s.$$

Finally, in the completely balanced case, $\mathbf{m}_s = \mathbf{m}$ and from Eqs. (4.2.21) and (4.2.22) we obtain $\hat{T}_{\text{SP}} = N\bar{y}_s$.

(Q.E.D.)

Example 4.2.2. Condition L is met by the the normal regression model where $\mathbf{X}$ and $\mathbf{W}$ are as in model SM2. Therefore, according to Theorem 4.2.1, the RMLP of T is the ratio predictor.

Example 4.2.3. Suppose that the normal regression model is such that $\mathbf{X} = (x_1, \ldots, x_N)'$ and $\mathbf{W} = \mathbf{I}_N$. It follows then from Eq. (4.2.13) that

$$\hat{T}_{\text{RML}} = N\bar{y}_s + (N-n)\hat{\beta}_1(\bar{x}_r - \bar{x}_s), \tag{4.2.23}$$

where $\hat{\beta}_1$ is the WLSE of β_1 under model SM3 with $g = 0$ (see Eq. 2.1.17). Therefore, in this case $\hat{T}_{\text{RML}}$ is not the BUP of T. In fact, $\hat{T}_{\text{RML}}$ in Eq. (4.2.23) is the BUP of T under model SM3 with $g = 0$ (see Example 2.2.3). This was expected, since, in the above regression model, condition L is not satisfied. Thus, when condition L is not satisfied, the RMLP approach may yield unreasonable predictors.

4.3. Maximum–Likelihood Predictors of the Population Variance S_y^2 Under the Normal Regression Model

4.3.1. Estimative Likelihood Predictors

According to Eq. (4.1.2) the EMLP of $\mathbf{y}_r$ is $\mathbf{X}_r\hat{\boldsymbol{\beta}}_s$. Accordingly, the EMLP of S_y^2 is

$$\begin{aligned}\hat{S}^2_{\text{EML}} = {} & \frac{n}{N}s_y^2 + (1-\frac{n}{N})[\hat{\boldsymbol{\beta}}_s'\mathbf{X}_r'\mathbf{A}_r\mathbf{X}_r\hat{\boldsymbol{\beta}}_s \\ & + \frac{n}{N}(\bar{y}_s - \frac{1}{N-n}\mathbf{1}_r'\mathbf{X}_r\hat{\boldsymbol{\beta}}_s)^2],\end{aligned} \tag{4.3.1}$$

where $\mathbf{A}_r = \frac{1}{N-n}(\mathbf{I}_{N-n} - \frac{1}{N-n}\mathbf{J}_{N-n})$. This predictor, however, is *not* BEP. The BEP is $E_{\hat{\psi}}[S_y^2]$. This is given by

Theorem 4.3.1. *The BEP of S_y^2 under the normal regression model $N(\mathbf{X}\boldsymbol{\beta}, \sigma^2 diag\{\mathbf{W}_s, \mathbf{W}_r\})$ is:*

$$\begin{aligned}\hat{S}^2_{\text{BE}} = {} & \frac{n}{N}s_y^2 + (1-\frac{n}{N})\Big\{\hat{\sigma}_s^2 tr\{\mathbf{A}_r\mathbf{W}_r\} + \hat{\boldsymbol{\beta}}_s\mathbf{X}_r'\mathbf{A}_r\mathbf{X}_r\hat{\boldsymbol{\beta}}_s \\ & + \frac{n}{N}\Big[\hat{\sigma}_s^2\mathbf{1}_r'\mathbf{W}_r\mathbf{1}_r + (\frac{1}{N-n}\mathbf{1}_r'\mathbf{X}_r\hat{\boldsymbol{\beta}}_s - \bar{y}_s)^2\Big]\Big\},\end{aligned} \tag{4.3.2}$$

where $\hat{\sigma}_s^2$ is given in Eq. (2.2.5).

Proof. Applying formulae (3.1.8), (3.1.9), and (3.1.24), we obtain, for given ψ,

$$E_\psi[S_{ry}^2 \mid \mathbf{y}_s] = \sigma^2 \mathrm{tr}\{\mathbf{A}_r\mathbf{W}_r\} + \boldsymbol{\beta}'\mathbf{X}_r'\mathbf{A}_r\mathbf{X}_r\boldsymbol{\beta}.$$

Moreover,

$$E_\psi[(\bar{y}_s - \bar{y}_r)^2 \mid \bar{y}_s] = \frac{\sigma^2}{(N-n)^2}\mathbf{1}_r'\mathbf{W}_r\mathbf{1}_r + (\frac{1}{N-n}\mathbf{1}_r'\mathbf{X}_r\boldsymbol{\beta} - \bar{y}_s)^2.$$

Substituting $\hat{\boldsymbol{\beta}}_s$ and $\hat{\sigma}_s^2$ for $\boldsymbol{\beta}$ and σ^2 we obtain Eq. (4.3.2).

(Q.E.D.)

It is interesting to note that

$$\hat{S}_{\mathrm{BE}}^2 = \hat{S}_{\mathrm{EML}}^2 + \left(1 - \frac{n}{N}\right)\hat{\sigma}_s^2[\mathrm{tr}\{\mathbf{A}_r\mathbf{W}_r\} + \frac{n}{N}\mathbf{1}_r'\mathbf{W}_r\mathbf{1}_r]. \tag{4.3.3}$$

Example 4.3.1. Consider model Eq. (4.2.1) with the assumptions that $\mathbf{X} = \mathbf{1}_N$ and $\mathbf{V} = \sigma^2\mathbf{I}_N$. The BEP of S_y^2 is, according to Eq. (4.3.2),

$$\hat{S}_{\mathrm{BE}}^2 = \frac{n}{N}s_y^2 + \left(1-\frac{n}{N}\right)\left(1-\frac{1}{N}\right)s_y^2 = s_y^2\left[1 - \frac{1}{N}\left(1-\frac{n}{N}\right)\right]. \tag{4.3.4}$$

The BUP of S_y^2 (Rodrigues et al., 1985) is

$$\hat{S}_{\mathrm{BU}}^2 = \frac{n}{N}s_y^2 + \left(1-\frac{n}{N}\right)\frac{n}{n-1}s_y^2 = s_y^2\left[1 + \frac{1}{n-1}\left(1-\frac{n}{N}\right)\right].$$

Both $\hat{S}_{\mathrm{BE}}^2$ and $\hat{S}_{\mathrm{BU}}^2$ are special cases of Eq. (2.3.41) and their predictive MSE can be obtained from Eq. (2.3.42).

The table below presents values of the efficiency ratio $E_\psi[\hat{S}_{\mathrm{BE}}^2 - S_y^2]^2/E_\psi[\hat{S}_{\mathrm{BU}}^2 - S_y^2]^2$ for $N = 100$ and several values of n.

Table 4.3.1. Efficiency of $\hat{S}_{\mathrm{BU}}^2$ Relative to $\hat{S}_{\mathrm{BE}}^2$.

n	10	20	50
$N = 100$	0.8432	0.92538	0.9717

4.3.2. Profile Likelihood Predictors

In the following theorem we prove that the MPLP of $\mathbf{y}_r$ is $X_r\hat{\boldsymbol{\beta}}_s$. This implies that the MPLP of S_y^2 is also given by formula (4.3.1).

Theorem 4.3.2. *Under the normal regression model, the MPLP of* $\mathbf{y}_r$ *is* $\mathbf{X}_r\hat{\boldsymbol{\beta}}_s$.

Proof. It is immediately obtained from the model that

$$\begin{aligned}\text{Lik}_P(\mathbf{y}_r \mid \mathbf{y}_s) \propto \exp\Big\{-\frac{1}{2}[(\mathbf{y}_s - \mathbf{X}_s\boldsymbol{\beta}_N)'\mathbf{W}_s^{-1}(\mathbf{y}_s - \mathbf{X}_s\boldsymbol{\beta}_N) \\ + (\mathbf{y}_r - \mathbf{X}_r\boldsymbol{\beta}_N)'\mathbf{W}_r^{-1}(\mathbf{y}_r - \mathbf{X}_r\boldsymbol{\beta}_N)]\Big\}, \quad \mathbf{y} \in \mathcal{Y}^{(r)},\end{aligned} \tag{4.3.5}$$

where $\boldsymbol{\beta}_N = \mathbf{P}_{sr}^{-1}(\mathbf{X}_s'\mathbf{W}_s^{-1}\mathbf{y}_s + \mathbf{X}_r'\mathbf{W}_r^{-1}\mathbf{y}_r)$.

Let

$$\begin{aligned}\mathbf{P}_{11} &= \mathbf{X}_s\mathbf{P}_{sr}^{-1}\mathbf{X}_s'\mathbf{W}_s^{-1}, & \mathbf{Q}_{11} &= \mathbf{I} - \mathbf{P}_{11}, \\ \mathbf{P}_{12} &= \mathbf{X}_s\mathbf{P}_{sr}^{-1}\mathbf{X}_r'\mathbf{W}_r^{-1}, & \mathbf{Q}_{12} &= -\mathbf{P}_{12}, \\ \mathbf{P}_{21} &= \mathbf{X}_r\mathbf{P}_{sr}^{-1}\mathbf{X}_s'\mathbf{W}_s^{-1}, & \mathbf{Q}_{21} &= -\mathbf{Q}_{21}, \\ \mathbf{P}_{22} &= \mathbf{X}_r\mathbf{P}_{sr}^{-1}\mathbf{X}_r'\mathbf{W}_r^{-1}, & \mathbf{Q}_{22} &= \mathbf{I} - \mathbf{P}_{22}.\end{aligned}$$

Let $\mathbf{P} = \begin{bmatrix}\mathbf{P}_{11} & \mathbf{P}_{12} \\ \mathbf{P}_{21} & \mathbf{P}_{22}\end{bmatrix}$ and $\mathbf{Q} = \begin{bmatrix}\mathbf{Q}_{11} & \mathbf{Q}_{12} \\ \mathbf{Q}_{21} & \mathbf{Q}_{22}\end{bmatrix}$.

It is easy to verify that $\mathbf{P}^2 = \mathbf{P}$ (idempotent), $\mathbf{Q}^2 = \mathbf{Q}$, and that $\mathbf{Q}_{12}\mathbf{X}_s + \mathbf{Q}_{12}\mathbf{X}_r = \mathbf{0}$, $\mathbf{Q}_{21}\mathbf{X}_s + \mathbf{Q}_{22}\mathbf{X}_r = \mathbf{0}$. Additional algebraic manipulations yield that

$$\begin{aligned}\text{Lik}_P(\mathbf{y}_r \mid \mathbf{y}_s) \propto \exp\Big\{-\frac{1}{2}[\mathbf{y}_r + (\mathbf{Q}_{22}'\mathbf{W}_r^{-1}\mathbf{Q}_{22} \\ + \mathbf{Q}_{12}'\mathbf{W}_s^{-1}\mathbf{Q}_{12})^{-1}(\mathbf{Q}_{12}'\mathbf{W}_s^{-1}\mathbf{Q}_{11} \\ + \mathbf{Q}_{22}'\mathbf{W}_r^{-1}\mathbf{Q}_{21})\mathbf{y}_s]'(\mathbf{Q}_{22}'\mathbf{W}_r^{-1}\mathbf{Q}_{22} \\ + \mathbf{Q}_{12}'\mathbf{W}_s^{-1}\mathbf{Q}_{12})[\mathbf{y}_r + (\mathbf{Q}_{22}'\mathbf{W}_r^{-1}\mathbf{Q}_{22} + \mathbf{Q}_{12}'\mathbf{W}_s^{-1}\mathbf{Q}_{12})^{-1} \\ (\mathbf{Q}_{12}'\mathbf{W}_s^{-1}\mathbf{Q}_{11} + \mathbf{Q}_{22}'\mathbf{W}_r^{-1}\mathbf{Q}_{12})\mathbf{y}_s]\Big\}; \quad \mathbf{y}_r \in \mathcal{Y}^{(r)}.\end{aligned} \tag{4.3.6}$$

Accordingly, the MPLP of $\mathbf{y}_r$ is

$$\begin{aligned}\hat{\mathbf{y}}_r = -(\mathbf{Q}_{22}'\mathbf{W}_r^{-1}\mathbf{Q}_{22} + \mathbf{Q}_{12}'\mathbf{W}_s^{-1}\mathbf{Q}_{12})^{-1}(\mathbf{Q}_{12}'\mathbf{W}_s^{-1}\mathbf{Q}_{11} \\ + \mathbf{Q}_{22}'\mathbf{W}_r^{-1}\mathbf{Q}_{21})\mathbf{y}_s.\end{aligned} \tag{4.3.7}$$

Moreover, one can verify that

$$\mathbf{Q}_{22}'\mathbf{W}_r^{-1}\mathbf{Q}_{22} + \mathbf{Q}_{12}'\mathbf{W}_s^{-1}\mathbf{Q}_{12} = \mathbf{W}_r^{-1}\mathbf{Q}_{22},$$

and

$$\mathbf{Q}_{12}'\mathbf{W}_s^{-1}\mathbf{Q}_{11} + \mathbf{Q}_{22}'\mathbf{W}_r^{-1}\mathbf{Q}_{21} = -\mathbf{W}_r^{-1}\mathbf{X}_r\mathbf{P}_{sr}^{-1}\mathbf{X}_s'\mathbf{W}_s^{-1}.$$

Substituting these results in Eq. (4.3.7), we obtain

$$\hat{\mathbf{y}}_r = \mathbf{Q}_{22}^{-1}\mathbf{X}_r\mathbf{P}_{sr}^{-1}(\mathbf{X}_s'\mathbf{W}_s^{-1}\mathbf{X}_s)\hat{\boldsymbol{\beta}}_s. \tag{4.3.8}$$

Finally, since $\mathbf{Q}_{21}\mathbf{X}_s + \mathbf{Q}_{22}\mathbf{X}_r = \mathbf{0}$,

$$\begin{aligned}\mathbf{X}_r\mathbf{P}_{sr}^{-1}(\mathbf{X}_s'\mathbf{W}_s\mathbf{X}_s) &= (\mathbf{I} - \mathbf{X}_r\mathbf{P}_{sr}^{-1}\mathbf{X}_r'\mathbf{W}_r^{-1})\mathbf{X}_r \\ &= \mathbf{Q}_{22}\mathbf{X}_r.\end{aligned}$$

Substituting this in Eq. (4.3.8), we obtain the result.

(Q.E.D.)

4.3.3. LH Likelihood Predictors

The conditional distribution of $\mathbf{y}_r$, given $\boldsymbol{\beta}_r = (\mathbf{X}_r'\mathbf{W}_r^{-1}\mathbf{X}_r)^{-1}\mathbf{X}_r'\mathbf{W}_r^{-1}\mathbf{y}_r$, is normal with mean $\mathbf{X}_r\boldsymbol{\beta}_r$ and covariance matrix:

$$\mathbf{W}_r - \mathbf{X}_r(\mathbf{X}_r'\mathbf{W}_r^{-1}\mathbf{X}_r)^{-1}\mathbf{X}_r'.$$

Hence, according to Eqs. (4.1.12) and (4.2.7) the LH predictive likelihood of $\mathbf{y}_r$, given $\hat{\boldsymbol{\beta}}_s$, is

$$\begin{aligned}\text{Lik}_{\text{LH}}(\mathbf{y}_r \mid \hat{\boldsymbol{\beta}}_s) \propto \exp\{&-\frac{1}{2}(\mathbf{y}_r - \mathbf{X}_r\boldsymbol{\beta}_r)' \\ &[\mathbf{W}_r - \mathbf{X}_r(\mathbf{X}_r'\mathbf{W}_r^{-1}\mathbf{X}_r)^{-1}\mathbf{X}_r']^{-1}(\mathbf{y}_r - \mathbf{X}_r\boldsymbol{\beta}_r) \\ &-\frac{1}{2}(\boldsymbol{\beta}_r - \hat{\boldsymbol{\beta}}_s)'\mathbf{E}_r'(\mathbf{I} - \mathbf{P}_{sr}^{-1})^{-1}\mathbf{E}_r(\boldsymbol{\beta}_r - \hat{\boldsymbol{\beta}}_s)\}.\end{aligned} \tag{4.3.9}$$

The MLHP of $\boldsymbol{\beta}_r$ is $\hat{\boldsymbol{\beta}}_s$ and hence the MLHP of $\mathbf{y}_r$ is $\mathbf{X}_r\hat{\boldsymbol{\beta}}_s$. This implies the following theorem.

Theorem 4.3.3. *Under the above normal regression model the MLHP of S_y^2 is*

$$\hat{S}^2_{\text{MLH}} = \frac{n}{N}s_y^2 + (1-\frac{n}{N})[\hat{\boldsymbol{\beta}}_s'\mathbf{X}_r'\mathbf{A}_r\mathbf{X}_r\hat{\boldsymbol{\beta}}_s + \frac{n}{N}(\frac{1}{N-n}\mathbf{1}_r'\mathbf{X}_r\hat{\boldsymbol{\beta}}_s - \bar{y}_s)^2]. \tag{4.3.10}$$

Proof. A function of an MLHP is an MLHP.

(Q.E.D.)

It is interesting to note that $\hat{S}^2_{\text{EML}} = \hat{S}^2_{\text{MPL}} = \hat{S}^2_{\text{MLH}}$. A maximum-likelihood predictor according to Royall's approach is unavailable.

4.4. Exercises

[4.1] Write the estimative likelihood $\text{Lik}_E(\mathbf{y}_r \mid \mathbf{y}_s)$ for model SM2 under normality.

[4.2] Let θ_L be any linear quantity of $\mathbf{y}$. Show that under the normal regression model, the EMLP of θ_L is the BUP of θ_L.

[4.3] Let $\theta_Q = \mathbf{y}_s' Q_s \mathbf{y}_s + \mathbf{y}_r' Q_r \mathbf{y}_r$. Find the bias of the EMLP of θ_Q under the general normal regression model.

[4.4] Consider the Poisson superpopulation model, according to which $y_1, \cdots, y_N$ are independent, $y_i \sim \text{Pois}(\beta x_i)$, $i = 1, \cdots, N$; where β is unknown. $0 < \beta < \infty$, and $x_1, \cdots, x_N$ are known positive quantities.
(i) Derive the MLE of β;
(ii) Write the estimative likelihood $\text{Lik}_E(\mathbf{y}_r \mid \mathbf{y}_s)$.
(iii) Show that the EMLP of $\mathbf{y}_r$ is $\hat{\beta}_s \mathbf{x}_r$.
(iv) Prove that the BUP of T is equal to the BEP of T, with respect to the squared-error prediction loss.

[4.5] Consider the Poisson model of Exercise 4.4.
(i) What are the minimal sufficient statistics U_s, U_r, and U_N?
(ii) Derive the conditional p.d.f. of U_s given U_N.
(iii) What is the Lauritzen–Hinkley likelihood of U_r, given U_N?
(iv) Prove that the MLHP of the population total T is the ratio predictor $\hat{T}_R = 1_s' \mathbf{y}_s + \dfrac{\bar{y}_s}{\bar{x}_s} \sum_{i \in r} x_i$. [Hint: $\hat{T}_{\text{MLH}}$ = smallest positive integer t, such that $\text{Lik}_{\text{LH}}(t \mid U_s)/\text{Lik}_{\text{LH}}(t-1 \mid U_s) \geq 1$.]

[4.6] Consider the location parameter exponential model, that is, $y_1, \cdots, y_N$ are i.i.d. and $y_i \sim \mu + G(1,1)$, $i = 1, \cdots, N$, where μ is unknown, $-\infty < \mu < \infty$.
(i) What is the EMLP of the population total T?
(ii) Is there an MLHP of T?

[4.7] Consider model SM6 under normality. Derive the RMLP of T.

[4.8] Prove that the RMLP given in formula (4.2.20) is equivalent to the BUP of T, in the case of SM3 under normality, with $g = 1$.

[4.9] What is the prediction bias of the BEP of S_y^2, given by Eq. (4.3.2)?

[4.10] Derive the BEP of S_y^2 for the case of model SM1 under normality. What are the bias and the prediction MSE of this predictor?

[4.11] Consider the general normal regression model. Show that $\hat{\boldsymbol{\beta}}_s$ is the BEP of $\boldsymbol{\beta}_N$. What is the MLHP of $\boldsymbol{\beta}_N$?

[4.12] (Zacks and Rodrigues, 1984) Consider the normal regression model $N(\mathbf{X}\boldsymbol{\beta}, \mathbf{V})$, and the following two phases of an iterative algorithm which is called the EM algorithm for estimating $\boldsymbol{\beta}$, where $\boldsymbol{\beta}_q$ is the value of $\boldsymbol{\beta}$ after q iterations of the algorithm:
E phase: Estimate the complete sufficient statistics for the population model, $S(\mathbf{y}) = \mathbf{X}'\mathbf{V}^{-1}\mathbf{y}$, according to

$$E[S(\mathbf{y}) \mid \mathbf{y}_s, \boldsymbol{\beta}_q] = \mathbf{X}'\mathbf{V}^{-1} \begin{pmatrix} \mathbf{y}_s \\ \mathbf{A}\mathbf{y}_s + \mathbf{B}\boldsymbol{\beta}_q \end{pmatrix},$$

where

$$\mathbf{A} = \mathbf{V}_{rs}\mathbf{V}_s^{-1} \text{ and } \mathbf{B} = \mathbf{X}_r - \mathbf{V}_{rs}\mathbf{V}_s^{-1}\mathbf{X}_s.$$

M phase: Determine

$$\beta_{q+1} = (\mathbf{X}'\mathbf{V}^{-1}\mathbf{X})^{-1}\mathbf{X}'\mathbf{V}^{-1}\begin{pmatrix} \mathbf{y}_s \\ \mathbf{A}\mathbf{y}_s + \mathbf{B}\boldsymbol{\beta}_q \end{pmatrix}.$$

Let

$$\mathbf{H}(\boldsymbol{\beta}) = (\mathbf{X}'\mathbf{V}^{-1}\mathbf{X})^{-1}\mathbf{X}'\mathbf{V}^{-1}\begin{pmatrix} \mathbf{y}_s \\ \mathbf{A}\mathbf{y}_s + \mathbf{B}\boldsymbol{\beta} \end{pmatrix}.$$

Show that $\hat{\boldsymbol{\beta}}_s$ given in Eq. (1.3.14) is the unique fixed point of the EM algorithm, that is,

$$\mathbf{H}(\hat{\boldsymbol{\beta}}_s) = \hat{\boldsymbol{\beta}}_s.$$

[4.13] Find the prediction MSE of the predictor given in Eq. (4.1.6).

[4.14] Consider the superpopulation model of Example 4.1.1(ii) with $\mu = 0$. Find the BUP of T and its prediction variance.

[4.15] Consider the superpopulation model defined in Example 4.1.1(ii). Show that the BEP of T is as given in Eq. (4.1.7). Find its prediction of MSE.

[4.16] Consider the superpopulation model of Example 4.1.4 with $\mu = 0$. Show that the MLHP of $\boldsymbol{\beta}_N$ is

$$\hat{\boldsymbol{\beta}}_{\text{MLH}} = \left(1 - \frac{1}{N}\right)\hat{\boldsymbol{\beta}}_s. \tag{4.4.1}$$

Find the prediction MSE of this predictor.

[4.17] Consider the normal regression model of Eq. (4.2.1). Conclude from Eq. (4.2.7) that $\hat{\boldsymbol{\beta}}_s$ is the MLHP of $\boldsymbol{\beta}_N$.

[4.18] Consider model SM6. Use Eq. (4.2.17) to derive the RMLP of T. Show that if condition L is satisfied, then, the RMLP = BUP.

[4.19] Consider the normal regression model (4.2.1). Show that the MLEP of $\boldsymbol{\beta}_N$ is $\hat{\boldsymbol{\beta}}_s$.

[4.20] Consider model SM1, with $e_i \sim N(0, \sigma^2)$, where σ^2 is known. Find the EMLP and BEP predictors of S_y^2.

[4.21] Consider model SM1 as in Exercise 4.20, with known σ^2. Find the MLHP of S_y^2 and its prediction MSE.

[4.22] Consider model SM2, with errors normally distributed and σ^2 known. Find the EMLP and BEP of S_y^2.

[4.23] Consider the normal regression model (4.2.1). Verify the details that lead to expression (4.3.3).

5 Classical and Bayesian Prediction Intervals

Prediction intervals (or sets) having certain characteristics fulfill the role of confidence and tolerance intervals in estimation theory. We start with some general concepts and two examples. For further reading on prediction and tolerance intervals, the reader is referred to the book by Aitchison and Dunsmore (1975) and the review paper by Bjørnstad (1990).

In the framework of predicting finite population quantities, consider a superpopulation model characterized by an element (parameter) ψ, which belongs to a space Ψ. Let $\theta(\mathbf{y})$ be a population quantity having a range $\mathcal{T}$. Let $\mathbf{y}_s$ be the sample vector. A set $\Theta(\mathbf{y}_s) \subset \mathcal{T}$, which depends on $\mathbf{y}_s$ only, is called a *prediction set.* The coverage of a prediction set is the probability $P_\psi[\theta \in \Theta(\mathbf{y}_s)]$. The coverage is generally a function of ψ. In certain cases, we can construct prediction sets having coverage with a lower bound uniformly greater than zero. Such prediction sets are discussed in the following sections.

5.1. Confidence Prediction Intervals

Definition 5.1.1. *A prediction set* $\Theta(\mathbf{y}_s)$*, such that, for* $0 < \alpha < 1$,

$$P_\psi[\theta \in \Theta(\mathbf{y}_s)] \geq (1-\alpha), \quad \textit{for all} \ \ \psi \in \Psi, \tag{5.1.1}$$

is called confidence prediction set of level $(1-\alpha)$.

We present two examples of confidence prediction intervals (CPI). The examples are followed by a general theorem on the CPI of the population total under the general normal regression model.

Example 5.1.1. Consider model SM1 under normality. Both β and σ are unknown.

(i) We show that $\hat{T}_E \pm [(N-n)/(n-1)]^{1/2} s_y t_{1-\alpha/2}[n-1]$ are the lower and upper limits of a CPI of level $(1-\alpha)$, for T; where $t_p[\nu]$ denotes the pth fractile of the t distribution with ν degrees of freedom, and $\hat{T}_E = N\bar{y}_s$, is the expansion predictor.

Indeed, $T - \hat{T}_E = (N-n)(\bar{y}_r - \bar{y}_s)$. Furthermore,

$$(N-n)(\bar{y}_r - \bar{y}_s) \sim N\left(0, \sigma^2 \frac{N-n}{n}\right),$$

and

$$s_y^2 \sim \frac{\sigma^2}{n} \chi^2[n-1].$$

Hence,

$$\frac{(N-n)(\bar{y}_r - \bar{y}_s)}{s_y} \sim \left(\frac{N-n}{n}\right)^{1/2} t[n-1].$$

Therefore,

$$P_\psi \left\{ \hat{T}_E - \sqrt{\frac{N-n}{n-1}} s_y t_{1-\alpha/2}[n-1] \le T \le \hat{T}_E + \sqrt{\frac{N-n}{n-1}} s_y t_{1-\alpha/2}[n-1] \right\}$$

$$= P_\psi \left\{ \frac{|T - \hat{T}_E|}{\sqrt{\frac{n-1}{N-n}} s_y} \le t_{1-\alpha/2}[n-1] \right\} = 1 - \alpha, \quad \text{for all } \psi = (\beta, \sigma).$$

(ii) We show now that $(s_y^2 A_{n,\alpha}, s_y^2 B_{n,\alpha})$ is a CPI for S_y^2 at level $1-\alpha$, where

$$A_{n,\alpha} = \frac{n}{N} + \left(1 - \frac{n}{N}\right) \frac{n}{n-1} \frac{1}{F_{1-\alpha/2}[n-1, N-n]},$$

$$B_{n,\alpha} = \frac{n}{N} + \left(1 - \frac{n}{N}\right) \frac{n}{n-1} F_{1-\alpha/2}[N-n, n-1],$$

and where $F_p[\nu_1, \nu_2]$ denotes the pth fractile of the central F distribution, with ν_1 and ν_2 degrees of freedom.

Notice first that,

$$\frac{S_y^2}{s_y^2} = \frac{n}{N} + (1 - \frac{n}{N}) \frac{S_{ry}^2 + \frac{n}{N}(\bar{y}_r - \bar{y}_s)^2}{s_y^2}.$$

Moreover, s_y^2, S_{ry}^2, and $\bar{y}_r - \bar{y}_s$ are mutually independent;

$$S_{ry}^2 \sim \frac{\sigma^2}{N-n} \chi^2[N-n-1]$$

and

$$\frac{n}{N}(\bar{y}_r - \bar{y}_s)^2 \sim \frac{\sigma^2}{N-n} \chi^2[1].$$

Hence,

$$\left[S_{ry}^2 + \frac{n}{N}(\bar{y}_r - \bar{y}_s)^2\right] \Big/ s_y^2 \sim \frac{n}{n-1}F[N-n, n-1].$$

From this, we obtain that

$$P_\psi\left\{s_y^2\left[\frac{n}{N} + \frac{n}{n-1}(1-\frac{n}{N})\frac{1}{F_{1-\alpha/2}[n-1, N-n]}\right] \leq \right.$$
$$\left. S_y^2 \leq s_y^2\left[\frac{n}{N} + \frac{n}{n-1}(1-\frac{n}{N})F_{1-\alpha/2}[N-n, n-1]\right]\right\} = 1-\alpha,$$

for all $\psi = (\mu, \sigma)$.

Example 5.1.2. Consider model SM2 under normality, that is, $\mathbf{y} \sim N(\beta\mathbf{x}, \sigma^2 \text{ diag}\{\mathbf{x}\})$, where $-\infty < \beta < \infty$, $0 < \sigma < \infty$. Both β and σ are unknown and $\psi = (\beta, \sigma)$.

(i) We derive the CPI of level $1-\alpha$ for T. Recall that

$$\hat{T}_R = n\bar{y}_s + (N-n)\bar{x}_r\frac{\bar{y}_s}{\bar{x}_s} \text{ and } \hat{\sigma}_s^2 = \frac{1}{n-1}\sum_{i\in s}\frac{1}{x_i}(y_i - x_i\frac{\bar{y}_s}{\bar{x}_s})^2.$$

Let $\hat{\sigma}_s^2 = Q_s/(n-1)$ and $Q_s = \mathbf{y}_s'\mathbf{H}_s\mathbf{y}_s$, where $\mathbf{H}_s = \text{diag}\{1/x_i;\ i = 1,\dots,n\} - [1/n\bar{x}_s]\mathbf{J}_s$. We assume here, without loss of generality that $\mathbf{s} = \{1, 2, \dots, n\}$. The covariance matrix of $\mathbf{y}_s$ is $\sigma^2\mathbf{W}_s$, where $\mathbf{W}_s = \text{diag}\{\mathbf{x}_s\}$. Thus, we obtain that $\mathbf{H}_s\mathbf{W}_s = \mathbf{I}_s - [1/n\bar{x}_s]\mathbf{1}_s\mathbf{x}_s'$, which is idempotent of rank $n-1$. Hence, $\hat{\sigma}_s^2 \sim [\sigma^2/(n-1)]\chi^2[n-1]$. It is also straightforward to show that $\hat{T}_R$ and $\hat{\sigma}_s^2$ are independent. Now,

$$T - \hat{T}_R = (N-n)(\bar{y}_r - \frac{\bar{x}_r}{\bar{x}_s}\bar{y}_s) \sim N(0, \sigma^2 N\frac{N-n}{n}\bar{x}\frac{\bar{x}_r}{\bar{x}_s}).$$

Therefore,

$$\frac{T - \hat{T}_R}{\hat{\sigma}_s} \sim (N\frac{N-n}{n}\bar{x}\frac{\bar{x}_r}{\bar{x}_s})^{1/2}t[n-1]$$

and the end points of the CPI for T, at level $(1-\alpha)$ are

$$\hat{T}_R \pm (N\frac{N-n}{n}\bar{x}\frac{\bar{x}_r}{\bar{x}_s})^{1/2}\hat{\sigma}_s t_{1-\alpha/2}[n-1].$$

(ii) Under the present model an exact $(1-\alpha)$ coverage CPI for S_y^2 does not exist, although some approximations might be tried.

It is more appealing to consider under SM2 the alternative population variance quantity

$$D_N^2 = \frac{1}{N-1}\sum_{i=1}^N \frac{1}{x_i}(y_i - x_i\frac{\bar{y}}{\bar{x}})^2 = \frac{1}{N-1}\mathbf{y}'\mathbf{H}_N\mathbf{y},$$

where $\mathbf{H}_N = \text{diag}\{1/x_i; i = 1, \ldots, N\} - (1/N\bar{x})\mathbf{J}_N$. If $D_s^2 = \hat{\sigma}_s^2$ and $D_r^2 = [1/(N-n)]\mathbf{y}_r'\mathbf{H}_r\mathbf{y}_r$, we obtain

$$D_N^2 = \frac{n-1}{N-1}D_s^2 + (1 - \frac{n-1}{N-1})[D_r^2 + \frac{n}{N\bar{x}}(\bar{y}_s\sqrt{\frac{\bar{x}_r}{\bar{x}_s}} - \bar{y}_r\sqrt{\frac{\bar{x}_s}{\bar{x}_r}})^2].$$

Let $n' = n - 1$ and $N' = N - 1$. One can verify that

$$\frac{D_N^2}{D_s^2} \sim \frac{n'}{N'} + (1 - \frac{n'}{N'})F[N-n, n-1].$$

Thus, a $(1-\alpha)$ CPI for D_N^2 is given by the lower and upper limits

$$D_s^2\left[\frac{n'}{N'} + (1 - \frac{n'}{N'})\frac{1}{F_{1-\alpha/2}[n-1, N-1]}\right],$$

$$D_s^2\left[\frac{n'}{N'} + (1 - \frac{n'}{N'})F_{1-\alpha/2}(N-n, n-1)\right].$$

The following theorem establishes the CPI for T, under the general normal regression model.

Theorem 5.1.1. *Under the normal regression model (2.2.3), with $\boldsymbol{\beta}$ and σ unknown, the CPI of T, of level $(1-\alpha)$, has the limit points*

$$\hat{T}_{\text{BU}} \pm g_{rs}\hat{\sigma}_s t_{1-\alpha/2}[n-p], \tag{5.1.2}$$

where $\hat{T}_{\text{BU}}$ is the BUP; $\hat{\sigma}_s^2$ is given by Eq. (2.2.5), p is the rank of $\mathbf{X}$, and

$$g_{rs} = \{\mathbf{1}_r'[\mathbf{W}_r + \mathbf{X}_r(\mathbf{X}_s'\mathbf{W}_s^{-1}\mathbf{X}_s)^{-1}\mathbf{X}_r']\mathbf{1}_r\}^{1/2}. \tag{5.1.3}$$

Proof. $T - \hat{T}_r = \mathbf{1}_r'(\mathbf{y}_r - \mathbf{X}_r\mathbf{B}_s\mathbf{y}_s) \sim N(0, \sigma^2 g_{rs}^2)$. Also, $\hat{\sigma}_s^2 \sim \frac{\sigma^2}{n-p}\chi^2[n-p]$ and $T - \hat{T}_R$ is independent of $\hat{\sigma}_s^2$, since $\mathbf{W}_{sr} = 0$. This implies Eq. (5.1.2). (Q.E.D.)

5.2. Tolerance Prediction Intervals for T

There are various concepts associated with tolerance intervals (see Guttman, 1970, and Aitchison and Dunsmore, 1975). We develop here prediction sets for the population total, T, which are *tolerance intervals*, providing $(1-\gamma)$ coverage at level of confidence $(1-\alpha)$.

We consider again the normal regression model (2.2.3), with $\boldsymbol{\psi} = (\boldsymbol{\beta}, \sigma)$ unknown. The conditional distribution of T, given $\mathbf{y}_s$, under $\boldsymbol{\psi}$, is $N(n\bar{y}_s + \mathbf{1}_r'\mathbf{X}_r\boldsymbol{\beta}, \sigma^2\mathbf{1}_r'\mathbf{W}_r\mathbf{1}_r)$. Hence, a $(1-\gamma)$ coverage interval for T is the interval having limit-points $(n\bar{y}_s + \mathbf{1}_r'\mathbf{X}_r\boldsymbol{\beta}) \pm z_{1-\gamma/2}\sigma(\mathbf{1}_r'\mathbf{W}_r\mathbf{1}_r)^{1/2}$. Here, z_p denotes the pth fractile of $N(0,1)$. These limit-points depend on $\boldsymbol{\psi}$ and are thus unknown. We construct an interval $T_{\alpha,\gamma}(\mathbf{y}_s)$, which contains the above $(1-\gamma)$ coverage interval, with confidence probability $(1-\alpha)$. Let $\Theta_\gamma(\mathbf{y}_s; \boldsymbol{\psi})$ denote a prediction set with coverage $1-\gamma$, under $\boldsymbol{\psi}$.

Definition 5.2.1. *A prediction set $T_{\alpha,\gamma}(\mathbf{y}_s)$ is called a $(1-\gamma, 1-\alpha)$ tolerance interval for T, if*

$$P_{\psi}[\Theta_{\gamma}(\mathbf{y}_s; \psi) \subset T_{\alpha,\gamma}(\mathbf{y}_s)] \geq 1-\alpha, \tag{5.2.1}$$

for all $\psi \in \Psi$.

Let $t_p[\nu; \delta]$ denote the pth fractile of the noncentral t distribution, with ν degrees of freedom and parameter of noncentrality δ. For computing formulae of these fractiles, see Johnson and Kotz (1970, Vol. 2, p. 207). Also notice that $t_p[\nu; \delta] = -t_{1-p}[\nu; -\delta]$.

Theorem 5.2.1. *Under the normal regression model (2.2.3), a $(1-\gamma, 1-\alpha)$ tolerance interval for T has limit points*

$$n\bar{y}_s + \mathbf{1}_r'\mathbf{X}_r\hat{\boldsymbol{\beta}}_s \pm t_{1-\alpha/2}(n-p; \delta_{\gamma})\hat{\sigma}_s h_{rs}, \tag{5.2.2}$$

where $\hat{\sigma}_s^2$ is given by Eq. (2.2.5),

$$\delta_{\gamma} = z_{1-\gamma/2}(\mathbf{1}_r'\mathbf{W}_r\mathbf{1}_r)^{1/2}h_{rs}^{-1}, \tag{5.2.3}$$

and

$$h_{rs} = [\mathbf{1}_r'\mathbf{X}_r(\mathbf{X}_s'\mathbf{W}_s^{-1}\mathbf{X}_s)^{-1}\mathbf{X}_r'\mathbf{1}_r]^{1/2}. \tag{5.2.4}$$

Proof. Let $n\bar{y}_s + \mathbf{1}_r'\mathbf{X}_r\hat{\boldsymbol{\beta}}_s + k_{\alpha,\gamma}\hat{\sigma}_s(\mathbf{1}_r'\mathbf{W}_r\mathbf{1}_r)^{1/2}$ be a $(1-\alpha/2)$ upper confidence limit for the upper limit point of $\Theta_{\gamma}(\mathbf{y}_s; \psi)$, which is $n\bar{y}_s + \mathbf{1}_r'\mathbf{X}_r\boldsymbol{\beta} + z_{1-\gamma/2}\sigma(\mathbf{1}_r'\mathbf{W}_r\mathbf{1}_r)^{1/2}$. To find $k_{\alpha,\gamma}$ we notice that

$$n\bar{y}_s + \mathbf{1}_r'\mathbf{X}_r\boldsymbol{\beta} + z_{1-\gamma/2}\sigma(\mathbf{1}_r'\mathbf{W}_r\mathbf{1}_r)^{1/2} \leq$$
$$n\bar{y}_s + \mathbf{1}_r'\mathbf{X}_r\hat{\boldsymbol{\beta}}_s + k_{\alpha,\gamma}\hat{\sigma}(\mathbf{1}_r'\mathbf{W}_r\mathbf{1}_r)^{1/2}$$

if, and only if,

$$\frac{\mathbf{1}_r'\mathbf{X}_r(\boldsymbol{\beta} - \hat{\boldsymbol{\beta}}_s) + z_{1-\gamma/2}\sigma(\mathbf{1}_r'\mathbf{W}_r\mathbf{1}_r')^{1/2}}{\hat{\sigma}_s} \leq k_{\alpha,\gamma}(\mathbf{1}_r'\mathbf{W}_r\mathbf{1}_r)^{1/2}.$$

Moreover, since $\mathbf{1}_r'\mathbf{X}_r(\boldsymbol{\beta} - \hat{\boldsymbol{\beta}}_s) \sim N(0, \sigma^2 h_{rs}^2)$, we obtain that

$$\frac{1}{\hat{\sigma}_s}[\mathbf{1}_r'\mathbf{X}_r(\boldsymbol{\beta} - \hat{\boldsymbol{\beta}}_s) + z_{1-\gamma/2}\sigma(\mathbf{1}_r'\mathbf{W}_r\mathbf{1}_r)^{1/2}] \sim h_{rs}t[n-p; \delta_{\gamma}].$$

Hence,

$$k_{\alpha,\gamma} = t_{1-\alpha/2}(n-p; \delta_{\gamma})h_{rs}(\mathbf{1}_r'\mathbf{W}_r\mathbf{1}_r)^{-1/2}.$$

Substituting this expression above, we obtain the upper limit of $T_{\alpha,\gamma}(\mathbf{y}_s)$. The lower limit of this tolerance interval is

$$n\bar{y}_s + \mathbf{1}_r'\mathbf{X}_r\hat{\boldsymbol{\beta}}_s + t_{\alpha/2}(n-p; -\delta_{\gamma})h_{rs}\hat{\sigma}_s = n\bar{y}_s + \mathbf{1}_r'\mathbf{X}_r\hat{\boldsymbol{\beta}}_s - t_{1-\alpha/2}(n-p; \delta_{\gamma})h_{rs}\hat{\sigma}.$$

This establishes Eq. (5.2.2).

(Q.E.D.)

If the sample size, n, is not too small, one can approximate $t_{1-\alpha/2}(n-p;\delta_\gamma)$ by

$$t_{1-\alpha/2}(n-p;\delta_\gamma) \cong \left[\delta_\gamma + z_{1-\alpha/2}\left(1+\frac{\delta_\gamma^2 - z_{1-\alpha/2}^2}{2(n-p)}\right)^{1/2}\right] \div \left[1-\frac{z_{1-\alpha/2}^2}{2(n-p)}\right], \tag{5.2.5}$$

as given in Johnson and Kotz (1970, Vol. 2, p. 270). Applying this approximation, one can show that $t_{1-\alpha/2}[n-p;\delta_\gamma]h_{rs} > t_{1-\alpha/2}[n-p]g_{rs}$. This means that the $(1-\gamma, 1-\alpha)$ tolerance intervals are wider than the $(1-\alpha)$ confidence prediction intervals.

5.3. Bayesian Prediction Intervals

In the Bayesian framework, we consider the predictive distribution of $\mathbf{y}_r$ given $\mathbf{y}_s$. This distribution depends on the prior distribution $\zeta(\psi)$ on Ψ. The Bayes predictive distribution induces a Bayes predictive distribution of $\theta(\mathbf{y})$. A set of θ values for which the Bayes predictive probability given $\mathbf{y}_s$ under ζ, is greater or equal to $(1-\alpha)$ is called a *Bayes prediction interval* (BPI) of coverage $(1-\alpha)$. Notice that, in contrast to the $(1-\alpha)$ CPI, or $(1-\gamma, 1-\alpha)$ tolerance interval, a $(1-\alpha)$ BPI depends on a particular prior. The following theorem characterizes the BPI, in the normal regression Bayesian model.

Theorem 5.3.1. *Under the Bayes normal regression model ψ_B, given by Eqs. (3.1.1) and (3.1.2), the $(1-\alpha)$ coverage BPI for T has the limit points*

$$\begin{aligned} &n\bar{y}_s + \mathbf{1}_r'\mathbf{X}_r\hat{\boldsymbol{\beta}}_B + \mathbf{1}_r'\mathbf{V}_{rs}\mathbf{V}_s^{-1}(\mathbf{y}_s - \mathbf{X}_s\hat{\boldsymbol{\beta}}_B) \\ &\pm z_{1-\alpha/2}\{\mathbf{1}_r'[\mathbf{V}_r - \mathbf{V}_{rs}\mathbf{V}_s^{-1}\mathbf{V}_{sr} \\ &+ (\mathbf{X}_r - \mathbf{V}_{rs}\mathbf{V}_s^{-1}\mathbf{X}_s)(\mathbf{X}_s'\mathbf{V}_s^{-1}\mathbf{X}_s \\ &+ \mathbf{B}^{-1})^{-1}(\mathbf{X}_r - \mathbf{V}_{rs}\mathbf{V}_s^{-1}\mathbf{X}_s)']\mathbf{1}_r\}^{1/2}. \end{aligned} \tag{5.3.1}$$

Proof. The theorem follows immediately from Theorem 3.1.1.

(Q.E.D.)

Corollary 5.3.1. *Under the conditions of Theorem 3.1.2, with the non-informative prior (3.1.11), the $(1-\alpha)$ coverage BPI reduces to the $(1-\alpha)$ CPI given by Eqs. (5.1.2) and (5.1.3).*

In the special case of the normal SM1 model, one can derive the predictive distribution of the population variance, S_y^2, theoretically, and obtain a BPI of coverage $(1-\alpha)$ (see Exercise 5.5). However, if the model is more complicated, like a normal SM2 for example, an analytical expression for the predictive distribution of S_y^2 is too complicated. One could obtain empirical estimates of this distribution by simulation, as shown in the following example.

Example 5.3.1. We consider a finite population of $N = 50$ elements and a sample of size $n = 20$. Each element of the population is characterized by two random quantities (x, y). It is assumed that $x_1, \ldots, x_N$ are i.i.d., random variables and that x_i is the integer part of a $N(100, 20^2)$ random variable.

It is further assumed that, $\mathbf{y} \mid \mathbf{x} \sim N(\beta\mathbf{x}, \sigma^2 \operatorname{diag}\{\mathbf{x}\})$. That is, conditional on $\mathbf{x}$, $\mathbf{y}$ satisfies SM2 under normality. We are interested in a Bayes prediction interval for the population variance S_y^2. For this example, we choose the parameters $\beta = 10$ and $\sigma = 15$. A sample of 50 quantities was simulated. The obtained value of S_y^2 was 37,788.6. For prediction, we apply the normal diffused prior, as given in Theorem 3.1.11. The predictive distribution of $\mathbf{y}_r$ given $(\mathbf{y}_s, \sigma)$ is $N\{\hat{\beta}_s\mathbf{x}_r, \sigma^2[\mathbf{W}_r + \frac{1}{n\bar{x}_s}\mathbf{x}_r\mathbf{x}_r']\}$, where $\mathbf{W}_r = \operatorname{diag}\{\mathbf{x}_r\}$ and $\mathbf{W}_s = \operatorname{diag}\{\mathbf{x}_s\}$. Since the value of σ^2 is unknown we substitute instead $\hat{\sigma}_s^2 = \frac{1}{n-1}\sum_{i=1}^{n}\frac{1}{x_i}(y_i - \hat{\beta}_s x_i)^2$.

Given the values of $\mathbf{x}$, we generate a random vector $\mathbf{y}_r$ from $N(\hat{\beta}_s\mathbf{x}_r, \hat{\sigma}_s^2$ $[\mathbf{W}_r + \frac{1}{n\bar{x}_s}\mathbf{x}_r\mathbf{x}_r'])$. We then compute the corresponding simulated value of $S_y^2 = \frac{n}{N}s_y^2 + (1-\frac{n}{N})[S_{ry}^2 + \frac{n}{N}(\bar{y}_r - \bar{y}_s)^2]$. This simulation was repeated 100 times independently. Examination of the histogram of these simulated variance quantities, and a normal probability plotting of these simulated quantities, show that the simulated quantities of S_y^2 are approximately normal. The average of the 100 simulated values of S_y^2 is 37,685.13 and their standard deviation is 408.94. Accordingly, $\hat{S}_{.975}^2 = 38,486$ and $\hat{S}_{.025}^2 = 36,883$, are estimates of the upper and lower limits of a BPI of coverage 0.95.

The following example deals with SM2 under Poisson distributions.

Example 5.3.2. Consider a population of N orchards. The ith orchard extends over x_i acres. The values of $x_1, \ldots, x_N$ are known. Let y_i, $i = 1, \ldots, N$, denote the number of trees in the ith orchard, which are infested with a certain fungus. We are interested in predicting the total number of infested trees, T, in the population. The present model is SM2 under Poisson, that is, $y_1, \ldots, y_N$ are independent, and $y_i \sim \operatorname{Pois}(\lambda x_i)$, $i = 1, \ldots, N$. The parameter λ represents the average number of infested trees per acre. Let $\mathbf{s} = \{1, 2, \ldots, n\}$ be a sample of the n orchards having the largest values of x_i.

The Bayes model assumes that λ has a prior gamma distribution, $G(\Lambda, \nu)$, such that $E[\lambda] = \nu/\Lambda$.

The joint p.d.f. of $\mathbf{y}_r$, given λ and $\mathbf{x}_r$, is

$$p(y_{n+1}, \ldots, y_N \mid \lambda, \mathbf{x}_r) = \left(\prod_{i=n+1}^{N} \frac{x_i^{y_i}}{y_i!} \right) \lambda^{(N-n)\bar{y}_r} \exp\{-\lambda(N-n)\bar{x}_r\}.$$

It is straightforward to verify that, the posterior distribution of λ given $\mathbf{y}_s$ is $G(\Lambda + n\bar{x}_s, \nu + n\bar{y}_s)$. Accordingly, the predictive p.d.f. of $\bar{y}_r$ given $\bar{y}_s$ is

$$p(y_{n+1}, \ldots, y_N \mid \mathbf{y}_s) = \left(\prod_{i=n+1}^{N} \frac{x_i^{y_i}}{y_i!} \right) \frac{(\Lambda + n\bar{x}_s)^{\nu + n\bar{y}_s}}{\Gamma(\nu + n\bar{y}_s)}$$

$$\int_0^\infty \lambda^{N\bar{y}+\nu-1} \exp\{-\lambda(\Lambda + N\bar{x})\} d\lambda$$

$$= \left(\prod_{i=n+1}^{N} \frac{x_i^{y_i}}{y_i!} \right) \frac{(\Lambda + n\bar{x}_s)^{\nu+n\bar{y}_s}}{(\Lambda + N\bar{x})^{\nu+N\bar{y}}} \frac{\Gamma(\nu + N\bar{y})}{\Gamma(\nu + n\bar{y}_s)} =$$

$$= \frac{T_r!}{\prod_{i=n+1}^{N} y_i!} \prod_{i=n+1}^{N} \pi_i^{y_i} \frac{\Gamma(\nu + n\bar{y}_s + T_r)}{T_r!\Gamma(\nu + n\bar{y}_s)} (1-\psi)^{\nu+n\bar{y}_s} \psi^{T_r},$$

where

$$T_r = \sum_{i=n+1}^{N} y_i, \quad \pi_i = x_i \Big/ \sum_{i=n+1}^{N} x_i, \quad \text{and} \quad \psi = \frac{(N-n)\bar{x}_r}{\Lambda + N\bar{x}}.$$

It follows that $\mathbf{y}_r \mid T_r, y_s \sim MN(T_r, \boldsymbol{\pi}_r)$ and that $T_r \mid \mathbf{y}_s \sim \text{NB}(\psi, \nu + n\bar{y}_s)$; where $MN(T_r, \boldsymbol{\pi}_r)$ designates a multinomial distribution with parameters T_r and $\boldsymbol{\pi}_r$, and $\text{NB}(\psi, \nu + n\bar{y}_s)$ designates the negative binomial distribution with probability function

$$g(t \mid \psi, \nu + n\bar{y}_s) = \frac{\Gamma(\nu + \bar{y}_s n + t)}{t!\Gamma(\nu + n\bar{y}_s)} (1-\psi)^{n+\nu\bar{y}_s} \psi^t, \quad t = 0, 1, \ldots.$$

Thus, the predictive distribution of T_r, given $\mathbf{y}_s$, is the $\text{NB}(\psi, \nu + n\bar{y}_s)$.

The expected value and variance of this distribution are (see Zacks, 1981, p. 32)

$$E[T_r \mid \mathbf{y}_s] = (\nu + n\bar{y}_s) \frac{(N-n)\bar{x}_r}{\Lambda + n\bar{x}_s},$$

and

$$\text{Var}[T_r \mid \mathbf{y}_s] = (\nu + n\bar{y}_s) \frac{(N-n)\bar{x}_r}{(\Lambda + n\bar{x}_s)^2} (\Lambda + N\bar{x}).$$

A $(1-\alpha)$ BPI for T can be obtained by determining the $\alpha/2$ and $(1-\alpha/2)$ fractiles of this negative binomial distribution. For large values of $(\nu + n\bar{y}_s)$ one can apply the normal approximation to obtain an approximation to the limits of the BPI, that is,

$$n\bar{y}_s + (\nu + n\bar{y}_s) \frac{(N-n)\bar{x}_r}{(\Lambda + n\bar{x}_s)} \pm z_{1-\alpha/2} \frac{1}{\Lambda + n\bar{x}_s} [(\nu + n\bar{y}_s)(\Lambda + N\bar{x})(N-n)\bar{x}_r]^{1/2}.$$

5.4. Exercises

[5.1] Consider model SM1 with exponentially distributed random variables. Derive the $(1-\alpha)$ CPI for the population total T.

[5.2] Consider model SM3 with $g=0$, under normality. Derive the $(1-\alpha)$ CPI for the population total T.

[5.3] Consider model SM3 with $g=1$, under normality. Define the population quantity $\sigma_N^2 = \frac{1}{N-2}\mathbf{y}'\mathbf{W}^{-1}[\mathbf{W}-\mathbf{X}(\mathbf{X}'\mathbf{W}^{-1}\mathbf{X})^{-1}\mathbf{X}']\mathbf{W}^{-1}\mathbf{y}$, where $\text{Var}[\mathbf{y}] = \sigma^2\mathbf{W}$. Determine a $(1-\alpha)$ CPI for σ_N^2.

[5.4] Consider model SM2 under Poisson distributions, that is, $y_1, \ldots, y_N$ are independent and $y_i \sim \text{Pois}(\lambda x_i)$, $i = 1, \ldots, N$. Let $T_s = \sum_{i\in s} y_i$. Show that a $(1-\gamma, 1-\alpha)$ tolerance interval for T, of the form $(0, T_{\alpha\gamma})$ has the limit point $T_{\alpha\gamma}$ = *smallest integer* j, $j \geq 0$, *such that* $\chi_\gamma^2[2j+2] \geq \chi_{1-\alpha}^2[2T_s+2](N\bar{x}/n\bar{x}_s)$.

[5.5] Consider model SM1 under normality. Derive a $(1-\alpha)$ BPI for S_y^2, with respect to the prior model $\beta \mid \sigma^2 \sim N(\beta^*, \tau^2)$ and $1/2\sigma^2 \sim G(\xi, \nu)$.

[5.6] As in Example 5.1.2, consider model SM2 under normality and derive $(1-\alpha)$ BPI for D_N^2 with respect to the diffused prior (3.1.11).

[5.7] Consider model SM2 under normality. Derive a $(1-\alpha)$ level CPI for $\beta_N = \sum_{i=1}^N y_i \Big/ \sum_{i=1}^N x_i$.

[5.8] Consider model SM3 with $y=0$ and $\beta_0 = 0$. Derive a $(1-\gamma, 1-\alpha)$ tolerance prediction interval for β_N.

6
The Effects of Model Misspecification, Conditions for Robustness, and Bayesian Modeling

As seen in the previous chapters, the optimality of predictors of population quantities might strongly depend on the particular model under which they have been derived. For example, if one assumes model SM2 (model ψ), as the *working model*, the ψ-BLUP of the population total is the ratio predictor

$$\hat{T}_R = n\bar{y}_s + \frac{\bar{y}_s}{\bar{x}_s}(N-n)\bar{x}_r.$$

On the other hand, if the *correct model* is SM3 with $g = 1$ (model ψ^*), then the BLUP of T is

$$\hat{T}_{\text{BLU}} = n\bar{y}_s + (N-n)\hat{\beta}_o + (N-n)\bar{x}_r\hat{\beta}_1$$

where formulae for $\hat{\beta}_o$ and $\hat{\beta}_1$ are given in Example 4.2.1. The prediction variance of $\hat{T}_{\text{BLU}}$ under model Ψ^* is

$$\text{Var}_{\psi^*}[\hat{T}_{\text{BLU}} - T] = \sigma^2\Bigg\{(N-n)\bar{x}_r + \frac{(N-n)^2}{n[(\frac{1}{n}\sum_{i\in s}\frac{1}{x_i})\bar{x}_s - 1]}\left(\bar{x}_s - 2\bar{x}_r + \bar{x}_r^2\frac{1}{n}\sum_{i\in s}\frac{1}{x_i}\right)\Bigg\}. \tag{6.0.1}$$

The predictor $\hat{T}_R$ is not unbiased under ψ^*, and its prediction ψ^*-MSE is

$$\text{E}_{\psi^*}[\hat{T}_R - T]^2 = \sigma^2\sum_{j=1}^{N} x_j\frac{\sum_{i\in r} x_i}{\sum_{i\in s} x_i} + (N-n)^2\beta_o^2(\frac{\bar{x}_r}{\bar{x}_s} - 1)^2. \tag{6.0.2}$$

Notice that if the sample design is balanced, that is, $\bar{x}_s = \bar{x}_r$, then, $\hat{T}_R = \hat{T}_{\text{BLU}} = N\bar{y}_s$ and

$$\text{Var}_{\psi^*}[\hat{T}_{\text{BLU}} - T] = E_{\psi^*}[\hat{T}_R - T]^2. \tag{6.0.3}$$

Thus, if the sample is balanced then $\hat{T}_{\mathbf{R}}$ is optimal also under ψ^*.

Royall and Herson (1973 a,b) provided a sufficient condition for the expansion estimator to be the BLUP under a general polynomial regression superpopulation model. Tallis (1978) proved that the condition is also necessary and Scott, Brewer, and Ho (1978) extended Royall and Herson's results by using a general variance function. Tallis's (1978) results were extended by Pfeffermann (1984). Applying Zyskind's (1967) general characterization of best linear unbiased estimators for general linear models, Pereira and Rodrigues (1983) gave necessary and sufficient conditions for a general linear unbiased predictor to be the BLUP of T under a general linear model with a diagonal covariance matrix. Their result was extended to the case of nondiagonal covariance matrices by Tam (1986). Bayesian robustness of the expansion predictor has been considered by Royall and Pfeffermann (1982). A general study of robustness of Bayesian prediction of the population total T, where general conditions for robustness are established, is considered in Bolfarine et al. (1987). Instead of imposing restrictions on the samples to be selected, Bolfarine (1989d) introduced Bayes predictors that adaptively consider the possibility that each one out of a series of alternative models is the correct model.

In the following section, we consider the robustness of linear predictors against misspecified regression models. Robust estimation of the prediction variance $\text{Var}_\psi[\hat{T}_{\text{BLU}} - T]$ is considered in Section 6.2. Some numerical studies are presented in Section 6.3. This material is followed by discussion in Section 6.4 of Bayesian robustness. Finally, Bayes adaptive selection of models is considered in Section 6.5.

6.1. Robust Linear Prediction of T

Consider a regression superpopulation model $\psi = (\boldsymbol{\beta}, \mathbf{V}, \mathbf{X})$. In Chapter 2, we derived the general form of the BLUP of T under a model ψ. Let $\psi^* = (\boldsymbol{\beta}^*, \mathbf{V}^*, \mathbf{X}^*)$ be an alternative regression model in which $\boldsymbol{\beta}^* = (\boldsymbol{\beta}', \boldsymbol{\delta}')'$ and $\mathbf{X}^* = (\mathbf{X}, \mathbf{Z})$, where $\mathbf{Z}$ is an $N \times J$ matrix of additional regressors. Under model ψ^*,

$$(6.1.1) \qquad \begin{aligned} \mathbf{y} &= \mathbf{X}^*\boldsymbol{\beta}^* + \mathbf{e}^* \\ &= \mathbf{X}\boldsymbol{\beta} + \mathbf{Z}\boldsymbol{\delta} + \mathbf{e}^*, \end{aligned}$$

where $\text{E}_{\psi^*}[\mathbf{e}^*] = 0$ and $\text{Var}_{\psi^*}[\mathbf{e}^*] = \mathbf{V}^*$. As we have seen in the introductory example, if the correct model is ψ^* rather than ψ, the BLUP under ψ, might not be even unbiased. Let $\hat{T}_{\text{BLU}}(\psi)$ denote the BLUP of T under ψ. According to Eq. (2.1.5),

$$\hat{T}_{\text{BLU}}(\psi) = \mathbf{1}_s'\mathbf{y}_s + \mathbf{l}_{sr}'(\psi)\mathbf{y}_s,$$

where

$$(6.1.2) \qquad \mathbf{l}_{sr}'(\psi) = \mathbf{1}_r'(\mathbf{X}_r - \mathbf{V}_{rs}\mathbf{V}_s^{-1}\mathbf{X}_s)\mathbf{B}_s + \mathbf{1}_r'\mathbf{V}_{rs}\mathbf{V}_s^{-1},$$

and $\mathbf{B}_s = (\mathbf{X}_s'\mathbf{V}_s^{-1}\mathbf{X}_s)^{-1}\mathbf{X}_s'\mathbf{V}_s^{-1}$.

From condition (2.1.2) we obtain the following lemma.

Lemma 6.1.1. *$\hat{T}_{\text{BLU}}(\psi)$ is ψ^*-unbiased if, and only if,*

(6.1.3) $$\mathbf{1}_r'(\mathbf{Z}_r - \mathbf{X}_r\mathbf{B}_s\mathbf{Z}_s) = \mathbf{1}_r'\mathbf{V}_{rs}\mathbf{V}_s^{-1}(\mathbf{I}_s - \mathbf{X}_s\mathbf{B}_s)\mathbf{Z}_s.$$

Proof. According to condition (2.1.2), $\mathbf{l}_{sr}'(\psi)\mathbf{y}_s$ is a ψ^*-unbiased predictor of $\mathbf{1}_r'\mathbf{y}_r$ if, and only if,

(6.1.4) $$\mathbf{l}_{sr}'(\psi)\mathbf{X}_s^* = \mathbf{1}_r'\mathbf{X}_r^*.$$

Substituting Eq. (6.1.2) into Eq. (6.1.4), we obtain the equivalent conditions

$$\text{(i) } \mathbf{1}_r'(\mathbf{X}_r - \mathbf{V}_{rs}\mathbf{V}_s^{-1}\mathbf{X}_s)\mathbf{B}_s\mathbf{X}_s + \mathbf{1}_r'\mathbf{V}_{rs}\mathbf{V}_s^{-1}\mathbf{X}_s = \mathbf{1}_r'\mathbf{X}_r,$$

and

$$\text{(ii) } \mathbf{1}_r'(\mathbf{X}_r - \mathbf{V}_{rs}\mathbf{V}_s^{-1}\mathbf{X}_s)\mathbf{B}_s\mathbf{Z}_s + \mathbf{1}_r'\mathbf{V}_{rs}\mathbf{V}_s^{-1}\mathbf{Z}_s = \mathbf{1}_r'\mathbf{Z}_r.$$

Condition (i) is satisfied since $\hat{T}_{\text{BLU}}(\psi)$ is obviously ψ-unbiased. This also holds algebraically, since $\mathbf{B}_s\mathbf{X}_s = \mathbf{I}_p$, where $\mathbf{I}_p$ is the identity matrix of dimension p. Condition (ii) implies Eq. (6.1.3).

(Q.E.D.)

Example 6.1.1. The present example is taken from Pereira and Rodrigues (1983). Suppose that model ψ is such that

$$\mathbf{X} = \mathbf{1}_N \text{ and } \mathbf{V} = \text{diag}\{f(x_1), \ldots, f(x_N)\}.$$

Let

$$\mathbf{Z} = \begin{pmatrix} x_1 & x_1^2 & \cdots & x_1^J \\ \vdots & \vdots & \ddots & \vdots \\ x_N & x_N^2 & \cdots & x_N^J \end{pmatrix},$$

and

$$\mathbf{V}^* = \text{diag}\{f^*(x_1), \ldots, f^*(x_N)\}.$$

Since $\mathbf{V}$ is diagonal, $\mathbf{V}_{rs} = \mathbf{0}$, and the ψ-BLUP of T is

(6.1.5) $$\hat{T}_{\text{BLU}}(\psi) = n\bar{y}_s + \frac{N-n}{\sum_{i \in s} 1/f(x_i)} \sum_{i \in s} \frac{y_i}{f(x_i)}.$$

According to Eq. (6.1.4), $\hat{T}_{\text{BLU}}(\psi)$ is ψ^*-unbiased if, and only if,

(6.1.6) $$\mathbf{1}_r'\mathbf{X}_r\mathbf{B}_s\mathbf{Z}_s = \mathbf{1}_r'\mathbf{Z}_r.$$

Condition (6.1.6) is equivalent to the condition

(6.1.7) $$\frac{\sum_{i \in s} x_i^j / f(x_i)}{\sum_{i \in s} 1/f(x_i)} = \frac{\sum_{i \in r} x_i^j}{N-n},$$

for all $j = 1, \ldots, J$.

Notice that when $f(x_i) = 1$, $i = 1, \ldots, N$, then condition (6.1.7) extends the condition of balanced sample design from the first moment of $\{x_i, i = 1, \ldots, N\}$ to its first J moments, simultaneously. Pereira and Rodrigues (1983) called designs that satisfy Eq. (6.1.7) *extended balanced* ones.

Definition 6.1.1. *The predictor $\hat{T}_{\text{BLU}}(\psi)$ is called completely robust with respect to model ψ^* if*

$$(6.1.8) \qquad \hat{T}_{\text{BLU}}(\psi) = \hat{T}_{\text{BLU}}(\psi^*).$$

Thus, in order for $\hat{T}_{\text{BLU}}(\psi)$ to be completely robust with respect to ψ^*, it should satisfy in addition to Eq. (6.1.3) also a condition which will imply that it is the BLUP with respect to model ψ^*. This condition will impose restrictions on the model ψ^*.

By Theorem 3 of Zyskind (1967) and Lemma 2.1.1, a necessary and sufficient condition for the ψ-unbiased estimator $(\mathbf{l}_{sr} - \mathbf{V}_s^{-1}\mathbf{V}_{sr}\mathbf{1}_r)'\mathbf{y}_s$ of $(\mathbf{l}_{sr} - \mathbf{V}_s^{-1}\mathbf{V}_{sr}\mathbf{1}_r)'\mathbf{X}_r\boldsymbol{\beta}$ to be BLU estimator is that $\mathbf{V}_s(\mathbf{l}_{sr} - \mathbf{V}_s^{-1}\mathbf{V}_{sr}\mathbf{1}_r) \in \mathcal{M}(\mathbf{X}_s)$, that is,

$$(6.1.9) \qquad \mathbf{V}_s\mathbf{l}_{sr} - \mathbf{V}_{sr}\mathbf{1}_r \in \mathcal{M}(\mathbf{X}_s),$$

where $\mathcal{M}(\mathbf{X}_s)$, is as in Theorem 2.1.2, the linear space generated by the columns of $\mathbf{X}_s$. It is easy to check that $\mathbf{l}_{sr}(\psi)$ given by Eq. (6.1.2) satisfies Eq. (6.1.9). Thus, $\mathbf{l}'_{sr}(\psi)\mathbf{y}_s$ is a BLUP of $\mathbf{1}'_r\mathbf{y}_r$, under ψ^*, if in addition to Eq. (6.1.4) it satisfies the condition

$$(6.1.10) \qquad \mathbf{V}_s^*\mathbf{l}_{sr}(\psi) - \mathbf{V}_{sr}^*\mathbf{1}_r \in \mathcal{M}(\mathbf{X}_s, \mathbf{Z}_s).$$

Lemma 6.1.2. *If condition (6.1.3) is satisfied then*

$$(6.1.11) \quad \mathbf{V}_s^*\mathbf{B}_s'(\mathbf{X}_r' - \mathbf{X}_s'\mathbf{V}_s^{-1}\mathbf{V}_{sr})\mathbf{1}_r + \mathbf{V}_s^*\mathbf{V}_s^{-1}\mathbf{V}_{sr}\mathbf{1}_r - \mathbf{V}_{sr}^*\mathbf{1}_r \in \mathcal{M}(\mathbf{X}_s^*),$$

if and only if,

$$\hat{T}_{\text{BLU}}(\psi) = \hat{T}_{\text{BLU}}(\psi^*).$$

Proof. Under condition (6.1.3), $\hat{T}_{\text{BLU}}(\psi)$ is ψ^*-unbiased. According to Eq. (6.1.2), condition (6.1.10) is reduced to Eq. (6.1.11).

(Q.E.D.)

Corollary 6.1.1. *If $\mathbf{V}_s = \mathbf{V}^*$ and condition (6.1.3) is satisfied, then $\hat{T}_{\text{BLU}}(\psi)$ is completely robust with respect to model ψ^*.*

Example 6.1.2. We have seen in Example 6.1.1 that condition (6.1.7) implies the ψ^*-unbiasedness of $\hat{T}_{\text{BLU}}(\psi)$ given by Eq. (6.1.5). By using condition (6.1.11), the predictor becomes a ψ^*-BLUP if, and only if, the sample is balanced and model ψ^* is such that

$$f^*(x_k) = f(x_j)\sum_{j=0}^{J} c_j x_k^j,$$

for a vector $(c_0, c_1, \ldots, c_J)$ and for every $k \in \mathbf{s}$.

Example 6.1.3. (Pereira and Rodrigues, 1983) Suppose that the elements of the superpopulation models ψ and ψ^* are such that

$$\mathbf{X} = \begin{pmatrix} x_1 \\ \vdots \\ x_N \end{pmatrix}, \mathbf{Z} = \begin{pmatrix} 1 & x_1^2 & \dots & x_1^J \\ \vdots & \vdots & \ddots & \vdots \\ 1 & x_N^2 & \dots & x_N^J \end{pmatrix}$$

and $\mathbf{V}$ and $\mathbf{V}^*$ are as in Example 6.1.2. Thus, it follows that

$$\mathbf{B}_s = (\sum_{j\in s} \frac{x_j}{f(x_j)})^{-1}\mathbf{1}_s'\mathbf{V}_s^{-1}$$

and Eq. (6.1.3) becomes

$$(6.1.12) \qquad \left(\sum_{j\in s} \frac{x_j^2}{f(x_j)}\right)^{-1} \left(\sum_{j\in s} \frac{x_j^{k+1}}{f(x_j)}\right) = \left(\sum_{j\notin s} x_j\right)^{-1} \sum_{j\notin s} x_j^k,$$

$j = 0, 1, \dots, J$. Property (6.1.12) was named by Pereira and Rodrigues (1983) *the extended overbalanced sample property* which is a generalization of the overbalanced sample property of Scott, Brewer, and Ho (1978). With the extended overbalanced sample property, the linear predictor

$$(6.1.13) \qquad \hat{T}_{\mathrm{BLU}}(\psi) = n\bar{y}_s + \left(\sum_{j\in s} \frac{x_k^2}{f(x_k)}\right)^{-1} \sum_{j\in r} x_j \sum_{j\in s} \frac{x_j y_j}{f(x_j)},$$

besides being ψ-BLUP is also ψ^*-*unbiased* in predicting T. By applying Lemma 6.1.2, Eq. (6.1.13) becomes ψ^*-BLUP if, and only if, the sample is extended overbalanced and

$$f^*(x_k) = f(x_k) \sum_{j=0}^{J} c_j x_k^{j-1},$$

for a vector $(c_0, c_1, \dots, c_J)$ and all $k \in s$. Notice that if $f(x) = x$, then

$$\hat{T}(\psi^*) = \hat{T}_R$$

and Eq. (6.1.12) reduces to the balanced sample case.

Example 6.1.4. Using Eqs. (6.1.3) and (6.1.11), it follows that the expansion predictor $\hat{T}_E$ is optimal with respect to the regression model $\psi = (\boldsymbol{\beta}, \mathbf{V})$ if, and only if,

(i) $\frac{n}{N}\mathbf{1}_s'\mathbf{X}_s = \mathbf{1}_r'\mathbf{X}_r$, and
(ii) $[\frac{N-n}{n}\mathbf{V}_s\mathbf{1}_s - \mathbf{V}_{sr}\mathbf{1}_r] \in \mathcal{M}(\mathbf{X}_s)$.

Notice that Condition (i) requires the sampling design to be balanced. This result generalizes Theorem 2 in Royall (1988), which is stated with respect to a diagonal covariance matrix $\mathbf{V}$.

Let

$$\hat{T}_{\text{BLU}}(\psi) = \mathbf{1}_s'\mathbf{y}_s + \mathbf{1}_r'\mathbf{T}_{rs}\mathbf{y}_s,$$

where

$$\mathbf{T}_{rs} = (\mathbf{X}_r - \mathbf{V}_{rs}\mathbf{V}_s^{-1}\mathbf{X}_s)\mathbf{B}_s + \mathbf{V}_{rs}\mathbf{V}_s^{-1}.$$

The prediction MSE of $\hat{T}_{\text{BLU}}(\psi)$ under ψ^* is thus

(6.1.14)
$$E_{\psi^*}[\mathbf{1}_r'\mathbf{T}_{rs}\mathbf{y}_s - \mathbf{1}_r'\mathbf{y}_r]^2 = \mathbf{1}_r'\mathbf{T}_{rs}\mathbf{V}_s^*\mathbf{T}_{sr}\mathbf{1}_r + \mathbf{1}_r'\mathbf{V}_r^*\mathbf{1}_r - 2\mathbf{1}_r'\mathbf{T}_{rs}\mathbf{V}_{sr}^*\mathbf{1}_r + [\mathbf{1}_r'(\mathbf{Z}_r - \mathbf{T}_{rs}\mathbf{Z}_s)\delta]^2,$$

where $\mathbf{T}_{sr} = \mathbf{T}_{rs}'$.

According to Eq. (2.1.6), the prediction variance of $\hat{T}_{\text{BLU}}(\psi^*)$ under ψ^* is

(6.1.15)
$$E_{\psi^*}[\hat{T}_{\text{BLU}}^* - T]^2 = \mathbf{1}_r'\mathbf{V}_r^*\mathbf{1}_r - \mathbf{1}_r'\mathbf{V}_{rs}^*\mathbf{V}_s^{*^{-1}}\mathbf{V}_{sr}^*\mathbf{1}_r + \mathbf{1}_r'(\mathbf{T}_{rs}^* - \mathbf{V}_{rs}^*\mathbf{V}_s^{*^{-1}})(\mathbf{T}_{sr}^* - \mathbf{V}_s^{*^{-1}}\mathbf{V}_{sr}^*)\mathbf{1}_r.$$

Thus, if the assumed model is ψ but the correct one is ψ^*, the relative efficiency of $\hat{T}_{\text{BLU}}(\psi)$ is the ratio of Eq. (6.1.15) to Eq. (6.1.14).

6.2. Estimation of the Prediction Variance

The problem of robust estimation of the prediction variance of $\hat{T}$ has been studied by Royall and Eberhart (1975), Royall and Cumberland (1978, 1981a,b), and Cumberland and Royall (1981). We focus attention in this section on estimating the prediction variance of the ratio predictor, $\hat{T}_R$, when model SM2 might be erroneous.

We have established earlier that the prediction variance of $\hat{T}_R$, under model SM2 (ψ), is [see formula (1.5.1)]

(6.2.1)
$$v_R = \sigma^2 N \frac{(1-f)}{f}\bar{x}\frac{\bar{x}_r}{\bar{x}_s},$$

where $f = n/N$. The BUP of σ^2 under model ψ (SM2 with normality) is given by

(6.2.2)
$$\hat{v}_R = \hat{\sigma}_s^2 N \frac{(1-f)}{f}\bar{x}\frac{\bar{x}_r}{\bar{x}_s},$$

where $\hat{\sigma}_s^2$ is specified in Eq. (2.2.11).

We consider now the properties of the estimator $\hat{v}_R$ under the more general model, namely, Model ψ^*: $\mathbf{y} \sim N(\beta\mathbf{x}, \sigma^2\mathbf{W})$, where $\mathbf{W} = \text{diag}\{w_1, \cdots, w_N\}$, $0 < w_i < \infty$ $(i = 1, \ldots, N)$. Model ψ is a special case of model ψ^*, where $w_i = x_i$ for all $i = 1, \ldots, N$.

The prediction variance of $\hat{T}_R$ under model ψ^* is

$$v_R^* = \sigma^2 \left[\left(\frac{1-f}{f}\right)^2 \left(\frac{\bar{x}_r}{\bar{x}_s}\right)^2 \sum_{i \in s} w_i + \sum_{i \in r} w_i \right]. \tag{6.2.3}$$

The expected value of the estimator $\hat{v}_R$ under model ψ^* is

$$E_{\psi^*}[\hat{v}_R] = \sigma_N^2 \frac{1-f}{f} \bar{x} \frac{\bar{x}_r}{\bar{x}_s} \frac{1}{n-1} \left(\sum_{i \in s} \frac{w_i}{x_i} - \sum_{i \in s} w_i / n\bar{x}_s \right). \tag{6.2.4}$$

It is easy to check that under SM2 (ψ), Eq. (6.2.4) reduces to Eq. (6.2.1). Generally, however, $\hat{v}_R$ is a biased estimator. For example, if ψ^* is such that $w_i = x_i^2$ $(i = 1, \ldots, N)$, then

$$v_R^* = \sigma^2 \left[\left(\frac{1-f}{f}\right)^2 \left(\frac{\bar{x}_r}{\bar{x}_s}\right)^2 \sum_{i \in s} x_i^2 + \sum_{i \in r} x_i^2 \right] \tag{6.2.5}$$

and

$$E_{\psi^*}[\hat{v}_R] = v_R \left[\bar{x}_s - \frac{s_x^2}{(n-1)\bar{x}_s} \right], \tag{6.2.6}$$

where $s_x^2 = \frac{1}{n} \sum_{i \in s} (x_i - \bar{x}_s)^2$. An index of the relative bias of $\hat{v}_R$ under ψ^* is

$$RB(\hat{v}_R; \psi^*) = \frac{E_{\psi} \alpha[\hat{v}_R] - v_R^*}{v_R^*}. \tag{6.2.7}$$

Let s_x^2 and $S_{r,x}^2$ be the variances of the x-values in $\mathbf{s}$ and in $\mathbf{r}$, respectively. Replacing $\sum_{i \in s} x_i^2$ by $n(s_x^2 + \bar{x}_s^2)$ and $\sum_{i \in r} x_i^2$ by $(N-n)(S_{r,x}^2 + \bar{x}_r^2)$ in Eq. (6.2.5), we can write

$$\begin{aligned} v_R^* &= v_R \left[(1-f) \frac{s_x^2 + \bar{x}_s^2}{\bar{x}_s^2} \cdot \frac{\bar{x}_r \bar{x}_s}{\bar{x}} \right] \\ &\quad + \sigma_N^2 (1-f)[S_{r,x}^2 + \bar{x}_r^2] \\ &= v_R (1-f)(1 + \gamma_{x,s}^2) \frac{\bar{x}_r \bar{x}_s}{\bar{x}} \\ &\quad + \sigma^2 N (1-f) \bar{x}_r^2 (1 + \gamma_{x,r}^2). \end{aligned} \tag{6.2.8}$$

where $\gamma_{x,s} = s_x/\bar{x}_s$ and $\gamma_{x,r} = S_{r,x}/\bar{x}_r$ are the coefficients of variation of the x-values in **s** and in **r**, respectively. If the sample is balanced, that is, $\bar{x}_s = \bar{x}_r = \bar{x}_s$, then $v_R = \sigma^2 N(1-f)\bar{x}/f$ and

(6.2.9)
$$v_R^* = v_R(1-f)(1+\gamma_{x,s}^2) + v_R f\bar{x}(1+\gamma_{x,r}^2).$$

In this case of balanced sample,

(6.2.10)
$$RB(\hat{v}_R;\psi^*) = \frac{\bar{x}(1-\gamma_{x,s}^2/(n-1)) - (1-f)(1+\gamma_{x,s}^2) - f\bar{x}(1+\gamma_{x,r}^2)}{(1-f)(1+\gamma_{x,s}^2) + f\bar{x}(1+\gamma_{x,r}^2)}.$$

This relative bias might be considerably large, even in the balanced sample case. This shows that the estimation of σ^2 is sensitive to gross misspecification of the model. In the following section, we provide a simulation study of the sensitivity of $\hat{v}_R$ when the correct model is SM3 ($g = 1$) rather than SM2.

6.3. Simulation Estimates of the ψ^*-MSE of $\hat{T}_R$

In this section, we present the results of a simulation study of the prediction MSE of $\hat{T}_R$ when the actual model is SM3 with $g = 1$. Note that the true and adopted models have the same covariance matrix and β_o is the only additional parameter. According to Example 6.1.2 and Corollary 6.1.1, $\hat{T}_R$ is optimal with respect to the model ψ^* if the sample is balanced with respect to the variable x. In the simulation study, we fixed $\beta_1 = 1.5$, $\sigma^2 = 0.5$, and x was generated at random from a uniform distribution on the interval $(1,9)$. Since obtaining exactly balanced samples is ordinarily impossible, ϵ–balanced samples (Herson, 1976) were considered. These samples satisfy the inequality

$$\left|\frac{\sqrt{n}(\bar{x}_s - \bar{x})}{S_x}\right| \le \epsilon,$$

where $S_x^2 = \sum_{i=1}^N (x_i - \bar{x})^2/N$.

Table 6.1 presents estimates of the MSE of $\hat{T}_R$ for several values of n, ϵ and β_o. These estimates are based on 200 samples. In the case of $\epsilon =$ 100, the sample may be considered as a simple random sample without replacement. Notice that in this case the MSE is highly affected by the fact that $\beta_o \neq 0$. However, as $\epsilon \to 0.25$, that is, as the sample becomes more balanced, the effect of β_0 on the MSE is diminished.

Table 6.2 shows estimates of the MSE of $\hat{T}_R$ when ψ is assumed, but the correct model is ψ^*, with $\beta_o = 4.0$. It shows that the ψ–optimal strategy (maximize $\bar{x}_s$) is highly inefficient under ψ^*.

Table 6.1. Simulation Estimates of the MSE of $\hat{T}_R$ under Model SM3 with $g = 1$.

		β_0			
n	ϵ	0.0	1.5	2.0	4.0
10	100	62,132	76,666	85,847	164,482
	1.0	62,761	69,484	72,952	92,958
	0.5	55,347	57,045	57,996	63,728
	0.25	55,582	56,489	56,896	59,045
50	100	12,220	16,166	18,458	32,510
	1.0	12,034	13,014	13,630	17,544
	0.5	12,760	12,745	12,817	13,494
	0.25	12,482	12,368	12,351	12,384
100	100	5,346	6,560	8,422	13,165
	1.0	5,523	5,911	6,187	8,017
	0.5	5,078	5,111	5,160	5,550
	0.25	4,549	4,542	4,549	4,626
200	100	1,839	2,276	2,585	4,642
	1.0	1,844	2,117	2,277	3,264
	0.5	1,993	2,988	2,021	2,212
	0.25	2,030	2,006	2,003	2,013
300	100	840	1,106	1,267	2,270
	1.0	876	2,004	1,084	1,595
	0.5	962	1,011	1,040	1,225
	0.25	966	957	957	977

Table 6.2. MSE Estimates of $\hat{T}_R$ under Model ψ^* ($\beta_o = 4.0$) and Samples Maximizing $\bar{x}_s$.

n	10	50	100	200	300
$\text{MSE}_{\psi^*}[\hat{T}_R]$	653,673.87	822,117.40	643,762.47	429,948.13	249,822.82

An alternative to balanced sampling is stratification. Consider the following strategy:

(i) The finite population is stratified according to the auxiliary variable x (model SM4);

(ii) Optimal allocation is considered with the predictor $\hat{T}_{SR}$ (Exercise 1.11);

(iii) Within each stratum a "balanced" sample is considered.

One can prove (see Exercise 6.3) that this strategy is more efficient than the strategy $\hat{T}_R$ with balanced sample. We illustrate this in Table 6.3. For Table 6.3, the parameters of models ψ and ψ^* are the ones of Table 6.1. The sample size considered was $n = 50$ and 200 sample were generated in each case.

Table 6.3. MSE Estimates of $\hat{T}_{SR}$ under Models ψ and ψ^*; $n = 50$.

	ϵ			
β_0	0.25	0.5	1.0	100
0	11,469.60	10,391.21	9,596.35	10,298.21
1.5	11,539.09	10,347.32	9,915.49	10,379.38
2.0	11,566.72	10,348.07	10,087.06	10,643.22
4.0	11,699.66	10,427.96	11,054.32	12,855.81
6.0	11,868.43	10,630.89	12,471.15	16,919.97
10.0	12,313.45	11,405.85	16,653.46	30,603.05
15.0	13,071.25	13,066.61	24,410.12	58,122.02
20.0	14,052.96	15,496.35	34,976.51	97,213.35
30.0	16,688.15	22,662.73	64,538.49	210,113.09

By comparing Tables 6.1 and 6.3, one concludes that stratification with balanced samples (as described above) is indeed more efficient against the kind of model misspecification described above.

6.4. Bayesian Robustness

As in Chapter 3, let ψ_B denote the Bayesian regression model $\mathbf{y}|\boldsymbol{\beta} \sim N(\mathbf{X}\boldsymbol{\beta}, \mathbf{V})$, and $\boldsymbol{\beta} \sim N(\mathbf{b}, \mathbf{B})$. Let ψ_B^* denote an alternative model $\mathbf{y}|\boldsymbol{\beta}^* \sim N(\mathbf{X}^*\boldsymbol{\beta}^*, \mathbf{V})$, where $\boldsymbol{\beta}^* = (\boldsymbol{\beta}, \boldsymbol{\delta})$,

$$\boldsymbol{\beta}^* \sim N(\mathbf{b}^*, \mathbf{B}^*),$$

$$\mathbf{b}^* = \begin{pmatrix} \mathbf{b} \\ \mathbf{d} \end{pmatrix} \text{ and } \mathbf{B}^* = \begin{pmatrix} \mathbf{B} & \mathbf{D}_o \\ \mathbf{D}_o & \mathbf{D} \end{pmatrix}.$$

Notice that both, ψ_B and ψ_B^*, specify the same covariance matrix for $\mathbf{y}$.

A set of conditions R on the models ψ and ψ^* is a robustness set if under these conditions the Bayesian inference is the same under the two models. The formalization of this concept is included in the following definitions.

Definition 6.4.1. *(Weak Robustness) A set of conditions R is a weak robustness set if under R the conditional (posterior) expectation of $\mathbf{y}_r$ (or of T) given $\mathbf{y}_s$ is the same under ψ_B and ψ_B^*.*

Definition 6.4.2. *(Robustness) A set of conditions R is a strong robustness set if under R the conditional (posterior) distribution of $\mathbf{y}_r$ (or of T) given $\mathbf{y}_s$ is the same under ψ_B and ψ_B^*.*

The roles of the moments, the prior distribution, and the population quantity of interest are highlighted in the following results.

Theorem 6.4.1. *The following conditions*

(i) $\boldsymbol{\delta}$ and $\boldsymbol{\beta}$ are priorly independent and
(ii) $\mathbf{1}_r'\mathbf{Z}_r = \mathbf{1}_r'[(\mathbf{X}_r - \mathbf{V}_{rs}\mathbf{V}_s^{-1}\mathbf{X}_s)(\mathbf{X}_s'\mathbf{V}_s^{-1}\mathbf{X}_s + \mathbf{B}^{-1})^{-1}\mathbf{X}_s' + \mathbf{V}_{rs}]\mathbf{V}_s^{-1}\mathbf{Z}_s$,

constitute a weak robustness set.

Proof. First write

$$\mathbf{y}^{(\delta)} = \mathbf{y} - \mathbf{Z}\boldsymbol{\delta} = \mathbf{X}\boldsymbol{\beta} + \mathbf{e},$$

where $\mathbf{e} \sim N(\mathbf{0}, \mathbf{V})$. Since $\boldsymbol{\beta}$ is independent of $\boldsymbol{\delta}$, we obtain from Theorem 3.1.1 that

$$E_{\psi_B^*}[\mathbf{y}_r^{(\delta)}|\mathbf{y}_s, \boldsymbol{\delta}] = E_{\psi_B^*}[\mathbf{y}_r^{(\delta)}|\mathbf{y}_s^{(\delta)}, \boldsymbol{\delta}] = \mathbf{X}_r\hat{\boldsymbol{\beta}}_B^{(\delta)} + \mathbf{V}_{rs}\mathbf{V}_s^{-1}[\mathbf{y}_s^{(\delta)} - \mathbf{X}_s\hat{\boldsymbol{\beta}}_B^{(\delta)}],$$

where

$$\hat{\boldsymbol{\beta}}_B^{(\delta)} = \mathbf{C}\hat{\boldsymbol{\beta}}_s^{(\delta)} + \mathbf{C}_o\mathbf{b},$$
$$\mathbf{C} = (\mathbf{X}_s'\mathbf{V}_s^{-1}\mathbf{X}_s + \mathbf{B}^{-1})^{-1}\mathbf{X}_s'\mathbf{V}_s^{-1}\mathbf{X}_s, \ \mathbf{C}_o = \mathbf{I} - \mathbf{C},$$

and

$$\hat{\boldsymbol{\beta}}_s^{(\delta)} = \hat{\boldsymbol{\beta}}_s - (\mathbf{X}_s'\mathbf{V}_s^{-1}\mathbf{X}_s)^{-1}\mathbf{X}_s'\mathbf{V}_s^{-1}\mathbf{Z}_s\boldsymbol{\delta}.$$

As before, $\hat{\boldsymbol{\beta}}_s$ is the weighted least–squares estimator of $\boldsymbol{\beta}$ under model ψ. Hence,

$$E_{\psi_B^*}[\mathbf{y}_r^{\delta)}|\mathbf{y}_s, \boldsymbol{\delta}] = E_{\psi_B}[\mathbf{y}_r|\mathbf{y}_s] - \mathbf{X}_r'(\mathbf{X}_s'\mathbf{V}_s^{-1}\mathbf{X}_s + \mathbf{B}^{-1})^{-1}\mathbf{X}_s'\mathbf{V}_s^{-1}\mathbf{Z}_s\boldsymbol{\delta}$$
$$- \mathbf{V}_{rs}\mathbf{V}_s^{-1}\mathbf{Z}_s\boldsymbol{\delta} + \mathbf{V}_{rs}\mathbf{V}_s^{-1}\mathbf{X}_s(\mathbf{X}_s'\mathbf{V}_s^{-1}\mathbf{X}_s + \mathbf{B}^{-1})^{-1}\mathbf{X}_s'\mathbf{V}_s^{-1}\mathbf{Z}_s\boldsymbol{\delta}$$

and

$$E_{\psi_B^*}[\mathbf{y}_r|\mathbf{y}_s, \boldsymbol{\delta}] = E_{\psi_B}[\mathbf{y}_r|\mathbf{y}_s] + \mathbf{Z}_r\boldsymbol{\delta} - [(\mathbf{X}_r - \mathbf{V}_{rs}\mathbf{V}_s^{-1}\mathbf{X}_s)$$
$$(\mathbf{X}_s'\mathbf{V}_s^{-1}\mathbf{X}_s + \mathbf{B}^{-1})^{-1}\mathbf{X}_s' + \mathbf{V}_{rs}]\mathbf{V}_s^{-1}\mathbf{Z}_s\boldsymbol{\delta}.$$

The result follows by noticing that

$$E_{\psi_B^*}[\mathbf{y}_r|\mathbf{y}_s] = E_{\psi_B^*}\{E_{\psi_B^*}[\mathbf{y}_r|\mathbf{y}_s, \boldsymbol{\delta}]|\mathbf{y}_s\}.$$

Q.E.D.

It is important to emphasize that only prior independence between δ and β is required for Theorem 6.4.1. However, for the next theorem, where conditions for robustness are established, independence and normality are essential.

Theorem 6.4.2. *Under the normal Bayes superpopulation models ψ_B and ψ_B^* the set R of Theorem 6.4.1 is a strong robustness set.*

Proof. Since $\mathbf{y}_r$ given $\mathbf{y}_s$ is normally distributed under ψ_B and ψ_B^*, and since under R,

$$E_{\psi_B}[T|\mathbf{y}_s] = E_{\psi_B^*}[T|\mathbf{y}_s],$$

we need to prove that the posterior variances are equal. Notice that

$$\mathrm{Var}_{\psi_B^*}[T|\mathbf{y}_s] = \mathrm{Var}_{\psi_B^*}\{E_{\psi_B^*}[T|\mathbf{y}_s,\boldsymbol{\delta}]|\mathbf{y}_s\} + E_{\psi_B^*}\{\mathrm{Var}_{\psi_B^*}[T|\mathbf{y}_s,\boldsymbol{\delta}]|\mathbf{y}_s\}.$$

Moreover, under R,

$$E_{\psi_B^*}[T|\mathbf{y}_s,\boldsymbol{\delta}] = E_{\psi_B}[T|\mathbf{y}_s].$$

Hence,

$$\mathrm{Var}_{\psi_B^*}\{E_{\psi_B^*}[T|\mathbf{y}_s,\boldsymbol{\delta}]|\mathbf{y}_s\} = 0.$$

On the other hand, since

$$\mathrm{Var}_{\psi_B^*}[\mathbf{1}_r'\mathbf{y}_r|\mathbf{y}_s,\boldsymbol{\delta}] = \mathrm{Var}_{\psi_B^*}[\mathbf{1}_r'\mathbf{y}_r^*|\mathbf{y}_s,\boldsymbol{\delta}] = \mathrm{Var}_{\psi_B}[\mathbf{1}_r'\mathbf{y}_r|\mathbf{y}_s],$$

we obtain that

$$E_{\psi_B^*}\{\mathrm{Var}_{\psi_B^*}[\mathbf{1}_r'\mathbf{y}_r|\mathbf{y}_s,\boldsymbol{\delta}]|\mathbf{y}_s\} = \mathrm{Var}_{\psi_B}[\mathbf{1}_r'\mathbf{y}_r|\mathbf{y}_s].$$

This concludes the proof.

(Q.E.D.)

Example 6.4.1. Suppose that $\mathbf{X} = \mathbf{1}_N$, $\mathbf{V} = \sigma^2\mathbf{I}$, b is a finite real number, B is a positive real number

$$\mathbf{Z} = \begin{pmatrix} x_1 & x_1^2 & \dots & x_1^J \\ \vdots & \vdots & \ddots & \vdots \\ x_N & x_N^2 & \dots & x_N^J \end{pmatrix},$$

and $\mathbf{D}_o = (0,\dots,0)'$. The Bayes predictor of T under the working model ψ is

$$\hat{T}_B = E_{\psi_B}[T|\mathbf{y}_s] = n\bar{y}_s + \frac{N-n}{n}(1+\frac{\sigma^2}{nB})^{-1}(\sum_{j\in\mathbf{s}} y_j + \frac{\sigma^2 b}{B}).$$

Condition (ii) of Theorem 6.4.1. reduces to the following system of equations

$$(1+\frac{\sigma^2}{nB})\frac{1}{N-n}\sum_{j\in\mathbf{r}} x_j^k = \frac{1}{n}\sum_{j\in\mathbf{s}} x_j^k, \quad k = 1,\dots,J.$$

If σ^2/B is approximately equal to zero, then the condition above is close to the balanced sample property and $\hat{T}_B$ above reduces to $\hat{T}_E$, the usual expansion estimator. This example shows that balanced samples may not play any important roles for robustness when informative priors are considered.

The next result, established by Bolfarine et al. (1987), shows that, in some special cases, conditions (i) and (ii) of Theorem 6.4.2 are indeed necessary and sufficient conditions for robustness.

Corollary 6.4.2. *If δ is a scalar with finite nonnull prior mean, $E_{\psi^*}[\delta]$, then R is a robustness set in relation to T if, and only if, R contains (i) and (ii) of Theorem 6.4.2.*

The result that follows appears in Royall and Pfeffermann (1982) and it establishes a set of conditions R under which the expansion estimator, $\hat{T}_E$ is robust with respect to models ψ and ψ^*. It is a direct consequence of Theorem 6.4.2. and Exercise 6.5.

Corollary 6.4.3. *Consider a noninformative prior distribution on β and δ. If $\mathbf{V}$ satisfies condition L and if the sample is balanced on $\mathbf{X}$ and on $\mathbf{Z}$, then the posterior distribution of T given $\mathbf{y}_s$ is, under ψ_B or ψ_B^*, normal with mean $\hat{T}_E = N\bar{y}_s$ and variance*

$$(\frac{N-n}{n})^2\mathbf{1}_s'\mathbf{V}_s\mathbf{1}_s - 2\frac{(N-n)}{n}\mathbf{1}_s'\mathbf{V}_{rs}\mathbf{1}_r + \mathbf{1}_r'\mathbf{V}_r\mathbf{1}_r.$$

According to Exercise 6.8, for some working models, balanced samples are indeed optimal in the sense of minimizing the posterior variance.

6.5. Bayesian Modeling

The question of robustness of predictors of population quantities was studied in the previous sections from the point of view of model misspecification. From this point of view, we have determined conditions on the model structure and on the sample selection so that the optimal predictors under the adopted model remain optimal under an alternative model. In this section, we develop Bayesian predictors that are based on several alternative models and thus are robust with respect to misspecification. This line of approach, which follows that of Bolfarine (1989) avoids the need of imposing restrictions (sometimes unrealistic) on the models and on the samples to be selected. Applications are considered for normal superpopulation models.

6.5.1. The Framework

As in Smith (1986) let $M_1, \ldots, M_k$ be the range of models under consideration. Let

$$f(\mathbf{y}|\psi^{(j)}, M_j) \tag{6.5.1}$$

be the density function (model) of the vector $\mathbf{y}$ under model M_j and $\psi^{(j)}$ be the vector of unknown parameters appearing in M_j with prior density $\zeta(\psi^{(j)})$, $j = 1, \ldots, k$. Then, if $\pi_j = P(M_j)$ denotes the prior probability that M_j is the correct model, the posterior probability of model M_j given $\mathbf{y}_s$, is

$$\pi_j^* = \frac{p(\mathbf{y}_s|M_j)\pi_j}{\sum_{j=1}^k p(\mathbf{y}_s|M_j)\pi_j}, \quad j = 1, \ldots, k, \tag{6.5.2}$$

where

$$p(\mathbf{y}_s|M_j) = \int f(\mathbf{y}_s|\boldsymbol{\psi}^{(j)}, M_j)\zeta(\boldsymbol{\psi}^{(j)}|M_j)d\boldsymbol{\psi}^{(j)}$$

is the predictive density corresponding to the observed units $\mathbf{y}_s$. Inference about $\boldsymbol{\psi}^{(j)}$ is based on the posterior density

$$p(\boldsymbol{\psi}^{(j)}|\mathbf{y}_s, M_j) \propto f(\mathbf{y}_s|\boldsymbol{\psi}^{(j)}, M_j)\zeta(\boldsymbol{\psi}^{(j)}|M_j). \tag{6.5.3}$$

Model (6.5.1) is now rewritten under model M_j as

$$\mathbf{y} = \mathbf{X}^{(j)}\boldsymbol{\beta}^{(j)} + \mathbf{e}^{(j)}, \tag{6.5.4}$$

where $E[\mathbf{e}^{(j)}] = \mathbf{0}$, $\text{Var}[\mathbf{e}^{(j)}] = \sigma^2\mathbf{W}^{(j)}$, and $\mathbf{W}^{(j)}$ is known and diagonal, $j = 1, \ldots, k$. Similarly, we extend the notation $\mathbf{X}_s$, $\mathbf{X}_r$, $\mathbf{W}_s$, and $\mathbf{W}_r$ to $\mathbf{X}_s^{(j)}$, $\mathbf{X}_r^{(j)}$, $\mathbf{W}_s^{(j)}$, and $\mathbf{W}_r^{(j)}$ under Eq. (6.5.4). An optimal (MSE) predictor of T is then given by

$$\begin{aligned}\hat{T}_{BM} &= E_{\boldsymbol{\psi}_B}[T|\mathbf{y}_s] = \sum_{j=1}^{k} \pi_j^* E_{\boldsymbol{\psi}_B}[T|\mathbf{y}_s, M_j] \\ &= \mathbf{1}_s'\mathbf{y}_s + \sum_{j=1}^{k} \pi_j^* \mathbf{1}_N'\mathbf{X}_r^{(j)} E_{\boldsymbol{\psi}_B}[\boldsymbol{\beta}^{(j)}|\mathbf{y}_s, M_j],\end{aligned} \tag{6.5.5}$$

where $E_{\boldsymbol{\psi}_B}[\boldsymbol{\beta}^{(j)}|\mathbf{y}_s, M_j]$ may be computed from Eq. (6.5.3). Notice that expression (6.5.5) follows from

$$\begin{aligned}E_{\boldsymbol{\psi}_B}[\mathbf{y}_r|\mathbf{y}_s, M_j] &= E_{\boldsymbol{\psi}_B}\left\{E_{\boldsymbol{\psi}}\left[\mathbf{y}_r|\mathbf{y}_s, \boldsymbol{\beta}^{(j)}, M_j\right]|\mathbf{y}_s, M_j\right\} \\ &= \mathbf{X}_r^{(j)} E_{\boldsymbol{\psi}_B}[\boldsymbol{\beta}^{(j)}|\mathbf{y}_s, M_j].\end{aligned}$$

Similarly, for predicting the population variance S_y^2, we may write

$$\widehat{S}^2_{BM} = E_{\boldsymbol{\psi}_B}[S_y^2|\mathbf{y}_s] = \sum_{j=1}^{k} \pi_j^* E_{\boldsymbol{\psi}_B}[S_y^2|\mathbf{y}_s, M_j], \tag{6.5.6}$$

where, under normality, $E_{\boldsymbol{\psi}_B}[S_y^2|\mathbf{y}_s, M_j]$ may be computed from Eq. (3.1.25).

6.5.2. The Normal Linear Model

Suppose that under the model M_j, $\mathbf{e}^{(j)} \sim N(\mathbf{0}, \sigma^2\mathbf{W}^{(j)})$, $j = 1, \ldots, k$. Assume also the following informative prior structure for $\boldsymbol{\psi}^{(j)} = [\boldsymbol{\beta}^{(j)}, \sigma^2]$:

$$\begin{aligned}\zeta[\boldsymbol{\beta}^{(j)}, \sigma^2|M_j] &= \zeta(\boldsymbol{\beta}^{(j)}|\sigma^2)\zeta(\sigma^2) \\ &\propto \frac{1}{\sigma^{m_j}} exp\{-h(\boldsymbol{\beta}^{(j)} - \mathbf{b}_o^{(j)})'(\boldsymbol{\beta}^{(j)} - \mathbf{b}_o)/2\sigma^2\}\times \\ &\quad (\sigma^2)^{-(\nu_0-1)/2} exp\{-\nu_o S_o^2/2\sigma^2\},\end{aligned} \tag{6.5.7}$$

where m_j is equal to the dimension of $\beta^{(j)}$ and h, ν_o, S_o^2, and $\mathbf{b}_o$ are all known. Accordingly,

(6.5.8)

$$\begin{aligned} p(\beta^{(j)}, \sigma^2|M_j) &\propto f(\mathbf{y}_s|\beta^{(j)}, \sigma^2, M_j)\zeta(\beta^{(j)}|\sigma^2, M_j)\zeta(\sigma^2) \\ &\propto \sigma^{-n} exp\{-[\mathbf{y}_s - \mathbf{X}_s^{(j)}\beta^{(j)}]'\mathbf{W}_s^{(j)}[\mathbf{y}_s - \mathbf{X}_s^{(j)}\beta^{(j)}]/2\sigma^2\} \\ &\quad \sigma^{-m_j} exp\{-h[\beta^{(j)} - \mathbf{b}_o^{(j)}]'[\beta^{(j)} - \mathbf{b}_o^{(j)}]/2\sigma^2\} \\ &\quad \sigma^{-(\nu_o - 1)} exp\{-\nu_o S_o^2/2\sigma^2\}. \end{aligned}$$

Integrating $\beta^{(j)}$ and σ^2, Eq. (6.5.8) reduces to

$$(6.5.9) \quad p(\mathbf{y}_s|M_j) = \frac{1}{\left|\mathbf{X}'^{(j)}_s \mathbf{W}_s^{(j)^{-1}} \mathbf{X}_s^{(j)} + h\mathbf{I}^{(j)}\right|^{1/2}} \frac{(\sqrt{2})^{a_j - 2}\Gamma(\frac{a_j}{2} - 1)}{(\sqrt{2\pi})^{n-m_j}[S_d^{(j)}]^{\frac{a_j}{2}-1}},$$

where $a_j = n + \nu_o - m_j$, $\mathbf{I}^{(j)}$ is the identity matrix of dimension m_j and

$$\begin{aligned} S_d^{(j)} &= [\mathbf{y}_s - \mathbf{X}_s^{(j)}\widehat{\beta}^{(j)}]'\mathbf{W}_s^{(j)^{-1}}[\mathbf{y}_s - \mathbf{X}_s^{(j)}\widehat{\beta}^{(j)}] + \nu_o S_o^2 \\ &\quad + [\widehat{\beta}^{(j)} - \mathbf{b}_o^{(j)}]' \left\{[\mathbf{X}'^{(j)}_s \mathbf{W}_s^{(j)} \mathbf{X}_s^{(j)}]^{-1} + h^{-1}\mathbf{I}^{(j)}\right\}^{-1} [\widehat{\beta}^{(j)} - \mathbf{b}_o^{(j)}], \end{aligned}$$

in which $\widehat{\beta}^{(j)}$ is the least–squares estimator of $\beta^{(j)}$. Furthermore, from Eq. (6.5.8), it follows that the marginal posterior density of $\beta^{(j)}$ is

$$(6.5.10) \quad \begin{aligned} p(\beta^{(j)}|\mathbf{y}_s, M_j) \propto \Bigg\{1 + \frac{1}{S_d^{(j)}}[\beta^{(j)} - \widehat{\beta}_B^{(j)}]'[\mathbf{X}'^{(j)}_s \mathbf{W}_s^{(j)} \mathbf{X}_s^{(j)} \\ + h\mathbf{I}^{(j)}][\beta^{(j)} - \widehat{\beta}_B^{(j)}]\Bigg\}^{-\frac{(n-m_j)}{2}}, \end{aligned}$$

where

$$\widehat{\beta}_B^{(j)} = [\mathbf{X}'^{(j)}_s \mathbf{W}_s^{(j)^{-1}} \mathbf{X}_s^{(j)} + h\mathbf{I}^{(j)}]^{-1}[\mathbf{X}'^{(j)}_s \mathbf{W}_s^{(j)^{-1}} \mathbf{X}_s^{(j)}\widehat{\beta}^{(j)} + h\mathbf{b}_o(j)].$$

Under the noninformative prior distribution

$$\zeta(\beta^{(j)}, \sigma^2) \propto 1/\sigma,$$

one obtains from Eq. (6.5.9) that

$$(6.5.11) \quad p(\mathbf{y}_s|M_j) = \frac{1}{\left|\mathbf{X}'^{(j)}_s \mathbf{W}_s^{(j)^{-1}} \mathbf{X}_s^{(j)}\right|^{1/2}} \frac{(\sqrt{2})^{a_j - 2}\Gamma(\frac{a_j}{2} - 1)}{(\sqrt{2})^{n-m_j}[S_d^{(j)}]^{\frac{a_j}{2}-1}},$$

where

$$S_d^{(j)} = [\mathbf{y}_s - \mathbf{X}_s^{(j)}\widehat{\beta}^{(j)}]'\mathbf{W}_s^{(j)^{-1}}[\mathbf{y}_s - \mathbf{X}_s^{(j)}\widehat{\beta}^{(j)}]$$

and $a_j = n+1-m_j$, $j = 1, \ldots, k$. The posterior density for $\boldsymbol{\beta}^{(j)}$ is obtained from Eq. (6.5.10).

Example 6.5.1. Consider the four models

$$y_i = \beta_o\delta_{1j} + x_i\beta_1\delta_{2j} + x_i^2\beta_2\delta_{3j} + e_i^{(j)}, \tag{6.5.12}$$

$i = 1, \ldots, N$, $j = 1, \ldots, 4$, where $\mathbf{e}^{(j)} \sim N(\mathbf{0}, \sigma^2\mathbf{W}^{(j)})$,

$$\mathbf{W}^{(j)} = \text{diag}\{x_1^{\delta_{2j}}, \ldots, x_N^{\delta_{2j}}\},$$

and

$$\delta_{1j} = \begin{cases} 1, & j = 1,3,4, \\ 0, & j = 2; \end{cases} \quad \delta_{2j} = \begin{cases} 1, & j = 2,3,4, \\ 0, & j = 1; \end{cases} \quad \delta_{3j} = \begin{cases} 1, & j = 4, \\ 0, & j = 1,2,3. \end{cases}$$

It follows from Eq. (6.5.11) that, under the noninformative prior for $[\boldsymbol{\beta}^{(j)}, \sigma^2]$,

$$p(\mathbf{y}_s|M_j) = C^{(j)}[S_d^{(j)}]^{-\frac{(a_j-2)}{2}}, \tag{6.5.13}$$

where $m_1 = m_2 = 1$, $m_3 = 2$, $m_4 = 3$,

$$C^{(1)} = \frac{2^{\frac{n-2}{2}}\Gamma(\frac{n}{2}-1)}{\sqrt{n}(2\pi)^{\frac{n-1}{2}}}, \quad S_d^{(1)} = \sum_{i\in s}(y_i - \bar{y}_s)^2, \quad a_1 = n;$$

$$C^{(2)} = \frac{2^{\frac{(n-2)}{2}}\Gamma(\frac{n}{2}-1)}{\sqrt{\sum_{i\in s} x_i}(2\pi)^{\frac{(n-1)}{2}}}, \quad S_d^{(2)} = \sum_{i\in s}\frac{1}{x_i}(y_i - \frac{\bar{y}_s}{\bar{x}_s}x_i)^2, \quad a_2 = n;$$

$$C^{(3)} = \frac{2^{\frac{(n-3)}{2}}\Gamma(\frac{n-1}{2}-1)}{\sqrt{D^{(3)}}(2\pi)^{\frac{n-2}{2}}}, \quad S_d^{(3)} = \sum_{i\in s}\frac{1}{x_i}[y_i - \widehat{\beta}_o^{(3)} - \widehat{\beta}_1^{(3)}x_i]^2, \quad a_3 = n-1;$$

$$C^{(4)} = \frac{2^{\frac{(n-4)}{2}}\Gamma(\frac{n-2}{2}-1)}{\sqrt{D^{(4)}}(2\pi)^{\frac{n-3}{2}}}, \quad S_d^{(4)} = \sum_{i\in s}\frac{1}{x_i}[y_i - \widehat{\beta}_o^{(4)} - \widehat{\beta}_1^{(4)}x_i - \widehat{\beta}_2^{(4)}x_i^2]^2,$$

$$a_4 = n-2,$$

and

$$D^{(3)} = \sum_{i\in s}x_i^{-1}\sum_{i\in s}x_i - n^2,$$

$$D^{(4)} = \sum_{i\in s}x_i^{-1}\sum_{i\in s}x_i\sum_{i\in s}x_i^3 + 2n\sum_{i\in s}x_i\sum_{i\in s}x_i^2 - (\sum_{i\in s}x_i)^3$$
$$- \left(\sum_{i\in s}x_i^2\right)^2\sum_{i\in s}x_i^{-1} - n^2\sum_{i\in s}x_i^3,$$

$\widehat{\beta}_i^{(j)}$ is the least-squares estimator of $\beta_i^{(j)}$, $i = 0,1,2$, $j = 1,2,3,4$. The Bayes predictor of $T = \sum_{i=1}^N y_i$, that adaptively considers each one of the possible models $M_1, \ldots, M_4$ to be the correct model, is obtained from Eq. (6.5.5), with $E_{\psi_B}[\boldsymbol{\beta}^{(j)}|\mathbf{y}_s, M_j] = \widehat{\boldsymbol{\beta}}^{(j)}$.

Example 6.5.2. Suppose that there are two models and

$$M_j\colon y_i = \beta_o\delta_{1j} + \beta_1 x_i \delta_{2j} + e_i^{(j)},$$

$j = 1, 2$. Assume that $\mathbf{e}^{(j)} \sim N(\mathbf{0}, \sigma^2\mathbf{W}^{(j)})$, and $\mathbf{W}^{(j)} = \text{diag}\{x_1^{\delta_{2j}}, \ldots, x_N^{\delta_{2j}}\}$,

$$\delta_{1j} = \begin{cases} 1, & j = 1, \\ 0, & j = 2, \end{cases} \quad \text{and} \quad \delta_{2j} = \begin{cases} 0, & j = 1, \\ 1, & j = 2. \end{cases}$$

Accordingly, a Bayes predictor of S_y^2, is given by

$$\widehat{S}^2_{BM} = \pi_1^* E_{\psi_B}[S_y^2|\mathbf{y}_s, M_1] + \pi_2^* E_{\psi_B}[S_y^2|\mathbf{y}_s, M_2], \tag{6.5.14}$$

where $E_{\psi_B}[S_y^2|\mathbf{y}_s, M_j]$, $j = 1, 2$, are given in Examples 2.1.1 and 2.1.2.

The efficiency of the above procedure in protecting the Bayes predictor against possible model misspecification is illustrated in some simulation studies presented below.

Example 6.5.3. To illustrate the behavior of predictor (6.5.5) some numerical comparisons are made with the expansion and the ratio estimators. For several values of β_o, populations of size $N = 100$ were generated according to the model $y_i = \beta_o + 1.2x_i + e_i$, where the e_i are independent and distributed as $N(0, \sigma^2 x_i)$, $i = 1, \ldots, N$ and $\sigma^2 = 1.0$. The x_i were generated at random independently, according to the uniform distribution $U(1, 10)$. The following range of alternative models was considered:

$$y_i = \beta_o\delta_{1j} + x_i\beta_1\delta_{2j} + e_i^{(j)},$$

$i = 1, \ldots, N$ and $j = 1, 2, 3$, where $\text{Var}[e_i] = \sigma^2 x_i^{\delta_{2j}}$,

$$\delta_{1j} = \begin{cases} 1, & j = 1, 3, \\ 0, & j = 2, \end{cases} \quad \text{and} \quad \delta_{2j} = \begin{cases} 1, & j = 2, 3, \\ 0, & j = 1. \end{cases}$$

As shown in Example 3.1.1, the ratio estimator $\widehat{T}_R$ is the Bayes predictor of $T = \sum_{i=1}^N y_i$ under model M_2, with noninformative prior on (β, σ^2). Similar results hold for the expansion predictor, $\widehat{T}_E$, under model M_1.

From each generated population, 1000 simple random samples of size $n = 10$ were selected without replacement. The mean squared error (MSE) of $\widehat{T}_E$, $\widehat{T}_R$ and $\widehat{T}_{BM}$ of (6.5.5), were estimated by $\sum(\widehat{T} - T)^2/1000$, where the summation extends over the 1000 selected samples. Table 6.5.1 shows the result of the simulation study. According to this table, the predictor $\widehat{T}_{BM}$ is considerably more efficient than the expansion predictor $\widehat{T}_E$. This efficiency does not change with β_0. On the other hand, the ratio predictor, $\widehat{T}_R$ is slightly more efficient than $\widehat{T}_{BM}$ when $\beta_0 = 0$. As β_0 increases $\widehat{T}_{BM}$ becomes more efficient.

Table 6.5.1. Efficiency of $\hat{T}_{BM}$ with Respect to $\hat{T}_R$ and $\hat{T}_E$.

β_0	$\frac{\text{MSE}\{\widehat{T}_E\}}{\text{MSE}\{\widehat{T}_{BM}\}}$	$\frac{\text{MSE}\{\widehat{T}_R\}}{\text{MSE}\{\widehat{T}_{BM}\}}$
0	13.072	0.928
15	12.970	1.686
20	13.004	2.211

Example 6.5.4. A population of size $N = 100$ was generated according to the model

$$y_i = 0.5 + 0.7x_i + 0.01x_i^2 + e_i,$$

where the e_i are independent and $N(0, \sigma^2 x_i)$, $i = 1, \ldots, N$, where $\sigma^2 = 5.0$ and the x_i were generated as $U(1; 10)$. The alternative models are the models in Eq. (6.5.12) in which the predictors $\widehat{T}_R$, $\widehat{T}_E$, and $\widehat{T}_{BM}$ are those of Example 6.5.3. As in that case, 1000 simple random samples of size $n = 10$ were selected and estimators for MSE($\widehat{T}$) were computed. The results are summarized in Table 6.5.2, which shows the efficiency of $\widehat{T}_{BM}$ with respect to the other two predictors.

Table 6.5.2. Efficiency of $\hat{T}_{BM}$ with respect to $\hat{T}_R$ and $\hat{T}_E$.

$\frac{\text{MSE}(\widehat{T}_R)}{\text{MSE}(\widehat{T}_{BM})}$	$\frac{\text{MSE}(\widehat{T}_E)}{\text{MSE}(\widehat{T}_{BM})}$
5.310	34.482

6.6. Exercises

[6.1] (Tallis, 1978) The elements of the models ψ and ψ^* are such that $\mathbf{X} = \mathbf{1}_N$, $\mathbf{V} = \sigma^2 \mathbf{I}_N$,

$$\mathbf{V}^* = \text{diag}\{f^*(x_1), \ldots, f^*(x_N)\}, \text{ and } \mathbf{Z} = \begin{pmatrix} x_1 & x_1^2 & \cdots & x_1^J \\ \vdots & \vdots & \ddots & \vdots \\ x_N & x_N^2 & \cdots & x_N^J \end{pmatrix}.$$

Find conditions under which the ψ-BUP of T is also the ψ^*-BUP.

[6.2] (Tam, 1986) Consider the superpopulation model ψ where $E[\mathbf{y}] = \mathbf{X}\boldsymbol{\beta}$ and $\text{Var}[\mathbf{y}] = \mathbf{V}$. Find conditions under which the predictors

$$\hat{T}_c = N\bar{y}_s + [\mathbf{1}_N'\mathbf{X} - \frac{N}{n}\mathbf{1}_s'\mathbf{X}_s]\hat{\boldsymbol{\beta}}_s$$

and

$$\hat{T}_{SP} = \mathbf{1}'_N \mathbf{X} \hat{\boldsymbol{\beta}}_s,$$

where $\hat{\boldsymbol{\beta}}_s$ is as in Eq. (1.3.14) are robust (optimal) with respect to the regression model $\psi = \psi(\boldsymbol{\beta}, \mathbf{V})$, where $\mathbf{V}$ is a general positive definite covariance matrix.

[6.3] (Royall and Herson, 1973b) Consider the superpopulation models

$$\psi : \mathbf{X} = \begin{pmatrix} x_1 \\ \vdots \\ x_N \end{pmatrix}; \quad \mathbf{V} = \sigma^2 \text{diag}\{x_1, \ldots, x_N\}$$

and

$$\psi^* : \mathbf{X}^* = \begin{pmatrix} 1 & x_1 \\ \vdots & \vdots \\ 1 & x_N \end{pmatrix}; \quad \mathbf{V}^* = \sigma^2 \text{diag}\{x_1, \ldots, x_N\}.$$

Suppose that the finite population is divided into K strata as in Example 1.2.4. Consider the following strategies:

(i) $\hat{T}_R$ is used with a simple balanced sample, that is,

$$\bar{x}_s = \bar{x}_r;$$

(ii) $\hat{T}_{SR}$ of Exercise 1.11 is used with stratified balanced sampling, that is, with a sample satisfying

$$\bar{x}_{sh} = \bar{x}_h,$$

$h = 1, \ldots, K$. Show that with strategies (i) and (ii) above

$$E_{\psi^*}[\hat{T}_{SR} - T]^2 \leq E_{\psi^*}[\hat{T}_R - T]^2.$$

[6.4] (Royall and Herson, 1973b) Let MSE_1 and MSE_2 be, respectively, the mean squared error of the ratio estimator of T with respect to (1) a balanced sample and (2) the optimal purposive sample discussed in Example 2.1.2. Show that

$$\frac{\text{MSE}_1}{\text{MSE}_2} = \min_{s \in \mathcal{S}} \left\{ \frac{\bar{x}_r}{\bar{x}_s} \right\}.$$

[6.5] (Pfeffermann, 1984) Let $\hat{T}_{BUP}$ be the BUP of T under the superpopulation model ψ specifying that

$$E[\mathbf{y}] = \mathbf{X}\boldsymbol{\beta} \text{ and } \text{Var}[\mathbf{y}] = \mathbf{V}.$$

Also, let

$$\mathbf{W}_s' = [\frac{1}{N-n}\mathbf{1}_r'\mathbf{V}_{rs} - \frac{1}{n}\mathbf{1}_s'\mathbf{V}_s].$$

Prove that if the sample is balanced on $\mathbf{X}$ and

$$\lim_{(n,N)\to\infty} \mathbf{W}_s'\mathbf{V}_s^{-1}\mathbf{W}_s = 0,$$

then

$$\hat{T}_{BLU} - \hat{T}_E \xrightarrow{P[\psi]} 0,$$

where $P[\psi]$ denotes convergence in probability under model ψ.

[6.6] Suppose that the ψ-model is such that $\mathbf{X} = [\tilde{\mathbf{X}}, \mathbf{X}^*]$, that is, unnecessary regression variables have been included in the adopted model. Show that the ψ-BLUP is also a ψ^*-BLUP if, and only if, Eqs. (6.1.3) and (6.1.11) hold.

[6.7] (Royall and Pfeffermann, 1982) Consider the superpopulation model ψ specified in Eq. (6.4.1). Let

$$\mathbf{D} = (\mathbf{X}_s'\mathbf{V}_s^{-1}\mathbf{X}_s)^{-1}, \ \mathbf{P}_s = \mathbf{X}_s\mathbf{D}\mathbf{X}_s'\mathbf{V}_s^{-1},$$

and, as before, $\mathbf{I}_s$ is the n–dimensional identity matrix. Condition L given in Eq. (2.1.11) implies that

(i) $\mathbf{1}_s'(\mathbf{I}_s - \mathbf{P}_s) = \mathbf{0}$,
(ii) $\mathbf{1}_s'\mathbf{V}_{rs}(\mathbf{I}_s - \mathbf{P}_s) = \mathbf{0}$,
(iii) $\mathbf{1}_r'\mathbf{V}_{rs}\mathbf{V}_s^{-1}(\mathbf{I}_s - \mathbf{P}_s) = \mathbf{0}$.

[6.8] (Royall and Pfeffermann, 1982) Consider the superpopulation model ψ of Eq. (6.4.1), where $\mathbf{V} = \sigma^2\mathbf{I}_N$ and with noninformative prior on $\boldsymbol{\beta}$. If one column of $\mathbf{X}$ is the vector $\mathbf{1}_N$ then, $\mathrm{Var}_{\psi_B}[T|\mathbf{y}_s]$ is minimized by taking a balanced sample on $\mathbf{X}$.

[6.9] Consider the superpopulation models ψ and ψ^* of Eqs. (6.4.1) and (6.4.2) with

$$\mathbf{X}' = \mathbf{1}_N, \ \mathbf{V} = \sigma^2\mathbf{I}_N, \ \boldsymbol{\beta} \sim N(\mathbf{b}, \mathbf{B}),$$

and

$$\mathbf{Z} = \begin{pmatrix} x_1 & x_1^2 & \dots & x_1^J \\ \vdots & \vdots & \ddots & \vdots \\ x_N & x_N^2 & \dots & x_N^J \end{pmatrix}.$$

Find the Bayes predictor of T under ψ and use Theorem 6.4.2 to find the set R of conditions under which it is weakly robust with respect to the alternative model ψ^*.

[6.10] Consider the superpopulation models ψ and ψ^* in Exercise 6.3. Suppose that $\boldsymbol{\beta} \sim N(\mathbf{b}, \mathbf{B})$ and noninformative prior is considered for $\boldsymbol{\delta}$, which is also independent of $\boldsymbol{\beta}$. Find a set R of conditions under which

$$E_{\psi}[T|\mathbf{y}_s] = E_{\psi^*}[T|\mathbf{y}_s].$$

[6.11] Consider the superpopulation models ψ and ψ^* of Eqs. (6.4.1) and (6.4.2), where noninformative prior is considered on $(\boldsymbol{\beta};\boldsymbol{\delta})$. Find the robust Bayes predictor of T and its posterior variance.

[6.12] Verify expression (6.0.3).

[6.13] Suppose that the elements of the superpopulation models ψ and ψ^* defined in Section 6.1 are such that

$$\mathbf{X} = \begin{pmatrix} 1 & x_1 \\ \vdots & \vdots \\ 1 & x_N \end{pmatrix} \text{ and } \mathbf{Z} = \begin{pmatrix} x_1^2 & \cdots & x_1^J \\ \vdots & \ddots & \vdots \\ x_N^2 & \cdots & x_N^J \end{pmatrix}.$$

Find conditions under which $\hat{T}_{BLU}(\psi) = \hat{T}_{BLU}(\psi^*)$.

[6.14] Verify expressions (6.2.3) and (6.2.4).

[6.15] Verify expressions (6.2.7) and (6.2.8).

[6.16] Show that $\hat{v}_{RM}$ in Eq. (6.2.9) is a consistent estimator of v_R.

[6.17] Consider predictor $\hat{T}_R$ and suppose that the correct model ψ^* specifies that $\mathbf{X}^* = (x_1, \ldots, x_N)'$ and $\mathbf{V}^* = \sigma^2\mathbf{I}_N$. Find an approximation to the relative bias [left side of Eq. (6.2.7)] of $\hat{v}_R$, when n is large and n/N is small.

[6.18] Prove Corollary 6.4.2.

[6.19] Prove Corollary 6.4.3.

[6.20] As an extension of Definition 6.4.1, we say that a set R of conditions is a weak robustness set if the Bayes predictor of T is the same under models ψ and ψ^* defined in Eqs. (6.4.1) and (6.4.2). Show that the set R of conditions in Theorem 6.4.1 constitute a weak robustness set for the Bayes predictor of T under the regression model ψ_B of Eq. (6.4.1), with respect to the linex loss function [see Eq. (3.5.1)].

[6.21] Verify expressions (6.5.9) and (6.5.10).

[6.22] Show that (6.2.10) is an approximate unbiased estimator of the prediction variance (6.2.11).

[6.23] (Royall and Cumberland, 1978) Consider models $\psi = (\boldsymbol{\beta}, \mathbf{V}, \mathbf{X})$ (adopted model) and $\psi^* = (\boldsymbol{\beta}^*, \mathbf{V}, \mathbf{X}^*)$ (true model). Let $\hat{T}_u(\psi)$ be any unbiased predictor of T under model ψ, that is, $E_{\psi}[\hat{T}_u - T] = 0$. Consider

$$\hat{v} = \sum_{i \in s} k_i r_i^2$$

as an estimator of $\text{Var}_{\psi}[\hat{T}_u(\psi) - T]$, where k_i are positive weights and r_i are linear functions of $\mathbf{y}_s$, with $E_{\psi}[r_i] = 0$. Notice that many estimators of the prediction variance may be written in the above form.

(i) Show that $E_{\psi^*}[\hat{T}_u(\psi) - T] \neq 0$ and that

$$\text{Var}_{\psi}[\hat{T}_u(\psi) - T] = \text{Var}_{\psi^*}[\hat{T}_u(\psi) - T].$$

(ii) Show that

$$E_{\psi^*}[\hat{v}] = \sum_{i \in s} k_i \text{Var}_{\psi^*}[r_i] + \sum_{i \in s} E_{\psi^*}[r_i^2].$$

(iii) Show that if $E_{\psi}[\hat{v}] = \text{Var}_{\psi}[\hat{T}_u(\psi) - T]$, then $\hat{v}$ has a positive bias with respect to ψ^*.

(iv) If the sample is balanced in the columns of $\mathbf{X}$ and $\mathbf{Z}$ and $\mathbf{V1}_N = \mathbf{X}\delta$, for some $p \times 1$ vector δ, then

$$E_{\psi^*}[\hat{v}] \geq E_{\psi^*}[\hat{T}_{BLU}(\psi) - T]^2.$$

7 Models with Measurement Errors

Superpopulation models with measurement errors are different from the superpopulation models considered previously. The variables involved may not be observed directly, but are mixed with measurement errors. Sprent (1966) proposed a method based on the generalized least squares approach for estimating linear regression coefficients. Lindley (1966) and Lindley and El–Sayad (1968) pioneered the Bayesian approach to that problem. Further Bayesian works for this problem can be found in Zellner (1971) and Reilly and Patino–Leal (1981). Some results with unequal probability sampling can be found in Chandhook (1982). A Bayesian approach for predicting T and S_y^2 when the variables involved in the regression model are measured with error is considered in Bolfarine and Rodrigues (1990). The estimation of regression coefficients based on asymptotic results is discussed in Fuller (1975). Fuller (1987) presented a general treatment of the inference problem for the error in variables superpopulation models in infinite populations.

In this chapter, we investigate properties of some predictors of the population total under error in variables models that depend on at most one auxiliary variable. Properties of the expansion predictor are investigated. Model SM3, with normally distributed errors, $g = 0$, where both variables x and y are measured with error is presented. The regression estimator, computed by using the observed variables, is considered as an element of a general class of regression estimators. Properties like bias and prediction variance of those predictors are investigated. A Bayesian approach to the prediction problem, when the variables in the model are measured with error, is considered in Section 7.2. By using Lindley's (1966) approach, approximate predictors are obtained for T and S_y^2 for the normal regression model.

7.1. The Location and Simple Regression Models

Model SM1 with measurement errors is considered first. Properties (like bias and MSE) of the expansion predictor, computed by using the observed variables, are investigated. Following this, the simple regression model with measurement errors is introduced. The regression predictor is considered as an element of a general class of predictors. An index of the relative excess in prediction risk is introduced.

7.1.1. Model SM1

The location error–in–variables model ψ_{E}, is the following. For $i = 1, \ldots, N$, let the observed variables be

$$Y_i = y_i + w_i, \tag{7.1.1}$$

where

$$y_i = \mu + e_i. \tag{7.1.2}$$

It is assumed that $E[\mathbf{e}] = E[\mathbf{w}] = 0$, $\text{Var}[\mathbf{e}] = \sigma_e^2\mathbf{I}_N$, and $\text{Var}[\mathbf{w}] = \sigma_w^2\mathbf{I}_N$, and that $\mathbf{e}$ and $\mathbf{w}$ are independent. Situations like the ones described by Eqs. (7.1.1) and (7.1.2) may occur, for example, when one is measuring the systolic blood pressure. The recorded value is then an estimate of the true value. Therefore, the y_is are not observed directly and the observed sample is $\mathbf{Y}_s = (Y_i, i \in \mathbf{s})$, which are estimates of the true ys. Let $\bar{Y}_s = \sum_{i\in s} Y_i/n$, be the observed sample mean. Since

$$E_{\psi_{\text{E}}}[\bar{Y}_s] = \mu,$$

it follows that an unbiased estimator of μ is

$$\hat{\mu} = \bar{Y}_s.$$

Consider the population total $T = \sum_{i\in \mathbf{s}} y_i + \sum_{i\in \mathbf{r}} y_i$. Notice that both components of T remain unknown even after $\mathbf{s}$ has been selected. Therefore, a possible estimative predictor of T is

$$\hat{T}_{\text{EE}} = n\hat{\mu} + (N-n)\hat{\mu} = N\bar{Y}_s. \tag{7.1.3}$$

This predictor is the usual expansion predictor, based on $\mathbf{Y}_s$.

Notice that $\hat{T}_{\text{EE}}$ is unbiased. Indeed,

$$\text{E}_{\psi_{\text{E}}}[\hat{T}_{\text{EE}} - T] = \text{E}_{\psi_{\text{E}}}[N\bar{Y}_s - \sum_{i=1}^{N} y_i] = N\mu - N\mu = 0.$$

Moreover, since

$$\mathrm{Cov}_{\psi_{\mathrm{E}}}[y_j, Y_i|\mu] = \begin{cases} \sigma_e^2; & j = i \\ 0; & j \neq i, \end{cases}$$

the prediction variance of $\hat{T}_{\mathrm{EE}}$ is

$$\begin{aligned}\mathrm{Var}_{\psi_{\mathrm{E}}}[\hat{T}_{\mathrm{EE}} - T] &= \mathrm{Var}_{\psi_{\mathrm{E}}}[N\bar{Y}_s - \sum_{j=1}^{N} y_j] \\ &= N\frac{(1-f)}{f}\sigma_e^2 + \frac{N^2}{n}\sigma_w^2,\end{aligned}$$

where $f = n/N$. This can be summarized in the following theorem.

Theorem 7.1.1. *$\hat{T}_{\mathrm{EE}}$ is a ψ_{E}–unbiased predictor of T, having prediction variance*

$$\mathrm{Var}_{\psi_{\mathrm{E}}}[\hat{T}_{\mathrm{EE}} - T] = \frac{N}{f}[\sigma_w^2 + (1-f)\sigma_e^2]. \tag{7.1.4}$$

If y_i is measured without error, then $\hat{T}_{\mathrm{EE}} = \hat{T}_{\mathrm{E}} = N\bar{y}_s$.

Notice that the form of the expansion predictor $N\bar{Y}_s$ is the same, whether the y_i values are measured with error or without. The prediction variance of the predictor increases by the quantity $N\sigma_w^2/f$, if the values of y are measured with error. The question is, *how can we determine prediction confidence intervals for* T?

In the *normal* model,

$$Y_i \mid y_i \sim N(y_i, \sigma_w^2), \quad i = 1, \ldots, N$$

and

$$y_i \mid \mu \sim N(\mu, \sigma_e^2), \quad i = 1, \ldots, N.$$

Hence the sample variance $s_y^2 = \frac{1}{n-1}\sum_{i \in s}(Y_i - \bar{Y}_s)^2$ is distributed like $(\sigma_e^2 + \sigma_w^2)\chi^2[n-1]/(n-1)$. It follows that

$$\frac{\hat{T}_{\mathrm{EE}} - T}{(\frac{N}{f})^{1/2} s_y} \sim (1 - \rho f)^{1/2} t[n-1],$$

where

$$\rho = \sigma_e^2/(\sigma_e^2 + \sigma_w^2). \tag{7.1.5}$$

Thus, if ρ is known, the limits of a $(1-\alpha)$ CPI for T are

$$\hat{T}_{\mathrm{EE}} \pm (1-\rho f)^{1/2} t_{1-\alpha/2}[n-1]\sqrt{\frac{N}{f}}\, s_y. \tag{7.1.6}$$

Obviously, if the ys are measured without error $\rho = 1$ and the limits of the CPI (7.1.6) reduce to the ones discussed in Chapter 5. If the value of ρ is unknown, one can use conservative intervals with $\rho = 0$. Such intervals will have coverage probability greater than $(1-\alpha)$, but their expected length will be larger than those with known ρ. The ratio of the expected length of the conservative intervals to those that can be applied when ρ is known is

$$\mathrm{ELR}(\rho) = \frac{1}{(1-\rho f)^{1/2}}, \qquad 0 \le \rho \le 1. \tag{7.1.7}$$

$\mathrm{ELR}(\rho)$ is a measure of relative additional penalty, due to ignorance of the ρ value. Notice that the parameter ρ is invariant with respect to affine transformations, and we cannot therefore eliminate it by considering transformations as change of scale or location. One could consider, however, Bayes equivariant prediction intervals, by assuming a prior distribution on ρ.

Another index of the effect of measurement error is the *relative excess of prediction risk* (REPR), that is,

$$\begin{aligned} \mathrm{REPR}(\rho) &= \frac{\mathrm{Var}_{\psi_{\mathrm{E}}}[\hat{T}_{\mathrm{EE}} - T]}{\mathrm{Var}_{\psi}[\hat{T}_{\mathrm{E}} - T]} - 1 \\ &= \frac{1-\rho}{\rho(1-f)}, \qquad 0 < \rho < 1. \end{aligned} \tag{7.1.8}$$

Notice that $\mathrm{REPR}(1) = 0$ while $\mathrm{REPR}(0) = \infty$.

7.1.2. Simple Regression Model

Suppose that the jth population element is associated with a pair (y_j, x_j), which is assumed to be a random sample from a bivariate normal distribution, $j = 1, \ldots, N$. Relating the two sets of variables, the model is

$$\begin{cases} y_j = \beta_o + \beta_1 x_j + e_j, \\ X_j = x_j + u_j, \\ Y_j = y_j + w_j. \end{cases} \qquad j = 1, \ldots, N, \tag{7.1.9}$$

The model characterized by model (7.1.9) is called the ψ_{RE}-model. The pair (y_j, x_j) is not directly observable. The observed data is $(Y_j, X_j), j \in \mathrm{s}$, where $\mathbf{s}$ indicates the observed sample. It is also assumed that the X_j are available for $j = 1, \ldots, N$, and that (x_j, e_j, u_j, w_j), are independent and

$$\begin{pmatrix} x_j \\ e_j \\ u_j \\ w_j \end{pmatrix} \sim N\left[\begin{pmatrix} \mu_x \\ 0 \\ 0 \\ 0 \end{pmatrix}, \begin{pmatrix} \sigma_x^2 & 0 & 0 & 0 \\ 0 & \sigma_e^2 & 0 & 0 \\ 0 & 0 & \sigma_u^2 & 0 \\ 0 & 0 & 0 & \sigma_w^2 \end{pmatrix}\right],$$

$j = 1, \ldots, N$.

The conditional joint distribution of (y_j, Y_j) given X_j may be obtained by using well-known properties of the multivariate normal distribution. Accordingly,

$$\begin{pmatrix} y_j \\ Y_j \end{pmatrix} | X_j \sim N(\eta(X_j), V), \tag{7.1.10}$$

where

$$\eta(X) = \begin{bmatrix} \beta_o + \beta_1\mu_x + \beta_1 k_x(X_j - \mu_x) \\ \beta_o + \beta_1\mu_x + \beta_1 k_x(X_j - \mu_x) \end{bmatrix},$$

$$V = \begin{bmatrix} \beta_1^2\sigma_u^2 k_x + \sigma_e^2 & \beta_1^2\sigma_u^2 k_x + \sigma_e^2 \\ \beta_1^2\sigma_u^2 k_x + \sigma_e^2 & \beta_1^2\sigma_u^2 k_x + \sigma_e^2 + \sigma_w^2 \end{bmatrix},$$

$k_x = \sigma_x^2/\sigma_X^2$ and $\sigma_X^2 = \sigma_x^2 + \sigma_u^2$, $j = 1, \ldots, N$.

7.1.3. Regression Type Predictors

Consider the general regression predictor of T

$$\hat{T}_{\text{GRE}} = N\bar{Y}_s + \hat{\beta}_{1\text{GE}}(N - n)(\bar{X}_r - \bar{X}_s), \tag{7.1.11}$$

where $\hat{\beta}_{1\text{GE}}$ is an estimator of β_1, which is independent of $\bar{Y}_s$ and of $\bar{X}_s = \sum_{j\in \mathbf{s}} X_j/n$, and that $E_{\psi_{\text{RE}}}[\hat{\beta}_{1\text{GE}}|\mathbf{X}] = E_{\psi_{\text{RE}}}[\hat{\beta}_{1\text{GE}}]$. The regression predictor

$$\hat{T}_{\text{REE}} = N\bar{Y}_s + \hat{\beta}_{1\text{E}}(N - n)(\bar{X}_r - \bar{X}_s), \tag{7.1.12}$$

with

$$\hat{\beta}_{1\text{E}} = \frac{\sum_{j\in s}(X_j - \bar{X}_s)Y_j}{\sum_{j\in s}(X_j - \bar{X}_s)^2} \tag{7.1.13}$$

is a particular case of a $\hat{T}_{\text{GRE}}$. To verify this, write $\hat{\beta}_{1\text{E}} = \mathbf{Y}_s'\mathbf{C}_s$, where

$$\mathbf{C}_s = \left(\mathbf{I}_n - \frac{1}{n}\mathbf{J}_n\right)\mathbf{X}_s \Big/ \mathbf{X}_s'\left(\mathbf{I}_n - \frac{1}{n}\mathbf{J}_n\right)\mathbf{X}_s. \tag{7.1.14}$$

According to Eq. (7.1.10)

$$\mathbf{Y}_s \mid \mathbf{X}_s \sim N\big((\beta_0 + \beta_1\mu_x)\mathbf{1}_s + \beta_1 k_x(\mathbf{X}_s - \mu_x\mathbf{1}_s), \\ (\beta_1^2\sigma_u^2 k_x + \sigma_e^2 + \sigma_w^2)\mathbf{I}_n\big). \tag{7.1.15}$$

Hence,

$$E[\hat{\beta}_{1\mathrm{E}} \mid \mathbf{X}_s] = [\beta_0 + \beta_1\mu_x(1-k_x)]\mathbf{1}_s'\mathbf{C}_s + \beta_1 k_x \mathbf{X}_s'\mathbf{C}_s = \beta_1 k_x.$$

Furthermore, since $\mathbf{1}_s'\mathbf{C}_s = 0$, $\bar{Y}_s$ and $\hat{\beta}_{1\mathrm{E}}$ are conditionally independent given $\mathbf{X}_s$. Moreover, the conditional distribution of $\bar{Y}_s$ given $\mathbf{X}_s$ is normal with conditional expectation $(\beta_0 + \beta_1\mu_x) + \beta_1 k_x(\bar{X}_s - \mu_x)$ and conditional variance $(\beta_1^2\sigma_u^2 k_x + \sigma_e^2 + \sigma_w^2)/n$. Hence, it depends on $\mathbf{X}_s$ only through $\bar{X}_s$. The conditional distribution of $\hat{\beta}_{1\mathrm{E}}$ given $\mathbf{X}_s$ is normal with conditional mean $\beta_1 k_x$, and conditional variance

$$\mathrm{Var}_{\psi_{\mathrm{RE}}}[\hat{\beta}_{1\mathrm{E}} \mid \mathbf{X}_s] = \frac{\beta_1^2\sigma_u^2 k_x + \sigma_e^2 + \sigma_w^2}{\sum_{i\in s}(X_i - \bar{X}_s)^2}. \tag{7.1.16}$$

Finally, since $\bar{X}_s$ and $\sum_{i\in s}(X_i - \bar{X}_s)^2$ are independent, we obtain that $\hat{\beta}_{1\mathrm{E}}$ is independent of $\bar{Y}_s$. Similarly, since the conditional distribution of $\hat{\beta}_{1\mathrm{E}}$, given $\mathbf{X}_s$, is a function of $\sum_{i\in s}(X_i - \bar{X}_s)^2$, $\hat{\beta}_{1\mathrm{E}}$ and $\bar{X}_s$ are independent. Notice that, if y_j and x_j ($j = 1, \cdots, N$) are measured without error, then $\hat{T}_{\mathrm{REE}} = \hat{T}_{\mathrm{RE}}$, which is the regression prediction given in Eq. (2.2.17). The following theorem (Bolfarine, 1989e) presents some properties of $\hat{T}_{\mathrm{GRE}}$.

Theorem 7.1.2. *The predictor $\hat{T}_{\mathrm{GRE}}$ is ψ_{RE}-unbiased with prediction variance*

(7.1.17)

$$\mathrm{Var}_{\psi_{\mathrm{RE}}}[\hat{T}_{\mathrm{GRE}} - T] = N\frac{(1-f)}{f}\Big(\sigma_e^2 + \beta_1^2 k_x\sigma_u^2 + \sigma_X^2\,\mathrm{Var}_{\psi_{\mathrm{RE}}}[\hat{\beta}_{1\mathrm{GE}}] \\ + \sigma_X^2\{\beta_1^2 k_x - 2\beta_1 k_x E_{\psi_{\mathrm{RE}}}[\hat{\beta}_{1\mathrm{GE}}] + (E_{\psi_{\mathrm{RE}}}[\hat{\beta}_{1\mathrm{GE}}])^2\}\Big) + \frac{N}{f}\sigma_w^2.$$

Proof. Since

$$E_{\psi_{\mathrm{RE}}}[\bar{Y}_s] = E_{\psi_{\mathrm{E}}}[\bar{y}] = \beta_o + \beta_1\mu_x = \mu_y,$$

it follows that

$$E_{\psi_{\mathrm{RE}}}[\hat{T}_{\mathrm{GRE}} - T] = E_{\psi_{\mathrm{RE}}}[N\bar{Y}_s + (N-n)\hat{\beta}_{1\mathrm{GE}}(\bar{X}_r - \bar{X}_s) - N\bar{y}] \\ = N\mu_y + (N-n)E_{\psi_{\mathrm{RE}}}[\hat{\beta}_{1\mathrm{GE}}](\mu_x - \mu_x) - N\mu_y = 0.$$

Furthermore,

(7.1.18)
$$\begin{aligned}\mathrm{Var}_{\psi_{\mathrm{RE}}}[\hat{T}_{\mathrm{GRE}}-T] &= \mathrm{Var}_{\psi_{\mathrm{RE}}}\{n(\bar{Y}_s-\bar{y}_s) \\ &\quad + (N-n)[(\bar{Y}_s-\bar{y}_r)+\hat{\beta}_{1\mathrm{GE}}(\bar{X}_r-\bar{X}_s)]\} \\ &= (N-n)^2\mathrm{Var}_{\psi_{\mathrm{RE}}}[(\bar{Y}_s-\bar{y}_r) \\ &\quad + \hat{\beta}_{1\mathrm{GE}}(\bar{X}_r-\bar{X}_s)] + n^2\mathrm{Var}_{\psi_{\mathrm{RE}}}[\bar{Y}_s-\bar{y}_s] \\ &\quad + 2n(N-n)\mathrm{Cov}_{\psi_{\mathrm{RE}}}\{(\bar{Y}_s-\bar{y}_s),(\bar{Y}_s-\bar{y}_r) \\ &\quad + \hat{\beta}_{1\mathrm{GE}}(\bar{X}_r-\bar{X}_s)\}.\end{aligned}$$

(see Exercise 7.2). Moreover,

(7.1.19)
$$\begin{aligned}\mathrm{Var}_{\psi_{\mathrm{RE}}}[\bar{Y}_s-\bar{y}_r] &= \frac{N}{n(N-n)}(\beta_1^2\sigma_x^2+\sigma_e^2)+\frac{\sigma_w^2}{n}, \\ \mathrm{Var}_{\psi_{\mathrm{RE}}}[\hat{\beta}_{1\mathrm{GE}}(\bar{X}_r-\bar{X}_s)] &= \frac{N}{n(N-n)}\sigma_X^2\{(E_{\psi_{\mathrm{RE}}}[\hat{\beta}_{1\mathrm{GE}}])^2 \\ &\quad + \mathrm{Var}_{\psi_{\mathrm{RE}}}[\hat{\beta}_{1\mathrm{GE}}]\}, \\ E_{\psi_{\mathrm{RE}}}[(\bar{Y}_s-\bar{y}_r)(\bar{X}_r-\bar{X}_s)] &= -\frac{N}{n(N-n)}\beta_1 k_x\sigma_X^2, \\ \mathrm{Cov}_{\psi_{\mathrm{RE}}}[(\bar{Y}_s-\bar{y}_r),\hat{\beta}_{1\mathrm{GE}}(\bar{X}_r-\bar{X}_s)] &= -\frac{N}{n(N-n)}\beta_1 k_x\sigma_X^2 E_{\psi_{\mathrm{RE}}}[\hat{\beta}_{1\mathrm{GE}}], \\ \mathrm{Cov}_{\psi_{\mathrm{RE}}}[\bar{Y}_s-\bar{y}_s,\bar{Y}_s-\bar{y}_r] &= \frac{\sigma_w^2}{n}, \\ \mathrm{Cov}_{\psi_{\mathrm{RE}}}[\bar{Y}_s-\bar{y}_s,\hat{\beta}_{1\mathrm{GE}}(\bar{X}_r-\bar{X}_s)] &= 0.\end{aligned}$$

Hence, the result follows by substituting results (7.1.19) into Eqs. (7.1.18). (Q.E.D.)

Notice that, if $\beta_1 = 0$ then the prediction variance (7.1.17) reduces to the prediction variance (7.1.4).

From Theorem 7.1.2 and from the fact that $E_{\psi_{\mathrm{RE}}}[\hat{\beta}_{1\mathrm{E}}] = k_x\beta_1$, we obtain

Corollary 7.1.1. *The predictor $\hat{T}_{\mathrm{REE}}$, is ψ_{RE}-unbiased and*

(7.1.20)
$$\begin{aligned}\mathrm{Var}_{\psi_{\mathrm{RE}}}[\hat{T}_{\mathrm{REE}}-T] &= N\frac{(1-f)}{f}\{\sigma_e^2+\beta_1^2k_x\sigma_u^2 \\ &\quad + \sigma_X^2\,\mathrm{Var}_{\psi_{\mathrm{RE}}}[\hat{\beta}_{1\mathrm{E}}]\}+\frac{N}{f}\sigma_w^2,\end{aligned}$$

where

(7.1.21)
$$\mathrm{Var}_{\psi_{\mathrm{RE}}}[\hat{\beta}_{1\mathrm{E}}] = \frac{\beta_1^2\sigma_u^2k_x+\sigma_e^2+\sigma_w^2}{(n-3)\sigma_X^2}.$$

Proof. Formula (7.1.20) follows directly from Theorem 7.1.2 and from the fact that $E_{\psi}[\hat{\beta}_{1\mathrm{E}}] = k_x\beta_1$. Furthermore, formula (7.1.17) implies that

$$\mathrm{Var}_{\psi_{\mathrm{RE}}}[\hat{\beta}_{1\mathrm{E}}] = (\beta_1^2\sigma_u^2 k_x + \sigma_e^2 + \sigma_w^2)E\left\{\frac{1}{\sum_{i\in s}(X_i - \bar{X}_s)^2}\right\}.$$

This yields formula (7.1.21).

(Q.E.D.)

Another predictor is obtained by using

$$\hat{\gamma} = k_x^{-1}\hat{\beta}_{1\mathrm{E}}, \tag{7.1.22}$$

as an estimator of β_1. See for example Fuller (1987). Accordingly,

$$E_{\psi_{\mathrm{RE}}}[\hat{\gamma}] = \beta_1.$$

Notice that the variance of $\hat{\gamma}$ can be computed from Eq. (7.1.21). Define the predictor

$$\hat{T}_{\mathrm{REU}} = N\bar{Y}_s + (N-n)\hat{\gamma}(\bar{X}_r - \bar{X}_s). \tag{7.1.23}$$

Corollary 7.1.2. *The predictor $\hat{T}_{\mathrm{REU}}$, is ψ_{RE}-unbiased, and*

$$\begin{aligned}\mathrm{Var}_{\psi_{\mathrm{RE}}}[\hat{T}_{\mathrm{REU}} - T] = N\frac{(1-f)}{f}\{\sigma_e^2 + \beta_1^2\sigma_u^2(1+k_x)\\ + \sigma_X^2\,\mathrm{Var}_{\psi_{\mathrm{RE}}}[\hat{\gamma}]\} + \frac{N}{f}\sigma_w^2.\end{aligned} \tag{7.1.24}$$

$\hat{T}_{\mathrm{REE}}$ and $\hat{T}_{\mathrm{REU}}$ are both unbiased predictors of T. By comparing Eqs. (7.1.20) and (7.1.24), it follows that $\hat{T}_{\mathrm{REE}}$ has a smaller prediction variance than $\hat{T}_{\mathrm{REU}}$, since

$$\mathrm{Var}_{\psi_{\mathrm{RE}}}[\hat{\beta}_{1\mathrm{E}}] \le \mathrm{Var}[\hat{\gamma}].$$

As in Section 7.1.1, the relative excess in prediction risk due to errors of measurements both in x and in y, is

$$\begin{aligned}\mathrm{REPRD}(\rho,\zeta) &= \frac{\mathrm{Var}_{\psi_{\mathrm{RE}}}[\hat{T}_{\mathrm{REE}} - T]}{\mathrm{Var}_{\psi_{\mathrm{RE}}}[\hat{T}_{\mathrm{RE}} - T]} - 1\\ &= \frac{n-2}{N-3}(1+\zeta) + \left(\frac{1}{n-3} + \frac{1}{1-f}\right)\frac{\rho}{1-\rho} - 1,\end{aligned} \tag{7.1.25}$$

where ρ is given in Eq. (7.1.5) and

$$\zeta = \frac{\beta_1^2\sigma_u^2 k_x}{\sigma_e^2}. \tag{7.1.26}$$

For large values of n,

$$\text{REPRD}(\rho,\zeta) \approx \text{REPR}(1-\rho) + f\zeta - (1-f). \tag{7.1.27}$$

ζ is the relative contribution to the prediction risk of the measurement errors in x.

Why and when should one use a predictor like $\hat{T}_{\text{REE}}$? An alternative could be the expansion predictor $\hat{T}_{\text{EE}} = N\bar{Y}_s$. This predictor is also ψ_{RE}-unbiased. It is straightforward to verify that

$$\begin{aligned} \text{Var}_{\psi_{\text{RE}}}[\hat{T}_{\text{EE}} - T] = \frac{N(1-f)}{f}[\sigma_e^2 + \beta_1^2\sigma_u^2 k_x(1+k_x) \\ + \beta_1^2\sigma_x^2 k_x^2] + \frac{N}{f}\sigma_w^2. \end{aligned} \tag{7.1.28}$$

Thus, $\hat{T}_{\text{REE}}$ has a smaller prediction variance than $\hat{T}_{\text{EE}}$ whenever

$$\frac{\sigma_e^2 + \sigma_w^2}{n-3} < \beta_1^2\sigma_u^2 k_x(k_x - \frac{1}{n-3}) + \beta_1^2 k_x^2\sigma_x^2. \tag{7.1.29}$$

If the parameters μ_x and k_x are *known*, one could consider the predictor

$$\hat{T}_{\text{REC}} = N\bar{Y}_s + (N-n)\hat{\gamma}(\bar{Z}_r - \bar{Z}_s), \tag{7.1.30}$$

where $Z_i = \mu_x + k_x(X_i - \mu_x)$, $i = 1, \cdots, N$, and

$$\hat{\gamma} = \frac{\sum_{i\in s}(Z_i - \bar{Z}_s)Y_i}{\sum_{i\in s}(Z_i - \bar{Z}_s)^2}. \tag{7.1.31}$$

We find that $\hat{T}_{\text{REC}}$ is ψ_{RE}-unbiased and

$$\begin{aligned} \text{Var}_{\psi_{\text{RE}}}[\hat{T}_{\text{REC}} - T|\mathbf{X}_s] = N\frac{(1-f)}{f}\{\beta_1^2 k_x\sigma_u^2 + \sigma_e^2\} \\ \left\{1 + n\frac{(1-f)(\bar{Z}_r - \bar{Z}_s)^2}{\sum_{i\in s}(Z_i - \bar{Z}_s)^2}\right\} \\ + \sigma_w^2\left\{\frac{N}{f} + (N-n)^2\frac{(\bar{Z}_r - \bar{Z}_s)^2}{\sum_{i\in s}(Z_i - \bar{Z}_s)^2}\right\}. \end{aligned} \tag{7.1.32}$$

Thus, the prediction variance, under model ψ_{RE} of $\hat{T}_{\text{REC}} - T$ is

$$\begin{aligned} \text{Var}_{\psi_{\text{RE}}}[\hat{T}_{\text{REC}} - T] = N\frac{1-f}{f}(\beta_1^2 k_x\sigma_u^2 + \sigma_e^2) \\ \left\{1 + n(1-f)E_{\psi_{\text{RE}}}\left[\frac{(Z_r - \bar{Z}_s)^2}{\sum_{i\in s}(Z_i - \bar{Z}_s)^2}\right]\right\} \\ + \sigma_w^2\left\{\frac{N}{f} + (N-n)^2 E_{\psi_{\text{RE}}}\left[\frac{(\bar{Z}_r - \bar{Z}_s)^2}{\sum_{i\in s}(Z_i - \bar{Z}_s)^2}\right]\right\}. \end{aligned} \tag{7.1.33}$$

Furthermore,

$$\frac{(\bar{Z}_s - \bar{Z}_r)^2}{\sum_{i \in s}(Z_i - \bar{Z}_s)^2} \sim \frac{n(N-n)}{N(n-1)} F[1, n-1] \tag{7.1.34}$$

and $E[F[1,n-1]] = (n-1)/(n-3)$. Hence,

$$E_{\psi_{\mathrm{RE}}}\left[\frac{(\bar{Z}_s - \bar{Z}_r)^2}{\sum_{i \in s}(Z_i - \bar{Z}_s)^2}\right] = \frac{n(N-n)}{(n-3)N}. \tag{7.1.35}$$

Substituting Eq. (7.1.35) into Eq. (7.1.33) we obtain

$$\begin{aligned} \mathrm{Var}_{\psi_{\mathrm{RE}}}[\hat{T}_{\mathrm{REC}} - T] &= N\frac{1-f}{f}(\beta_1^2 \sigma_u^2 k_x + \sigma_e^2)\left[1 + \frac{n^2(1-f)^2}{n-3}\right] \\ &+ \sigma_w^2\left[\frac{N}{f} + \frac{n}{n-3}N^2(1-f)^3\right]. \end{aligned} \tag{7.1.36}$$

7.2. Bayesian Models with Measurement Errors

In this section, a Bayesian approach is considered for the error in variables superpopulation models considered in the previous section. First we consider the location model (model SM1) and then the two-stage sampling model (model SM6). The simple regression error model is considered later. The Bayesian model with measurement errors is denoted by ψ_{BE} in the sequel.

7.2.1. Model SM1

Consider the simple location superpopulation model with error in variables (7.1.1) and (7.1.2), and noninformative prior distribution for μ. Given the observed data $\mathbf{Y}_s$, it follows that the posterior density of μ is proportional to

$$\exp\left[-\frac{n}{2}\frac{(\mu - \bar{Y}_s)^2}{\sigma_e^2 + \sigma_w^2}\right].$$

Hence,

$$\mu | \mathbf{Y}_s \sim N(\bar{Y}_s, \frac{\sigma_e^2 + \sigma_w^2}{n}). \tag{7.2.1}$$

Moreover, since

$$E_{\psi_{\mathrm{E}}}[y_i | \mu, \mathbf{Y}_s] = \begin{cases} \mu + \dfrac{\sigma_e^2}{\sigma_e^2 + \sigma_w^2}(Y_i - \mu), & i \in \mathbf{s}, \\ \mu, & i \notin \mathbf{s}, \end{cases} \tag{7.2.2}$$

it follows that the Bayes predictor of T is given by

$$\begin{aligned}\hat{T}_{\mathrm{BE}} &= E_{\psi_{\mathrm{BE}}}[\sum_{i=1}^{N} y_i|\mathbf{Y}_s] = E_{\psi_{\mathrm{BE}}}[\sum_{i\in\mathbf{s}} y_i|\mathbf{Y}_s] + E_{\psi_{\mathrm{BE}}}[\sum_{i\in\mathbf{r}} y_i|\mathbf{Y}_s] \\ &= NE_{\psi_{\mathrm{BE}}}[\mu|\mathbf{Y}_s] + n\frac{\sigma_e^2}{\sigma_e^2+\sigma_w^2}(\bar{Y}_s - E_{\psi_{\mathrm{BE}}}[\mu|\mathbf{Y}_s]) \\ &= N\bar{Y}_s.\end{aligned} \tag{7.2.3}$$

Thus, the expansion predictor $\hat{T}_{\mathrm{EE}}$ is also a Bayes predictor of T. The posterior variance of T is given by

$$\begin{aligned}\mathrm{Var}_{\psi_{\mathrm{BE}}}[T|\mathbf{Y}_s] &= E_{\psi_{\mathrm{BE}}}\left\{\mathrm{Var}_{\psi_{\mathrm{E}}}\left[\sum_{i=1}^{N} y_i|\mu,\mathbf{Y}_s\right]|\mathbf{Y}_s\right\} \\ &\quad + \mathrm{Var}_{\psi_{\mathrm{BE}}}\left\{E_{\psi_{\mathrm{E}}}\left[\sum_{i=1}^{N} y_i|\mu,\mathbf{Y}_s\right]|\mathbf{Y}_s\right\} \\ &= n\frac{\sigma_e^2\sigma_w^2}{\sigma_e^2+\sigma_w^2} + (N-n)\sigma_e^2 \\ &\quad + \mathrm{Var}_{\psi_{\mathrm{BE}}}\left\{\left[\frac{n\sigma_w^2}{\sigma_e^2+\sigma_w^2} + (N-n)\right]\mu|\mathbf{Y}_s\right\} \\ &= N\frac{(1-f)}{f}\sigma_e^2 + \frac{N}{f}\sigma_w^2.\end{aligned}$$

The above results may be summarized in the following theorem.

Theorem 7.2.1. *The Bayes predictor of T with respect to noninformative prior distribution of μ is*

$$E_{\psi_{\mathrm{BE}}}[T|\mathbf{Y}_s] = N\bar{Y}_s = \hat{T}_{\mathrm{EE}},$$

with posterior variance

$$\mathrm{Var}_{\psi_{\mathrm{BE}}}[T|\mathbf{Y}_s] = \frac{N(1-f)}{f}\sigma_e^2 + \frac{N}{f}\sigma_w^2. \tag{7.2.4}$$

Notice that the posterior variance of T coincides with the prediction variance (7.1.4). This result is expected, since a noninformative prior distribution has been considered for μ. Using the posterior distribution of μ in Eq. (7.2.1), it follows from Eq. (7.2.2) and the law of the iterated expectation that

$$E_{\psi_{\mathrm{BE}}}[y_i|\mathbf{Y}_s] = \begin{cases}\bar{Y}_s, & i\in\mathbf{r},\\ \bar{Y}_s + \rho(Y_i-\bar{Y}_s), & i\in\mathbf{s}.\end{cases}$$

where, as in the previous section, $\rho = \sigma_e^2/(\sigma_e^2 + \sigma_w^2)$. Moreover, $\mathbf{\Sigma} = \text{Var}_{\psi_{\text{BE}}}[\mathbf{y}|\mathbf{Y}_s]$ can be computed by writing

$$\text{Var}_{\psi_{\text{BE}}}[\mathbf{y}|\mathbf{Y}_s] = E_{\psi_{\text{BE}}}\{\text{Var}_{\psi_{\text{BE}}}[\mathbf{y}|\mu;\mathbf{Y}_s]|\mathbf{Y}_s\} + \text{Var}_{\psi_{\text{BE}}}\{E_{\psi_{\text{BE}}}[\mathbf{y}|\mathbf{Y}_s;\mu]|\mathbf{Y}_s\},$$

where

$$\text{Var}_{\psi_{\text{BE}}}[y_i|\mathbf{Y}_s;\mu] = \begin{cases} \sigma_e^2, & i \in \mathbf{r}, \\ \rho\sigma_w^2, & i \in \mathbf{s}, \end{cases}$$

and $E_{\psi_{\text{BE}}}[y_i|\mu;\mathbf{Y}_s]$ is obtained from Eq. (7.2.2). Therefore, since $S_y^2 = \mathbf{y}'(\mathbf{I}_N - \mathbf{J}_N/N)\mathbf{y}/N$, we obtain that the Bayes predictor of S_y^2 is

$$\begin{aligned} E_{\psi_{\text{BE}}}[S_y^2|\mathbf{Y}_s] &= \frac{1}{N}E_{\psi_{\text{BE}}}\left[\mathbf{y}'\left(\mathbf{I}_N - \frac{1}{N}\mathbf{J}_N\right)\mathbf{y}|\mathbf{Y}_s\right] \\ &= \frac{1}{N}\text{tr}\left\{\left(\mathbf{I}_N - \frac{1}{N}\mathbf{J}_N\right)\mathbf{\Sigma}_r\right\} \\ &\quad + \frac{1}{N}E_{\psi_{\text{BE}}}[\mathbf{y}'|\mathbf{Y}_s]\left(\mathbf{I}_N - \frac{1}{N}\mathbf{J}_N\right)E_{\psi_{\text{BE}}}[\mathbf{y}|\mathbf{Y}_s]. \end{aligned}$$

Some algebraic manipulations yield

$$\begin{aligned} \hat{S}_{\text{BE}} &= E_{\psi_{\text{BE}}}[S_y^2|\mathbf{Y}_s] \\ &= \frac{n}{N}\left(\frac{\sigma_e^2}{\sigma_e^2+\sigma_w^2}\right)^2 s_Y^2 + \frac{(N-n)}{N}\left[\sigma_e^2 + \frac{(n-1)}{N-n}\frac{\sigma_w^2\sigma_e^2}{\sigma_e^2+\sigma_w^2}\right] \qquad (7.2.5) \\ &= f\rho^2 s_Y^2 + (1-f)\left[\sigma_e^2 + \frac{n-1}{N(1-f)}\rho\sigma_w^2\right], \end{aligned}$$

where

$$s_Y^2 = \frac{1}{n}\sum_{i\in\mathbf{s}}(Y_i - \bar{Y}_s)^2.$$

Notice that predictor (7.2.5) reduces to predictor (3.1.27) if $\sigma_w^2 = 0$.

7.2.2. Model SM6

In this section, we consider a Bayesian approach to the two-stage sampling with error in variables, by extending the approach considered by Scott and Smith (1969) to the ordinary two-stage sampling scheme (see Example 3.1.2 and Exercise 3.3). Considering the same partition of the population as in SM6, we specify that

$$y_{hj} = \mu_h + e_{hj},$$

$$\mu_h = \mu + u_h.$$

The observed variable is

$$Y_{hj} = y_{hj} + w_{hj},$$

where e_{hj}, u_h, and v_h are all independent, $e_{hj} \sim N(0, \sigma^2_{eh})$, $u_h \sim N(0, \sigma^2_v)$, and $w_{hj} \sim N(0, \sigma^2_{wh})$, $j = 1, \ldots, M_h$, $h = 1, \ldots, K$. The model specified by the above notations is denoted by ψ_{E}. Let $\boldsymbol{\mu}' = (\mu_1, \ldots, \mu_K)$ and $\mathbf{Y}_s = \{Y_{hj}; j \in \mathbf{s}_h \text{ and } h \in \mathbf{s}\}$ be the observed data corresponding to the observed sample $\mathbf{s}$. Since,

$$\mathrm{Cov}_{\psi_{\mathrm{E}}}[y_{hj}, y_{ql}|\mu_h] = \mathrm{Cov}_{\psi_{\mathrm{E}}}[y_{hj}, Y_{ql}|\boldsymbol{\mu}] = \begin{cases} \sigma^2_{eh}, & h = q, j = l \\ 0, & \text{otherwise,} \end{cases}$$

and $\mathrm{Var}_{\psi_{\mathrm{E}}}[Y_{hj}|\boldsymbol{\mu}] = \sigma^2_{eh} + \sigma^2_{wh}$, it follows that

$$E_{\psi_{\mathrm{E}}}[y_{hj}|\mathbf{Y}_s, \mu_h] = \begin{cases} \mu_h + \rho_h(Y_{hj} - \mu_h), & h \in \mathbf{s}, j \in \mathbf{s}_h, \\ \mu_h, & \text{otherwise,} \end{cases} \tag{7.2.6}$$

where $\rho_h = \sigma^2_{eh}/(\sigma^2_{eh} + \sigma^2_{wh})$. Moreover, by using Exercise 3.3, it follows that

$$\hat{\mu}_h = E_{\psi_{\mathrm{BE}}}[\mu_h|\mathbf{Y}_s] = \lambda_h \bar{Y}_{sh} + (1 - \lambda_h)\bar{Y}_s, \tag{7.2.7}$$

where

$$\bar{Y}_{sh} = \sum_{j \in \mathbf{s}_h} \frac{Y_{hj}}{m_h}, \quad \bar{Y}_{s.} = \sum_{h \in \mathbf{s}} \frac{\lambda_h \bar{Y}_{sh}}{\sum_{j \in \mathbf{s}} \lambda_j},$$

and

$$\lambda_h = \begin{cases} \dfrac{\sigma^2_v}{\sigma^2_v + \dfrac{\sigma^2_{eh} + \sigma^2_{wh}}{m_h}}, & h \in \mathbf{s}, \\ 0, & h \notin \mathbf{s}. \end{cases}$$

Therefore, from Eqs. (7.2.6) and (7.2.7) above, it can be shown that

$$\begin{aligned} \hat{T}_{\mathrm{BE}} = E_{\psi_{\mathrm{BE}}}[T|\mathbf{Y}_s] &= \sum_{h \in \mathbf{s}} m_h \rho_h \bar{Y}_{sh} \\ &+ \sum_{h \in \mathbf{s}} m_h(1 - \rho_h)[\lambda_h \bar{Y}_{sh} + (1 - \lambda_h)\bar{Y}_{s.}] \\ &+ \sum_{h \in \mathbf{s}} (M_h - m_h)[\lambda_h \bar{Y}_{sh} + (1 - \lambda_h)\bar{Y}_{s.}] + (\sum_{h \notin \mathbf{s}} M_h)\bar{Y}_{s.}. \end{aligned} \tag{7.2.8}$$

$\hat{T}_{\mathrm{BE}}$ is similar to predictor (4) in Scott and Smith (1969). In the particular case where $m_h = m$, $\sigma^2_{wh} = \sigma^2_w$, $\sigma^2_{eh} = \sigma^2_e$, for all h,

$$\lambda_h = \lambda = \frac{\sigma^2_v}{\sigma^2_v + (\sigma^2_e + \sigma^2_w)/m}, \quad \bar{Y}_{s.} = \sum_{h \in \mathbf{s}} \frac{\bar{Y}_{sh}}{n}$$

and, hence,

$$\hat{T}_{\mathrm{BE}} = \sum_{h \in \mathbf{s}} m_h \bar{Y}_{sh} + \sum_{h \in \mathbf{s}} (M_h - m_h)[\lambda \bar{Y}_{sh} + (1 - \lambda)\bar{Y}_{s.}] + (\sum_{h \notin \mathbf{s}} M_h)\bar{Y}_{s.}. \tag{7.2.9}$$

Thus, even in this simple situation, predictor $\hat{T}_{\mathrm{BE}}$ still depends on the model variances.

The posterior variance of T may also be derived. The results of Exercise 3.3 imply that

(7.2.10)
$$\mathrm{Cov}_{\psi_{\mathrm{BE}}}[\mu_h, \mu_q|\mathbf{Y}_s] = \begin{cases} (1-\lambda_h)^2 w^2 + (1-\lambda_h)\sigma_v^2 = c_{hh}, & h = q, \\ (1-\lambda_h)(1-\lambda_q)w^2 = c_{hq}, & h \neq q, \end{cases}$$

where

$$w^2 = \left[\sum_{h\in \mathbf{s}} \left(\sigma_v^2 + \frac{\sigma_{eh}^2 + \sigma_{wh}^2}{m_h}\right)^{-1}\right]^{-1}.$$

Therefore, it can be shown that

(7.2.11)
$$\begin{aligned}\mathrm{Var}_{\psi_{\mathrm{BE}}}[T|\mathbf{Y}_s] = &\sum_{h\in \mathbf{s}} B_h^2 c_{hh} + \sum_{h\in \mathbf{s}} \sum_{q\neq h\in \mathbf{s}} B_h B_q c_{hq} + \sum_{h\in \mathbf{s}} m_h \rho_h \sigma_{wh}^2 \\ &+ \sum_{h\in \mathbf{s}} (M_h - m_h)\sigma_{eh}^2 + \sum_{h\notin \mathbf{s}} M_h \sigma_{eh}^2 + \sum_{h\notin \mathbf{s}} c_{hh},\end{aligned}$$

where

$$B_h = \sum_{h\in s} [m_h \rho_h + (N_h - m_h)].$$

7.2.3. Simple Regression

In this section, a Bayesian approach is considered for predicting T and S_y^2 under the superpopulation model (7.1.6). For simplicity, we take $\beta_o = 0$. The case of $\beta_o \neq 0$ does not introduce additional theoretical difficulty. Since,

(7.2.12)
$$E_{\psi_{\mathrm{REC}}}[y_i|Y_i, \beta_1] = \begin{cases} \beta_1 Z_i + \frac{\beta_1^2 k_x \sigma_u^2 + \sigma_e^2}{\beta_1^2 k_x \sigma_u^2 + \sigma_e^2 + \sigma_w^2}(Y_i - \beta_1 Z_i), & i \in \mathbf{s} \\ \beta_1 Z_i, & i \in \mathbf{r}, \end{cases}$$

a Bayesian predictor of T is,

(7.2.13)
$$\begin{aligned}\hat{T}_{\mathrm{BE}} = E_{\psi_{\mathrm{BE}}}[T|\mathbf{Y}_s, \mathbf{X}_s] &= E_{\psi_{\mathrm{BE}}}\left[\sum_{i\in \mathbf{s}} y_i|\mathbf{Y}_s, \mathbf{X}_s\right] \\ &+ E_{\psi_{\mathrm{BE}}}\left[\sum_{i\in \mathbf{r}} y_i|\mathbf{Y}_s, \mathbf{X}_s\right] \\ &= \sum_{i\in \mathbf{s}} Y_i E_{\psi_{\mathrm{BE}}}\left[\frac{\beta_1^2 k_x \sigma_u^2 + \sigma_e^2}{\beta_1^2 k_x \sigma_u^2 + \sigma_e^2 + \sigma_w^2}|\mathbf{Y}_s, \mathbf{X}_s\right] \\ &+ \sigma_w^2 \sum_{i\in \mathbf{s}} Z_i E_{\psi_{\mathrm{BE}}}\left[\frac{\beta_1}{\beta_1^2 k_x \sigma_u^2 + \sigma_e^2 + \sigma_w^2}|\mathbf{Y}_s, \mathbf{X}_s\right] \\ &+ \sum_{i\in \mathbf{r}} Z_i E_{\psi_{\mathrm{BE}}}[\beta_1|\mathbf{Y}_s, \mathbf{X}_s].\end{aligned}$$

The prediction variance is given by

$$\begin{aligned}
\mathrm{Var}_{\psi_{\mathrm{BE}}}\left[\sum_{i=1}^{N} y_i \middle| \mathbf{Y}_s, \mathbf{X}_s\right] &= \mathrm{Var}_{\psi_{\mathrm{BE}}}\left\{\frac{\beta_1^2 k_x \sigma_u^2 + \sigma_e^2}{\beta_1^2 k_x \sigma_u^2 + \sigma_e^2 + \sigma_v^2}\sum_{i\in s} Y_i\right. \\
&\quad \left. + \beta_1\left[\frac{\sigma_v^2}{\beta_1^2 k_x \sigma_u^2 + \sigma_e^2 + \sigma_v^2} + \sum_{i\in r} Z_i\right] \middle| \mathbf{Y}_s, \mathbf{X}_s\right\} \\
&\quad + n\sigma_v^2 E_{\psi_{\mathrm{BE}}}\left[\frac{\beta_1^2 \sigma_u^2 k_x + \sigma_e^2}{\beta_1^2 \sigma_u^2 k_x + \sigma_e^2 + \sigma_v^2} \middle| \mathbf{Y}_s, \mathbf{X}_s\right] \\
&\quad + (N-n) E_{\psi_{\mathrm{BE}}}\left[\beta_1^2 k_x \sigma_u^2 + \sigma_e^2 | \mathbf{Y}_s, \mathbf{X}_s\right].
\end{aligned} \tag{7.2.14}$$

For computing the posterior mean and variance of T given above, the posterior distribution of β_1 given $\mathbf{Y}_s$ and $\mathbf{X}_s$ is required. It can be shown that the posterior density of β_1 given $\mathbf{Y}_s$ and $\mathbf{X}_s$ is proportional to

$$\frac{1}{(\beta_1^2 k_x \sigma_u^2 + \sigma_e^2 + \sigma_w^2)^{n/2}} \exp\left\{\frac{1}{2}\sum_{i\in s}\frac{(Y_i - \beta_1 Z_i)^2}{\beta_1^2 k_x \sigma_u^2 + \sigma_e^2 + \sigma_w^2}\right\}. \tag{7.2.15}$$

Hence, the posterior mean and variance of β_1 can be computed numerically. Approximations of posterior moments by the Laplace method for integrals (Tierney and Kadane, 1986) may also be used.

Using a normal approximation of Eq. (7.2.15), it is possible to obtain approximate analytical expressions for Eqs. (7.2.13) and (7.2.14) when y_i is measured without error, that is, $\sigma_w^2 = 0$. Following Lindley (1966), it can be shown that when $\sigma_w^2 = 0$, the posterior density (7.2.15) is approximately normal with mean

$$\hat{\beta}_1 = E_{\psi_{\mathrm{BE}}}[\beta_1 | \mathbf{Y}_s, \mathbf{X}_s] = \frac{S_{zy}}{S_{zz} - \lambda k_x \sigma_u^2} \tag{7.2.16}$$

and variance

$$V(\hat{\beta}_1) = \mathrm{Var}_{\psi_{\mathrm{BE}}}[\beta_1 | \mathbf{Y}_s, \mathbf{X}_s] = \frac{\sigma_e^2 + \hat{\beta}_1^2 k_x \sigma_u^2}{S_{zz} - \lambda k_x \sigma_u^2}, \tag{7.2.17}$$

where $S_{zy} = \sum_{i\in s} Z_i Y_i$, $S_{zz} = \sum_{i\in s} Z_i^2$, $S_{yy} = \sum_{i\in s} Y_i^2$, and

$$\lambda = \frac{k_x \sigma_u^2 S_{yy} + \sigma_e^2 S_{zz} - \sqrt{(k_x \sigma_u^2 S_{yy} + \sigma_e^2 S_{zz})^2 - 4(S_{zz}S_{yy} - S_{zy}^2) k_x \sigma_u^2 \sigma_e^2}}{2 k_x \sigma_u^2 \sigma_e^2}.$$

Thus, using the normal approximation given above, the posterior mean and variance given by Eqs. (7.2.13) and (7.2.14), respectively, reduce to

$$\hat{T}_{\mathrm{BE}} = n\bar{Y}_s + \frac{S_{zy}}{S_{zz} - \lambda k_x \sigma_u^2}\sum_{j\in r} Z_i \tag{7.2.18}$$

and

$$\text{Var}_{\psi_{\text{BE}}}[T|\mathbf{Y}_s, \mathbf{X}_s] = (N-n)\{\sigma_e^2 + k_x\sigma_u^2[\hat{\beta}_1^2 + V(\hat{\beta}_1)]\} + \left(\sum_{i\in\mathbf{r}} Z_i\right)^2 V(\hat{\beta}_1), \tag{7.2.19}$$

where $\hat{\beta}_1$ and $V(\hat{\beta}_1)$ are given above. If x_i is also measured without error, then predictor $\hat{T}_{\text{BE}}$ in Eq. (7.2.18) reduces to the ordinary Bayesian predictor of T, by making $\sigma_u^2 = 0$.

Furthermore, if y_i is measured without error ($\sigma_w^2 = 0$), a posterior mean for the population variance may be found by using the normal approximation (discussed above) to the posterior density (7.2.15). Using Eq. (3.1.25), it can be shown that

$$\begin{aligned}\hat{S}^2_{\text{BE}} = E_{\psi_{\text{BE}}}[S_y^2|\mathbf{Y}_s] = \frac{n}{N}s_Y^2 + \left(1-\frac{n}{N}\right)\Bigg\{\left(1-\frac{1}{N}\right) \\ \{\sigma_e^2 + k_x\sigma_u^2[\hat{\beta}_1^2 + V(\hat{\beta}_1)]\} \\ + S_{rz}^2[\hat{\beta}_1^2 + V(\hat{\beta}_1)] + \frac{n}{N}[(\bar{Y}_s - \hat{\beta}_1\bar{Z}_r)^2 + \bar{Z}_r^2 V(\hat{\beta}_1)]\Bigg\},\end{aligned} \tag{7.2.20}$$

where

$$S_{zr}^2 = \sum_{i\in\mathbf{r}} \frac{(Z_i - \bar{Z}_r)^2}{N-n}, \quad \bar{Z}_r = \sum_{i\in\mathbf{r}} \frac{Z_i}{N-n},$$

and $\hat{\beta}_1$ and $V(\hat{\beta}_1)$ are respectively the posterior mean and variance of β_1 given $\mathbf{Y}_s$, which, for large n, are given by Eqs. (7.2.16) and (7.2.17). Notice also that, if x_i is measured without error, the posterior mean of S_y^2 is still given by Eq. (7.2.20), but with

$$\hat{\beta}_1 = \frac{\sum_{i\in\mathbf{s}} X_i Y_i}{\sum_{i\in\mathbf{s}} X_i^2} \text{ and } V(\hat{\beta}_1) = \frac{\sigma_e^2}{\sum_{i\in\mathbf{s}} X_i^2},$$

as in Example 3.1.5.

7.2.4. Orthogonal Transformations

Another case that is considered in Bolfarine and Sandoval (1990), is when σ_x^2 and σ_u^2 are unknown and $k_x = \sigma_x^2/\sigma_X^2$ is known. As pointed out by Fuller (1987), there are a number of situations, particularly in psychology, sociology, and survey sampling, where k_x could be considered known. Thus, we may write $\sigma_u^2 = k\sigma_e^2$ where

$$k = \frac{(1-k_x)}{k_x}.$$

The joint distribution of (Y_k, X_k) is

$$\begin{pmatrix} X_k \\ Y_k \end{pmatrix} \sim N\left(\begin{pmatrix} \beta\mu_x \\ \mu_x \end{pmatrix}, \begin{bmatrix} \beta^2\sigma_x^2 + \sigma_e^2 & \beta\sigma_x^2 \\ \beta\sigma_x^2 & (k+1)\sigma_x^2 \end{bmatrix} \right),$$

$k = 1, \cdots, N$. Before computing the conditional distribution of Y_k given X_k, we transform the original parameter $(\beta, \sigma_e^2, \sigma_x^2)$ to a new parameter vector $(\beta, \lambda_0, \lambda_1)$, where β is orthogonal to (λ_0, λ_1) as defined by Cox and Reid (1987). After computing the elements of the Fisher information matrix needed for finding the orthogonal parametrization $(\beta, \lambda_0, \lambda_1)$, we arrive at the following set of differential equations

$$(7.2.21) \qquad (1+k^2)\sigma_x^4 \frac{\partial \sigma_e^2}{\partial \beta} + k(k+1)\beta^2\sigma_x^4 \frac{\partial \sigma_x^2}{\partial \beta} = -2k(k+1)\beta\sigma_x^6,$$

and

$$(7.2.22) \qquad k(k+1)\beta^2\sigma_x^4 \frac{\partial \sigma_e^2}{\partial \beta} + [2k\beta^2 D + \sigma_e^4(k=1)^2] \frac{\partial \sigma_x^2}{\partial \beta} = -2\beta^3 k^2 \sigma_x^2.$$

A solution to this set of differential equations yields the one to one transformation

$$(7.2.23) \qquad \lambda_0 = (k+1)\sigma_e^2 + k\lambda_1\beta^2$$

and

$$(7.2.24) \qquad \lambda_1 = \sigma_x^2.$$

In this new parameterization, it follows that the joint distribution of (Y_i, X_i) is bivariate normal with mean vector and covariance matrix given, respectively, by

$$\begin{pmatrix} \beta\mu_x \\ \mu_x \end{pmatrix} \quad \text{and} \quad \begin{pmatrix} (\lambda_0 + \lambda_1\beta^2)/(k+1) & \beta\lambda_1 \\ \beta\lambda_1 & (k+1)\lambda_1 \end{pmatrix}.$$

The conditional distribution of Y_i given X_i is, accordingly,

$$(7.2.25) \qquad Y_i \mid X_i \sim N(\beta Z_i, \frac{\lambda_0}{k+1}),$$

where

$$Z_i = \mu_x + \frac{1}{k+1}(X_i - \mu_x),$$

$i = 1, \cdots, N$. By considering the noninformative prior

$$p(\beta, \lambda_0) \propto \frac{1}{\lambda_0},$$

it can be shown that the posterior density of β given $\mathbf{Z}_s = (Z_i,\ i \in \mathbf{s})$ and $\mathbf{Y}_s$ is

$$(7.2.26) \qquad p(\beta \mid \mathbf{Y}_s, \mathbf{Z}_s) \propto \left(1 - \frac{(\beta - \hat{\beta}_s)^2}{(n-1)k_x\hat{\lambda}_0}\right)^{-\frac{n}{2}},$$

where

$$(7.2.27) \qquad \hat{\beta}_s = \frac{\sum_{i\in \mathbf{s}} Y_i Z_i}{\sum_{i\in \mathbf{s}} Z_i^2} \quad \text{and} \quad \hat{\lambda}_0 = \frac{\sum_{i=1}^{n}(Y_i - \hat{\beta}Z_i)^2}{(n-1)k_x \sum_{i=1}^{n} Z_i^2}.$$

Thus, as implied from Corollary 3.1.2, the Bayes predictor of T is given by

$$(7.2.28) \qquad \hat{T}_{\mathrm{BE}} = n\bar{y}_s + \hat{\beta}_s \sum_{i\in s} Z_i,$$

where $\hat{\beta}_s$ is the posterior mean, as follows from Eq. (7.2.26). Furthermore, the posterior variance of T is

$$(7.2.29) \qquad \mathrm{Var}[T|Y_s, Z_s] = \frac{n-1}{n-3}k_x\hat{\lambda}_0\left\{(N-n) + \frac{\left(\sum_{i\in \mathbf{r}} Z_i\right)^2}{\sum_{i\in \mathbf{s}} Z_i^2}\right\}.$$

7.3. Exercises

[7.1] Derive the prediction variance (7.1.4).
[7.2] Verify expression (7.1.13).
[7.3] Find the expected value of the conditional variance given in Eq. (7.1.17).
[7.4] Consider model ψ_{E} defined by Eqs. (7.1.1) and (7.1.2). Show that the posterior distribution of μ given $\mathbf{y}_s$, using a noninformative prior distribution for μ is as given in Eq. (7.2.1).
[7.5] Derive predictor (7.2.5).
[7.6] Using Lemma 3.1.1, derive expressions (7.2.8), (7.2.9), and (7.2.10), and the posterior variance (7.2.11).
[7.7] Derive the posterior distribution of β_1 given in Eq. (7.2.15).
[7.8] Using Lindley (1966), derive the approximations (7.2.16) and (7.2.17).
[7.9] Show that the approximate Bayes predictor of T and the posterior variance are given by Eqs. (7.2.18) and (7.2.19).

[7.10] Derive the Bayes predictor (7.2.20).

[7.11] Derive the posterior variance of predictor (7.2.5).

[7.12] Verify that the differential equations that lead to the transformations (7.2.23) and (7.2.24) are given by Eqs. (7.2.21) and (7.2.22).

[7.13] Verify that Eqs. (7.2.23) and (7.2.24) are a set of solutions to the differential equations (7.2.21) and (7.2.22).

[7.14] Verify that the conditional distribution of Y_i given X_i is as given in Eq. (7.2.25).

[7.15] Verify that the conditional distribution of β given $\mathbf{y}_s$ is as given in Eq. (7.2.26).

[7.16] Verify that the posterior variance of predictor (7.2.28) is as given in Eq. (7.2.29).

8
Asymptotic Properties in Finite Populations

In this chapter, we discuss limiting properties of predictors of finite population quantities. Both, the sample size and the population size should be large. Some limiting properties like consistency and asymptotic normality of certain predictors, under certain regression models, are considered. First we discuss conditions under which linear predictors of the population total T are asymptotically normal. Consistent estimators of the prediction variance are also considered. Next we investigate the limiting behavior of predictors of β_N.

8.1. Predictors of T

As seen in the previous chapters, to investigate the behavior of the predictors of population quantities of interest, it is often necessary to make distributional assumptions about the vector $\mathbf{y}$. Another approach is to use large sample approximations. In this section, we investigate the asymptotic properties of linear predictors of the population total T. The results extend straightforwardly to the case of predicting any linear quantity θ_L. Let "$\xrightarrow{D[\psi]}$" ("$\xrightarrow{P[\psi]}$") denote convergence in distribution (convergence in probability) under model ψ. In this chapter, we assume that the elements of of $\mathbf{y}$ are independent.

The asymptotic model is as follows:

Let $\{\mathcal{P}_k, k \geq 1\}$ be a sequence of finite populations, and let $\{\mathbf{s}_k, k \geq 1\}$ be the corresponding samples from these populations and $\mathbf{r}_k = \mathcal{P}_k - \mathbf{s}_k$. Let N_k be the size of $\mathcal{P}_k$ and n_k the size of $\mathbf{s}_k$. We assume that $N_{k+1} \geq N_k$, $n_{k+1} \geq n_k$, and that

(i) $\lim_{k\to\infty} N_k = \infty$, $\lim_{k\to\infty} n_k = \infty$, and $\lim_{k\to\infty} (N_k - n_k) = \infty$.

(ii) For each k, the vector of interest, $\mathbf{y}^{(k)}$ (of dimension N_k) satisfies the regression model

$$\mathbf{y}^{(k)} = \mathbf{X}^{(k)}\boldsymbol{\beta} + \mathbf{e}^{(k)}, \tag{8.1.1}$$

where $E[\mathbf{e}^{(k)}] = \mathbf{0}$ and $V[\mathbf{e}^{(k)}] = \sigma^2 W^{(k)}$ (diagonal).

Let $\mathbf{y}_{s_k}$ be the subvector of $\mathbf{y}^{(k)}$ corresponding to $\mathbf{s}_k$. Let $T^{(k)} = \mathbf{1}'_{N_k}\mathbf{y}^{(k)}$ be the total of $\mathcal{P}_k$ and let $\hat{T}^{(k)}_{LU} = \mathbf{l}'_{r_k s_k}\mathbf{y}_{s_k}$ be a linear unbiased predictor of $T^{(k)}$. Accordingly, $\hat{T}^{(k)}_{LU} - T^{(k)} = \sum_{i\in s_k} l_{r_k s_k i} y_i - \sum_{j\in r_k} y_j$. Moreover, since $\hat{T}^{(k)}_{LU}$ is unbiased, $\sum_{i\in s_k} l_{r_k s_k}\mathbf{X}_i^{(k)'}\beta = \sum_{j\in r_k}\mathbf{X}_j^{(k)'}\boldsymbol{\beta}$, where $\mathbf{X}_l^{(k)'}$ ($l = 1,\ldots,N$) denotes the lth row of the matrix $X^{(k)}$. It follows that

$$\frac{\hat{T}^{(k)}_{LU} - T^{(k)}}{\{\mathrm{Var}_{\psi(k)}[\hat{T}^{(k)}_{LU} - T^{(k)}]\}^{1/2}} = \sum_{j=1}^{N_k} a_j^{(k)}[y_j - \mathbf{X}_j^{(k)'}\boldsymbol{\beta}], \tag{8.1.2}$$

where

$$a_j^{(k)} = \begin{cases} l_{r_k s_k j}/\{\mathrm{Var}_{\psi(k)}[\hat{T}^{(k)}_{LU} - T_k]\}^{1/2}, & \text{if } j \in \mathbf{s}_k, \\ \\ -1/\{\mathrm{Var}_{\psi(k)}[\hat{T}^{(k)}_{LU} - T_k]\}^{1/2}, & \text{if } j \in \mathbf{r}_k. \end{cases} \tag{8.1.3}$$

We will use the symbols $o_p(\cdot)$ and $O_p(\cdot)$ to denote the rate of convergence in probability, according to the model. These symbols are defined in the following manner. Let $\{U_k, k \geq 1\}$ and $\{V_k, k \geq 1\}$ be two sequences of random variables. We say that $U_k = o_p(V_k)$, as $k \to \infty$, if $U_k/V_k \xrightarrow{p} 0$. We say that $U_k = O_p(V_k)$, as $k \to \infty$, if there exists a finite positive constant B, such that $P[|U_k/V_k| \leq B] \to 1$ as $k \to \infty$.

Theorem 8.1.1. *If for each k, $y_1, \ldots, y_{N_k}$ are independent and if $\hat{T}^{(k)}_{LU}$ satisfies the conditions*

(i) $\lim_{k\to\infty} \mathrm{Var}_{\psi(k)}[\hat{T}^{(k)}_{LU} - T^{(k)}] = \infty$, *and*

(ii) for any $\epsilon > 0$, $\lim_{k\to\infty} \sum_{j=1}^{N_k} \int_{|a_j^{(k)}[y_j - \mathbf{X}^{(k)'}\beta]| > \epsilon} [a_j^{(k)}]^2 [y_j - \mathbf{X}_j^{(k)'}\boldsymbol{\beta}]^2$

$dF_j^{(k)}(y_j) = 0$,

where $F_j^{(k)}(y)$ is the distribution function of y_j in $\mathcal{P}^{(k)}$, then

$$\frac{\hat{T}^{(k)}_{LU} - T^{(k)}}{\{\mathrm{Var}_{\psi(k)}[\hat{T}^{(k)}_{LU} - T^{(k)}]\}^{1/2}} \xrightarrow{D[\psi]} N(0,1). \tag{8.1.4}$$

Proof. Since, for each $k \geq 1$,

$$\mathrm{Var}_{\psi(k)}\left[\sum_{j=1}^{N_k} a_j^{(k)}[y_j - \mathbf{X}_j^{(k)'}\boldsymbol{\beta}]\right] = 1,$$

conditions (i) and (ii) are equivalent to the Lindeberg–Feller conditions (see Serfling, 1980, p. 29).

(Q.E.D.)

The superpopulation parameter σ^2 can be estimated consistently by

$$\hat{\sigma}_k^2 = \frac{1}{n_k - p}[\mathbf{y}_{s_k} - \mathbf{X}_{s_k}^{(k)}\hat{\boldsymbol{\beta}}_{s_k}]'\mathbf{W}_{s_k}^{-1}[\mathbf{y}_{s_k} - \mathbf{X}_{s_k}^{(k)}\hat{\boldsymbol{\beta}}_{s_k}], \tag{8.1.5}$$

provided $\hat{\boldsymbol{\beta}}_{s_k} \xrightarrow{P[\psi]} \boldsymbol{\beta}$, where $\hat{\boldsymbol{\beta}}_{s_k}$ is the weighted least-squares estimator of $\boldsymbol{\beta}$. The asymptotic distribution of $\hat{\boldsymbol{\beta}}_{s_k}$ will be studied in the next section. Presently we just mention that a sufficient condition for the ψ-consistency of $\hat{\boldsymbol{\beta}}_{s_k}$, as $k \to \infty$, is that $||[\mathbf{X}_{s_k}^{(k)\prime}\mathbf{W}_{s_k}^{-1}\mathbf{X}_{s_k}]^{-1}|| \to 0$ as $k \to \infty$.

Let $\hat{T}_{BLU}^{(k)}$ be the best linear unbiased predictor of T. Theorem 8.1.1 implies the following corollary.

Corollary 8.1.1. *Under the conditions of Theorem 8.1.1,*

$$\frac{\hat{T}_{BLU}^{(k)} - T^{(k)}}{\{\mathbf{1}_{r_k}'\mathbf{W}_{r_k}^{(k)}\mathbf{1}_k + \mathbf{1}_{r_k}'\mathbf{X}_{r_k}^{(k)}[\mathbf{X}_{s_k}^{(k)\prime}\mathbf{W}_{s_k}^{(k)-1}\mathbf{X}_{s_k}^{(k)}]^{-1}\mathbf{X}_{r_k}^{(k)\prime}\mathbf{1}_{r_k}\}^{1/2}} \xrightarrow{D[\psi]} N(0,1).$$

Example 8.1.1. In this example we omit the subscript (superscript) k, for the sake of simplifying notation. Consider model SM2 and assume that $y_1, \ldots, y_N$ are mutually independent. Furthermore, assume that for each element of the population, $0 \le x^* \le x_i \le x^{**} < \infty$ $(i = 1, \ldots, N)$. The best linear unbiased predictor of T is $\hat{T}_R$, the ratio predictor. Furthermore, the coefficients a_j specified by Eq. (8.1.3) assume the form

$$a_j = \begin{cases} \dfrac{(N-n)\bar{x}_r/n\bar{x}_s}{\sigma[(N-n)\bar{x}_r\frac{N\bar{x}_N}{n\bar{x}_s}]^{1/2}}, & \text{if } j \in s, \\[2ex] \dfrac{-1}{\sigma[(N-n)\bar{x}_r\frac{N\bar{x}_N}{n\bar{x}_s}]^{1/2}}, & \text{if } j \in r. \end{cases} \tag{8.1.6}$$

Let

$$\begin{aligned} b^* &= \sup_{1\le j\le N} |a_j| \\ &= \frac{1}{\sigma\sqrt{N\bar{x}_N}} \max\left\{\frac{(N-n)\bar{x}_r}{n\bar{x}_s}, \frac{n\bar{x}_s}{(N-n)\bar{x}_r}\right\}. \end{aligned} \tag{8.1.7}$$

Assume that $n \to \infty$, $N - n \to \infty$ and $\dfrac{n}{N} \to f$, where $0 < f < 1$. Accordingly, since the values of x_i belong to the interval $[x^*, x^{**}]$, $b^* =$

$O(\frac{1}{\sqrt{N}})$. Moreover, for any positive ϵ,

$$\begin{aligned}\sum_{j=1}^{N} a_j^2 \int_{|y-x_j\beta|>(\epsilon/|a_j|)} (y - x_j\beta)^2 dF_j(y) \\ \leq b^{*2} \sum_{j=1}^{N} \int_{|y-x_j\beta|>(\epsilon/b^*)} (y - x_j\beta)^2 dF_j(y).\end{aligned} \tag{8.1.8}$$

Let $b^* = c_N/\sqrt{N}$, where $\sup_{N\geq 1} c_N \leq c^* < \infty$. Since $\int (y - x_j\beta)^2 dF_j(y) = \sigma^2 x_j < \sigma^2 x^{**}$

$$\begin{aligned}b^{*2} \sum_{j=1}^{N} \int_{|y-x_j\beta|>(\epsilon/b^*)} (y - x_j\beta)^2 dF_j(y) \\ = c_N^2 \frac{1}{N} \sum_{j=1}^{N} \int_{|y-x_j\beta|>(\epsilon/c_N)\sqrt{N}} (y - x_j\beta)^2 dF_j(y) < \delta\end{aligned} \tag{8.1.9}$$

for arbitrarily small values of δ, if N is sufficiently large. Thus, conditions (i) and (ii) of Theorem 8.1.1 are satisfied, and $\hat{T}_R$ is asymptotically normal.

8.2. The Asymptotic Distribution of $\hat{\beta}_{s_k}$

In this section we investigate the asymptotic properties of the weighted least–squares predictor $\boldsymbol{\beta}_{s_k}$ of $\boldsymbol{\beta}_{N_k}$, as $k \to \infty$. We assume model (8.1.1) and the same asymptotic framework as in the previous chapter. In particular, $y_1, \ldots, y_{N_k}$ are mutually independent, for each k.

Theorem 8.2.1. *Let $\{\mathcal{P}_k, k \geq 1\}$ be a sequence of finite populations, generated according to the linear regression model $\psi^{(k)}$ (8.1.1). Assume*

(i) $N_k \to \infty$ *as* $k \to \infty$;

(ii) $f_{n_k} = \frac{n_k}{N_k} \to f$, $0 < f < 1$, *as* $k \to \infty$;

(iii) $\mathbf{M}_x^{(k)} = \frac{1}{N_k} \sum_{j=1}^{N_k} \frac{1}{w_j} \mathbf{X}_j^{(k)} \mathbf{X}_j^{(k)\prime} \to \boldsymbol{\Sigma}_{xx}$, *as* $k \to \infty$;

(iii) $\mathbf{M}_{s_k x}^{(k)} = \frac{1}{n_k} \sum_{j\in s_k} \frac{1}{w_j} \mathbf{X}_j^{(k)} \mathbf{X}_j^{(k)\prime} \to \boldsymbol{\Sigma}_{xx}$, *as* $k \to \infty$;

(v) $E_{\psi^{(k)}}[|e_j^{(k)}|^{2+\delta}] < \infty$, *for each* $j = 1, \cdots, N_k$,

where $w_1, \ldots, w_{N_k}$ are the diagonal elements of $\mathbf{W}^{(k)}$; $\mathbf{X}_j^{(k)}$ is the transpose of the j–th row of $\mathbf{X}^{(k)}$; $\boldsymbol{\Sigma}_{xx}$ is a $(p \times p)$ positive definite matrix; $\delta > 0$; and $e_j^{(k)} = y_j - \mathbf{X}_j^{(k)\prime} \boldsymbol{\beta}$ $(j = 1, \ldots, N_k)$. Then $\hat{\boldsymbol{\beta}}_{s_k}$ is a ψ–consistent estimator of $\boldsymbol{\beta}$; $||\hat{\boldsymbol{\beta}}_{s_k} - \boldsymbol{\beta}_{N_k}|| \xrightarrow{P[\psi]} 0$,

$$\sqrt{n_k}(\hat{\boldsymbol{\beta}}_{s_k} - \boldsymbol{\beta}_{N_k}) \underset{k\to\infty}{\xrightarrow{D[\psi]}} N[\mathbf{0}, \sigma^2(1-f)\boldsymbol{\Sigma}_{xx}^{-1}], \tag{8.2.1}$$

and

$$\sqrt{n_k}(\hat{\boldsymbol{\beta}}_{s_k} - \boldsymbol{\beta}) \underset{k\to\infty}{\xrightarrow{D[\psi]}} N(\mathbf{0}, \sigma^2\boldsymbol{\Sigma}_{xx}^{-1}). \tag{8.2.2}$$

Proof. Notice that under model $\psi^{(k)}$,

$$\hat{\boldsymbol{\beta}}_{s_k} = [\mathbf{M}_{sx}^{(k)}]^{-1} \sum_{j\in s_k} \frac{y_j}{w_j} \mathbf{X}_j^{(k)},$$

and

$$\boldsymbol{\beta}_{N_k} = [\mathbf{M}_x^{(k)}]^{-1} \frac{1}{N_k} \sum_{j=1}^{N_k} \frac{y_j}{w_j} \mathbf{X}_j^{(k)}.$$

Substituting $y_j = X_j^{(k)'}\boldsymbol{\beta} + e_j$ above, we obtain

$$\hat{\boldsymbol{\beta}}_{s_k} = \boldsymbol{\beta} + [\mathbf{M}_{sx}^{(k)}]^{-1} \frac{1}{n_k} \sum_{j\in s} \frac{e_j}{w_j} \mathbf{x}_j^{(k)}, \tag{8.2.3}$$

and

$$\boldsymbol{\beta}_{N_k} = \boldsymbol{\beta} + [\mathbf{M}_x^{(k)}]^{-1} \frac{1}{N_k} \sum_{j} \frac{e_j}{w_j} \mathbf{x}_j^{(k)}. \tag{8.2.4}$$

By Chebychev's inequality, for any $\delta > 0$,

$$\begin{aligned} &P_{\psi(k)}\left[\frac{1}{n_k}\left|\left|\sum_{j\in s_k} \frac{e_j}{w_j}\mathbf{x}_j^{(k)}\right|\right| > \delta\right] \\ &\leq \frac{1}{\delta^2 n_k^2}\left|\left|\sum_{j\in s_k} \mathbf{x}_j^{(k)}\mathbf{x}_j^{(k)'}\right|\right| = \frac{1}{\delta^2 n_k}||\mathbf{M}_{sx}^{(k)}|| \to 0, \end{aligned} \tag{8.2.5}$$

as $k \to \infty$, according to condition (iv). This proves that $\hat{\boldsymbol{\beta}}_{s_k}$ is a ψ-consistent estimator of $\boldsymbol{\beta}$ and that $||\hat{\boldsymbol{\beta}}_{s_k} - \boldsymbol{\beta}|| = O_p(\frac{1}{\sqrt{n_k}})$ as $k \to \infty$.

In a similar fashion we show that $||\beta_{N_k} - \beta|| = O_p(\frac{1}{\sqrt{N_k}})$ as $k \to \infty$. From Eq. (8.2.4), one obtains

(8.2.6)

$$\begin{aligned}\sqrt{n_k}(\hat{\beta}_{s_k} - \beta_{N_k}) &= \mathbf{\Sigma}_{xx}^{-1}\Bigg[\left(\frac{1}{n_k} - \frac{1}{N_k}\right)\sum_{j\in s_k}\frac{e_j}{w_j}\mathbf{x}_j^{(k)} \\ &\quad - \frac{1}{N_k}\sum_{j\in r_k}\frac{e_j}{w_j}\mathbf{x}_j^{(k)}\Bigg] + R_{xk} \\ &= \mathbf{\Sigma}_{xx}^{-1}\Bigg[(1 - f_{n_k})\frac{1}{\sqrt{n_k}}\sum_{j\in s_k}\frac{e_j}{w_j}\mathbf{x}_j^{(k)} \\ &\quad - f_{n_k}^{1/2}(1 - f_{n_k})^{1/2}\frac{1}{\sqrt{N_p - n_k}}\sum_{j\in r_k}\frac{e_j}{w_j}\mathbf{x}_j^{(k)}\Bigg] + \mathbf{R}_{xk},\end{aligned}$$

where

$$\begin{aligned}\mathbf{R}_{xk} &= \sqrt{n_k}\Bigg\{[(\mathbf{M}_{sx}^{(k)})^{-1} - \mathbf{\Sigma}_{xx}^{-1}]\frac{1}{n_k}\sum_{j\in s_k}\frac{y_j}{w_j}\mathbf{x}_j^{(k)} \\ &\quad - [(\mathbf{M}_{x}^{(k)})^{-1} - \mathbf{\Sigma}_{xx}^{-1}]\frac{1}{N_k}\sum_{j=1}^{N_k}\frac{y_j}{w_j}\mathbf{x}_j^{(k)}\Bigg\} \\ &= o_p(1), \qquad \text{as } k \to \infty.\end{aligned} \tag{8.2.7}$$

Finally, from the central limit theorem,

$$\frac{1}{\sqrt{n_k}}\sum_{j\in s_k}\frac{e_j}{w_j}\mathbf{x}_j^{(k)} \underset{k\to\infty}{\overset{D[\psi]}{\longrightarrow}} N(0, \sigma^2\mathbf{\Sigma}_{xx}) \tag{8.2.8}$$

and

$$\frac{1}{\sqrt{N_k - n_k}}\sum_{j\in r_k}\frac{e_j}{w_j}\mathbf{x}_j^{(k)} \underset{k\to\infty}{\overset{D[\psi]}{\longrightarrow}} N(0, \sigma^2\mathbf{\Sigma}_{xx}). \tag{8.2.9}$$

Hence, from Eqs. (8.2.6) – (8.2.9) and the independence of the terms on the left hand side of Eqs. (8.2.8) and (8.2.9), we obtain Eq. (8.2.1). From Eqs. (8.2.3) and (8.2.8), we imply Eq. (8.2.2).

(Q.E.D.)

A consistent estimator of the asymptotic variance in Eqs. (8.2.1) and (8.2.5) is obtained by replacing σ^2 by the consistent estimator given in Eq. (8.1.5).

Example 8.2.1. Consider the superpopulation model SM2. As seen in Example 2.1.1, $\beta_N = \bar{y}/\bar{x}$ and analogously, $\hat{\beta}_s = \bar{y}_s/\bar{x}_s$. To satisfy condition (iii) of Theorem 8.2.1, assume that

$$\sum_{j\in s_k}\frac{x_j}{n_k} \text{ and } \sum_{j\in r_k}\frac{x_j}{(N_k - n_k)} \longrightarrow \Sigma_{xx}. \tag{8.2.10}$$

Then, Theorem 8.2.1 implies that

$$(8.2.11) \qquad \sqrt{n_k}(\hat{\beta}_{s_k} - \beta_{N_k}) \xrightarrow{D[\psi]} N[0, (1-f)\frac{\sigma^2}{\Sigma_{xx}}].$$

Moreover, a consistent estimator of the asymptotic variance in Eq. (8.2.11) may be obtained by replacing σ^2 by the consistent estimator given in Eq. (2.2.11).

Fuller (1975a) considered a different superpopulation model for $\mathcal{P}$. Each element of $\mathcal{P}$ is associated with a random vector $(\mathbf{X}, \mathbf{Y})$, where, $E[\mathbf{X}] = \mathbf{0}$. The realization of $\mathbf{X}$ is known for all the elements of $\mathcal{P}$. The values of $\mathbf{y}$ are observed only on the elements of a sample $\mathbf{s}$. This model, described below in more detail, is called the ψ_F-model. Define the regression coefficient $\boldsymbol{\beta}$ to be

$$(8.2.12) \qquad \boldsymbol{\beta} = \boldsymbol{\Sigma}_{xx}^{-1}\boldsymbol{\Sigma}_{xy},$$

where

$$\boldsymbol{\Sigma}_{xx} = E_{\psi_F}[\mathbf{X}_j\mathbf{X}_j'] \text{ and } \boldsymbol{\Sigma}_{xy} = E_{\psi_F}[\mathbf{X}_j y_j],$$

for any $j = 1, \ldots, N$. It is also assumed that the entries of the matrix

$$\mathbf{G} = \text{Var}_{\psi_F}[\mathbf{X}_j e_j],$$

where $e_j = y_j - \mathbf{X}_j\boldsymbol{\beta}$, $j = 1, \ldots, N$, are all finite. As before, the first column of $\mathbf{X}$ may be a column of ones. The proof of the next theorem is similar to that of Theorem 8.2.1.

Theorem 8.2.2. *Consider a finite population model ψ_F satisfying conditions (i) – (iv) of Theorem 8.2.1 and the condition*

$$(v)' \qquad E[|X_{ij}e_j^{(k)}|^4] < \infty,$$

$j = 1, \ldots, N_k$, $i = 1, \ldots, p$, for each $k \geq 1$. Then as $k \to \infty$

$$(8.2.13) \qquad \sqrt{n_k}(\hat{\boldsymbol{\beta}}_{s_k} - \boldsymbol{\beta}_{N_k}) \xrightarrow{D[\psi_F]} N[\mathbf{0}, (1-f)\boldsymbol{\Sigma}_{xx}^{-1}\mathbf{G}\boldsymbol{\Sigma}_{xx}^{-1}].$$

Theorem 8.2.3. *Under the assumptions of Theorem 8.2.2, it follows that*

$$(8.2.14) \qquad \sqrt{n}(\hat{\boldsymbol{\beta}}_{s_k} - \boldsymbol{\beta}) \xrightarrow{D[\psi_F]} N(\mathbf{0}, \boldsymbol{\Sigma}_{xx}^{-1}\mathbf{G}\boldsymbol{\Sigma}_{xx}^{-1}).$$

The next result presents a consistent estimator of $\mathbf{G}$, which may be used in the above theorems.

Theorem 8.2.4. *Under the assumptions of Theorem 8.2.2,*

$$\hat{\mathbf{G}}^{(k)} = \frac{1}{n_k - p} \sum_{j \in s_k} \mathbf{X}_j^{(k)} \mathbf{X}_j^{(k)'} \hat{e}_j^2,$$

where $\hat{e}_j = y_j - \mathbf{X}_j^{(k)'} \hat{\boldsymbol{\beta}}_{s_k}$, *is a consistent estimator of* $\mathbf{G}$.

Example 8.2.2. Consider model ψ_F, where the pair (y_i, x_i) $i = 1, \ldots, N$ is such that

$$E_{\psi_F}[x_i^2] = \Sigma_{xx} \text{ and } E_{\psi_F}[x_i y_i] = \Sigma_{xy}.$$

Let $e_i = y_i - x_i\beta$, $i = 1, \ldots, N$ where $\beta = \Sigma_{xy}/\Sigma_{xx}$ and $G = \mathrm{Var}_{\psi_F}[x_i e_i]$. Thus,

$$\hat{\beta}_s = (\sum_{i \in s} \frac{x_i^2}{n})^{-1} \sum_{i \in s} \frac{x_i y_i}{n} \text{ and } \beta_N = (\sum_{i=1}^{N} \frac{x_i^2}{N})^{-1} \sum_{i=1}^{N} \frac{x_i y_i}{N}.$$

It follows from Theorem 8.2.4 that

$$\sqrt{n_k}(\hat{\beta}_{s_k} - \beta_{N_k}) \xrightarrow{D[\psi_F]} N[0, (1-f)G\Sigma_{xx}^{-2}].$$

8.3. The Linear Regression Model with Measurement Errors

In this section, model (8.1.1) is generalized to include measurement errors. The effect of such kind of errors upon estimated regression coefficients has long been recognized as a serious problem. Cochran (1968) and Fuller (1975a, 1987) report distortions that are introduced into the regression coefficient estimates when the variables in the regression equation are measured with error. The model we describe next generalizes the simple regression model with measurement errors introduced in Eq. (7.1.9).

Relating the two sets of variables $\mathbf{y}$ and $\mathbf{X}$, we consider the linear regression model where

$$y_i = \mathbf{X}_i'\boldsymbol{\beta} + e_i, \tag{8.3.1}$$

$\mathbf{X}_i' = (x_{i1}, \ldots, x_{ip})$, $\mathbf{V} = \sigma_e^2 \mathbf{I}_N$, $i = 1, \ldots, N$, and where y_i and x_{ij} are not observed directly. The observed variables are

$$Y_i = y_i + w_i \tag{8.3.2}$$

and

$$X_{ij} = x_{ij} + u_{ij}, \tag{8.3.3}$$

for $i = 1, \ldots, N$ and $j = 1, \ldots, p$ as in Section 7.1.2. Let $\mathbf{u}_i = (u_{i1}, \ldots, u_{ip})'$, $i = 1, \ldots, N$. We assume that $\mathbf{u}_i$, e_i, and x_{ij} are independent. Furthermore, $(w_i, \mathbf{u}_i)$ are also zero mean independent vectors, where $i = 1, \ldots, N$. The covariance matrix of $(Y_j, X_{j1}, \ldots, X_{jp})$ is given by the sum

$$\begin{pmatrix} \sigma_y^2 & \mathbf{\Sigma}_{yx} \\ \mathbf{\Sigma}_{xy} & \mathbf{\Sigma}_{xx} \end{pmatrix} + \begin{pmatrix} \sigma_w^2 & \mathbf{\Sigma}_{wu} \\ \mathbf{\Sigma}_{uw} & \mathbf{\Sigma}_{uu} \end{pmatrix},$$

where the second matrix is considered to be known and $\mathbf{\Sigma}_{xx}$ is known and nonsingular. Let $\tilde{\mathbf{X}} = (\tilde{\mathbf{X}}_1, \ldots, \tilde{\mathbf{X}}_N)'$, where $\tilde{\mathbf{X}}_i = (X_{i1}, \ldots, X_{ip})'$. Suppose also that

$$E[x_{jk}] = \mu_{xk},$$

which are assumed to be known for $j = 1, \ldots, N$ and $k = 1, \ldots, p$. Let

$$\boldsymbol{\mu}_x = (\mu_{x1}, \ldots, \mu_{xp})'.$$

After the sample s has been observed, consider the partition

$$\mathbf{Y} = \begin{pmatrix} \mathbf{Y}_s \\ \mathbf{Y}_r \end{pmatrix} \text{ and } \tilde{\mathbf{X}} = \begin{pmatrix} \tilde{\mathbf{X}}_s \\ \tilde{\mathbf{X}}_r \end{pmatrix}.$$

This superpopulation model is denoted by ψ_{GRE}.

The objective is to predict the finite population quantity

$$\tilde{\beta}_N = (\sum_{i=1}^{N} \frac{\tilde{\mathbf{X}}_i \tilde{\mathbf{X}}_i'}{N})^{-1} \sum_{i=1}^{N} \frac{\tilde{\mathbf{X}}_i y_i}{N}.$$

A natural way of correcting $\hat{\boldsymbol{\beta}}_s$ for measurement error is by defining

$$\hat{\beta}_{cs} = (\mathbf{M}_{XX} - \mathbf{\Sigma}_{uu})^{-1}(\mathbf{M}_{XY} - \mathbf{\Sigma}_{uw}), \tag{8.3.4}$$

whenever the inverse of the right-hand side of Eq. (8.3.4) exists, where

$$\mathbf{M}_{XX} = \sum_{i \in s} \frac{\tilde{\mathbf{X}}_i \tilde{\mathbf{X}}_i'}{n} \text{ and } \mathbf{M}_{XY} = \sum_{i \in s} \frac{\tilde{\mathbf{X}}_i Y_i}{n}.$$

To guarantee that the matrix to be inverted in Eq. (8.3.4) is positive definite, we replace $(\mathbf{M}_{XX} - \mathbf{\Sigma}_{uu})$ by the matrix

$$\mathbf{H} = \begin{cases} \mathbf{M}_{XX} - \mathbf{\Sigma}_{uu}, & g \geq 1, \\ \\ \mathbf{M}_{XX} - g\mathbf{\Sigma}_{uu}, & g < 1, \end{cases}$$

where g is the smallest root of the determinental equation

$$\left| \begin{pmatrix} \mathbf{M}_{YY} & \mathbf{M}_{XY} \\ \mathbf{M}_{YX} & \mathbf{M}_{XX} \end{pmatrix} - g \begin{pmatrix} \sigma_w^2 & \mathbf{\Sigma}_{wu} \\ \mathbf{\Sigma}_{uw} & \mathbf{\Sigma}_{uu} \end{pmatrix} \right| = 0,$$

where $\mathbf{M}_{YY} = \sum_{i \in s} Y_i^2/n$ and $\mathbf{M}_{YX} = \mathbf{M}'_{XY}$. Thus, after correcting for measurement errors, a predictor of $\tilde{\beta}_N$ may be defined as

$$\hat{\boldsymbol{\beta}}^*_{cs} = \mathbf{H}^{-1}(\mathbf{M}_{xy} - \mathbf{\Sigma}_{wu}). \tag{8.3.5}$$

Bock and Peterson (1975) and Fuller (1987) discuss procedures for constructing the positive definite matrix $\mathbf{H}$. Some large sample properties of the predictor $\hat{\boldsymbol{\beta}}^*_{cs}$ are presented next. As before, $f_{n_k} = n_k/N_k \to f$, as $k \to \infty$. Furthermore,

$$\begin{aligned} \mathbf{G} &= \mathrm{Var}_{\psi_{GRE}}[\tilde{\mathbf{X}}_i e_i] \\ &= \sigma_e^2 \mathrm{diag}\{\mu_{x1}^2 + \mathrm{Var}_{\psi_{GRE}}[X_{i1}], \ldots, \mu_{xp}^2 + \mathrm{Var}_{\psi_{GRE}}[X_{ip}]\}. \end{aligned} \tag{8.3.6}$$

Notice that, if $\mathbf{\Sigma}_{xx} + \mathbf{\Sigma}_{uu}$ is diagonal, then,

$$\mathbf{G} = \sigma_e^2(\mathrm{diag}\{\mu_{x1}^2, \ldots, \mu_{xp}^2\} + \mathbf{\Sigma}_{xx} + \mathbf{\Sigma}_{uu}).$$

Theorem 8.3.1. *Let $\{\mathcal{P}_k, k \geq 1\}$ be a sequence of finite populations as considered in Theorem 8.2.1, but where y_i and x_{ij} are measured with error according to Eqs. (8.3.2) and (8.3.3). Let $Y_i = \tilde{\mathbf{X}}_i\boldsymbol{\beta} + \xi_i$, $i = 1, \ldots, N$, and suppose that the $4 + \delta$ ($\delta > 0$) moments of $(w_i, u_{i1}, \ldots, u_{ip})$ are finite. Let*

$$n_k \,\mathrm{Var}_{\psi_{GRE}}[(\mathbf{M}_{X\xi}^{(k)} - \mathbf{\Sigma}_{u\xi})] = \mathbf{A},$$

where

$$E_{\psi_{GRE}}[\mathbf{u}_i \xi_i] = \mathbf{\Sigma}_{u\xi}$$

and

$$\mathbf{M}_{X\xi}^{(k)} = \sum_{j \in s_k} \frac{\tilde{\mathbf{X}}_j^{(k)} \xi_j^{(k)}}{n_k}.$$

Then,

$$\sqrt{n}(\hat{\boldsymbol{\beta}}^*_{cs_k} - \boldsymbol{\beta}_{N_k}) \xrightarrow[k\to\infty]{D[\psi_{GRE}]} N[\mathbf{0}, \mathbf{\Sigma}_{xx}^{-1}(\mathbf{A} - f\mathbf{G})\mathbf{\Sigma}_{xx}^{-1}], \tag{8.3.7}$$

where $\mathbf{G}$ is given in Eq. (8.3.6).

Proof. Because of the adjustment associated with the computation of $\hat{\beta}^*_{cs_k}$, we may write (see Fuller, 1987, p. 165)

$$\hat{\boldsymbol{\beta}}^*_{cs_k} - \tilde{\boldsymbol{\beta}}_{N_k} = \hat{\boldsymbol{\beta}}_{cs_k} - \tilde{\boldsymbol{\beta}}_{N_k} + O_p(n_k^{-1/2}).$$

From the assumptions of the theorem, it follows that

$$(8.3.8)\quad \sqrt{n_k}[\mathbf{M}_{X\xi}^{(k)} - \mathbf{\Sigma}_{u\xi}] = \frac{1}{\sqrt{n_k}}\sum_{i\in s_k}[\tilde{\mathbf{X}}_i^{(k)}\xi_i^{(k)} - \mathbf{\Sigma}_{u\xi}] \overset{D[\psi_{GRE}]}{\longrightarrow} N(\mathbf{0},\mathbf{A}).$$

Furthermore, from the moment assumptions, we may write

$$\mathbf{M}_{XX}^{(k)} = \mathbf{\Sigma}_{xx} + \mathbf{\Sigma}_{uu} + O_p(n_k^{-1/2})$$

and

$$\mathbf{M}_{XY}^{(k)} = \mathbf{\Sigma}_{xy} + \mathbf{\Sigma}_{uw} + O_p(n_k^{-1/2}).$$

Hence, by using the above results,

$$\sqrt{n_k}(\hat{\boldsymbol{\beta}}_{cs_k} - \tilde{\boldsymbol{\beta}}_{N_k}) = \sqrt{n_k}\mathbf{\Sigma}_{xx}^{-1}(\mathbf{M}_{X\xi}^{(k)} - \mathbf{\Sigma}_{u\xi} - \mathbf{R}_{N_k}) + O_p(n_k^{-1/2}),$$

where

$$\mathbf{R}_{N_k} = \frac{1}{N_k}\sum_{i=1}^{N_k}\mathbf{X}_i^{(k)}e_i^{(k)}.$$

Moreover,

$$(8.3.9)\quad \mathrm{Var}_{\psi_{GRE}}[\mathbf{R}_{N_k}] - 2Cov_{\psi_{GRE}}[\mathbf{R}_{N_k}, \mathbf{M}_{X\xi} - \mathbf{\Sigma}_{u\xi}] = -\frac{\mathbf{G}}{N_k}.$$

Finally, the assumptions of the theorem and the central limit theorem (Serfling, 1980) imply Eq. (8.3.7).

(Q.E.D.)

Theorem 8.3.2. *Under the assumptions of Theorem 8.3.1,*

$$\sqrt{n_k}(\hat{\boldsymbol{\beta}}_{cs_k}^* - \boldsymbol{\beta}) \overset{D[\psi_{GRE}]}{\longrightarrow} N(\mathbf{0}, \mathbf{\Sigma}_{xx}^{-1}\mathbf{A}\mathbf{\Sigma}_{xx}^{-1}).$$

Proof. The result follows from the presentation

$$(8.3.10)\quad \sqrt{n_k}(\hat{\boldsymbol{\beta}}_{s_k} - \boldsymbol{\beta}) = \mathbf{\Sigma}_{xx}^{-1}[\mathbf{M}_{X\xi}^{(k)} - \mathbf{\Sigma}_{u\xi}] + O_p(n_k^{-1/2}).$$

(Q.E.D.)

Theorem 8.3.3. *Under the assumptions of Theorem 8.3.1,*

$$(8.3.11)\quad \hat{\mathbf{A}}_k = \frac{1}{n_k - p}\sum_{i\in s_k}(\tilde{\mathbf{X}}_i^{(k)}\hat{\xi}_i - \mathbf{\Sigma}_{u\xi})(\tilde{\mathbf{X}}_i^{(k)}\hat{\xi}_i - \mathbf{\Sigma}_{u\xi})',$$

where $\hat{\xi}_i = Y_i - \tilde{\mathbf{X}}_i\hat{\boldsymbol{\beta}}_{cs}^*$, *is a consistent estimator of* $\mathbf{A}$.

Proof. The result follows by noticing that

$$\hat{\xi}_i = \xi_i + O_p(n^{-1/2})$$

and by applying the weak law of the large numbers.

(Q.E.D)

Model ψ_F considered in Section 8.2 can be extended to the situation discussed in this section, where both vectors $\mathbf{y}$ and $\mathbf{X}$ are measured with errors. Extensions of Theorems 8.3.1, 8.3.2, and 8.3.3 can be readily formulated.

8.4. Exercises

[8.1] Consider the asymptotic framework of Theorem 8.1.1. Prove that a sufficient condition for the consistency of $\hat{\sigma}_k^2$ defined in Eq. (8.1.5) is that $||(\mathbf{X}'_{s_k}\mathbf{W}^1_{s_k}\mathbf{X}_{s_k})^{-1}|| \to 0$ as $k \to \infty$.

[8.2] Verify formula (8.1.6).

[8.3] Consider model SM3 with $g = 0$.

(i) State the sufficient conditions and derive the asymptotic distribution of:

$$\text{(a) } \sqrt{n_k}(\hat{\boldsymbol{\beta}}_{s_k} - \boldsymbol{\beta}_{N_k}) \text{ and (b) } \sqrt{n_k}(\hat{\boldsymbol{\beta}}_{s_k} - \boldsymbol{\beta}), \text{ as } n_k, N_k \to \infty.$$

(ii) State sufficient conditions under which

$$\frac{\hat{T}^{(k)}_{BLU} - T^{(k)}}{\sqrt{\text{Var}_\psi[\hat{T}^{(k)}_{BLU} - T^{(k)}]}} \xrightarrow[k\to\infty]{D[\psi]} N(0,1).$$

(iii) Find a consistent estimator that could replace σ^2 in (i) and (ii).

[8.4] Consider model SM3 with $\beta_o = g = 0$. Do (i), (ii), and (iii) of Exercise 8.3.

[8.5] Consider model SM3 with $g = 1$. Do (i), (ii), and (iii) of Exercise 8.3.

[8.6] Consider model SM1 with $0 < \sigma^2 < \infty$.

(i) Show that

$$\sqrt{n_k}(\bar{y}_{s_k} - \beta) \xrightarrow[k\to\infty]{D[\psi]} N[0, (1-f)\sigma^2].$$

(ii) Show that

$$\sqrt{n_k}\begin{pmatrix} \bar{y}_{s_k} - \beta \\ \bar{y}_{s_k} - \bar{y} \end{pmatrix} \xrightarrow[k\to\infty]{D[\psi]} N\left[\begin{pmatrix} 0 \\ 0 \end{pmatrix}, \begin{pmatrix} 1 & 1-f \\ 1-f & 1-f \end{pmatrix}\sigma^2\right].$$

[8.7] Verify expressions (8.2.4), (8.2.5), (8.2.6), and (8.2.7).

[8.8] Prove Theorems 8.2.3, 8.2.4, and 8.2.5.

[8.9] Verify expressions (8.3.9) and (8.3.10).

[8.10] Consider model SM3 with $g = \beta_o = 0$. Suppose that

$$y_i = \beta x_i + e_i, \;\; X_i = x_i + u_i,$$

$i = 1, \ldots, N$, where e_i, u_i, and x_i are all independent, with mean zero and $\text{Var}[e_i] = \sigma_e^2$, $\text{Var}[u_i] = \sigma_u^2$, and $\text{Var}[x_i] = \sigma_x^2$.

(i) State sufficient conditions and find the asymptotic distribution of $\sqrt{n_k}(\hat{\beta}_{cs} - \tilde{\beta}_{N_k})$ and $\sqrt{n_n}(\hat{\beta}_{cs} - \beta)$.
(ii) Let

$$\hat{T}^{(k)} = n_k \bar{Y}_s + \hat{\beta}_{cs} \sum_{i \in r} \tilde{X}_i,$$

be a predictor of $T^{(k)}$. State the sufficient conditions and derive the asymptotic distribution of $\sqrt{n_n}(\hat{T}^{(k)} - T^{(k)})$.

[8.11] Consider the regression model $\psi = \psi(\beta, \mathbf{V})$, where $\mathbf{V} = \sigma^2 \mathbf{W}$ and $\mathbf{W}$ is known and general (not necessarily diagonal). Find sufficient conditions under which

$$\sqrt{n} \frac{\hat{T}_{BLU} - T}{\text{Var}_\psi[\hat{T}_{BLU} - T]} \xrightarrow{D[\psi]} N(0, 1),$$

where $\hat{T}_{BLU}$ is given in Eq. (2.1.5).

9
Design Characteristics of Predictors

In the present chapter we study the properties of predictors of the population total, T, from the sampling design point of view. In particular, we focus attention on the *asymptotic design unbiasedness* (ADU) of the predictors. This property has been suggested by Brewer (1979) for predictors with a single auxiliary variable. A generalization of the concept to the case of several regressors was given by Wright (1983). Further extensions of the models were considered by Rodrigues and Bolfarine (1988) and by Montanari (1988). See also the studies of Tam (1988a,b) and Rodrigues and Bolfarine (1989). The ψ-model considered in this chapter is the regression model (1.2.1) with a diagonal covariance matrix $\mathbf{V} = \sigma^2 \mathrm{diag}\{v_1, \ldots, v_N\}$.

9.1. The QR Class of Predictors

Let π_i, $i = 1, \ldots, N$ denote the first-order inclusion probabilities associated with a sampling design. Let $\mathbf{\Pi} = \mathrm{diag}\{\pi_1, \ldots, \pi_N\}$. Let q_i, $i = 1, \ldots, N$ be arbitrary positive numbers and $\mathbf{Q} = \mathrm{diag}\{q_1, \ldots, q_N\}$. Furthermore, let $\mathbf{Q}^* = \mathrm{diag}\{q_1\pi_1, \ldots, q_N\pi_N\}$ and $\mathbf{R} = \mathrm{diag}\{r_1, \ldots, r_N\}$, where $r_i \geq 0$. Assume that $\mathbf{X}_s'\mathbf{Q}_s\mathbf{X}_s$ and $\mathbf{X}_s'\mathbf{Q}_s^*\mathbf{X}_s$ are positive definite.

Definition 9.1.1. *A QR predictor for the population total T is*

$$\hat{T}_{QR} = \mathbf{1}_N'\mathbf{X}\hat{\beta}_Q + \mathbf{1}_s'\mathbf{R}_s\hat{\mathbf{e}}_s, \tag{9.1.1}$$

where

$$\hat{\beta}_Q = (\mathbf{X}_s'\mathbf{Q}_s\mathbf{X}_s)^{-1}\mathbf{X}_s'\mathbf{Q}_s\mathbf{y}_s \tag{9.1.2}$$

and $\hat{\mathbf{e}}_s = (\mathbf{y}_s - \mathbf{X}_s\hat{\beta}_Q)$.

As the next example shows, our interest in QR predictors comes from the fact that various choices of $\mathbf{Q}_s$ and $\mathbf{R}_s$ yield familiar predictors.

Example 9.1.1. Consider the regression model $\psi = \psi(\beta, \mathbf{V})$, where $\mathbf{X} = (x_1, \ldots, x_N)'$ and $\mathbf{V} = \sigma^2 \text{diag}\{v_1, \ldots, v_N\}$. Let $\hat{y}_s = \sum_{j \in s} y_j/\pi_j$ and $\hat{x}_s = \sum_{j \in s} x_j/\pi_j$. Notice that if $v_i = x_i$, $i = 1, \ldots, N$, then the model ψ reduces to model SM2. Five different predictors of T are considered next with the choices of r_j and q_j which make them QR predictors:

(i) Horvitz–Thompson Predictor (Hájek, 1971)

$$\hat{T}_{HTR} = \hat{\beta}_R \sum_{j=1}^{N} x_j, \tag{9.1.3}$$

where $\hat{\beta}_R = \hat{y}_s/\hat{x}_s$, $q_j = \pi_j/x_j$, $r_j = 0$.

(ii) Combined Regression Through the Origin (Wright, 1983)

$$\hat{T}_{CR} = N\hat{y}_s + N\hat{\beta}_{CR}(\bar{x} - \hat{x}_s), \tag{9.1.4}$$

where $\hat{\beta}_{CR} = (\sum_{j \in s} x_j^2/\pi_j)^{-1} \sum_{j \in s} x_j y_j/\pi_j$, $q_j = 1/\pi_j$ and $r_j = N/\pi_j$.

(iii) Generalized Regression Predictor (Cassel et al., 1977)

$$\hat{T}_{GREG} = N\hat{y}_s + N\hat{\beta}_s(\bar{x} - \hat{x}_s),$$

where $\hat{\beta}_s = (\sum_{j \in s} x_j^2/v_j)^{-1} \sum_{j \in s} x_j y_j/v_j$, $q_j = 1/v_j$, and $r_j = N/\pi_j$.

(iv) ψ-BLUP (Royall, 1970)

$$\hat{T}_{BLU} = n\bar{y}_s + \hat{\beta}_s \sum_{j \in \mathbf{r}} x_j,$$

where $q_j = 1/v_j$ and $r_j = 1$.

(v) Brewer's predictor (Brewer, 1979)

$$\hat{T}_{BR} = n\bar{y}_s + \hat{\beta}_{BR} \sum_{j \in r} x_j, \tag{9.1.5}$$

where $\hat{\beta}_{BR} = (\sum_{j \in s} q_j x_j^2)^{-1} \sum_{j \in s} q_j x_j y_j$, $q_j = (1 - \pi_j)/\pi_j x_j$ and $r_j = 1$.

As the next example shows, QR predictors also arise in the multiple regression case.

Example 9.1.2. Some multiple regression examples of QR predictors under the model $\psi = \psi(\boldsymbol{\beta}, \mathbf{V})$ are as follows:

(i) Simple Projection Predictors ($\mathbf{R} = \mathbf{0}$ and $\mathbf{Q} = \mathbf{V}^{-1}$)

$$\hat{T}_{SP} = \mathbf{1}'_N \mathbf{X} \hat{\boldsymbol{\beta}}_s$$

(ii) Linear Regression Predictors ($\mathbf{R} = \mathbf{I}_N$ and $\mathbf{Q} = \mathbf{V}^{-1}$)

$$\hat{T}_{BLU} = \mathbf{1}'_N \mathbf{X} \hat{\boldsymbol{\beta}}_s + \mathbf{1}'_s \hat{\mathbf{e}}_s$$

(iii) Generalized Regression Predictors ($\mathbf{R} = \mathbf{\Pi}^{-1}$)

(9.1.6) $$\hat{T}_{GREG} = \mathbf{1}'_N \mathbf{X}\hat{\beta}_Q + \mathbf{1}'_s \mathbf{\Pi}_s^{-1} \hat{\mathbf{e}}_s,$$

where $\hat{\beta}_Q$ and $\hat{\mathbf{e}}_s$ are as in Definition 7.4.1 above.

9.2. ADU Predictors

The asymptotic framework that we consider is similar to that of Theorem 8.1.1. Let $\{\mathcal{P}_k, k \geq 1\}$ be a sequence of finite populations, such that

(i) the population size N_k is increasing with k;

(ii) for each k, the vector of interest $\mathbf{y}^{(k)}$ of dimension N_k satisfies the regression model

$$\psi^{(k)}: \mathbf{y}^{(k)} = \mathbf{X}^{(k)}\boldsymbol{\beta} + \mathbf{e}^{(k)},$$

$E[\mathbf{e}^{(k)}] = \mathbf{0}$ and $\text{Var}[\mathbf{e}^{(k)}] = \sigma^2 \mathbf{W}^{(k)}$. Let $\mathbf{s}_k$ be a sample of size n_k from $\mathcal{P}_k$, where n_k is increasing in k.

Let $T^{(k)}$ denote the total corresponding to population $\mathcal{P}_k$ and $\hat{T}_{QR}^{(k)}$ the QR-predictor of $T^{(k)}$ corresponding to a sample $\mathbf{s}_k$ from $\mathcal{P}_k$.

Definition 9.2.1. *Predictor $\hat{T}_{QR}$ is ADU for the population total T if*

$$\lim_{k\to\infty} E_p[\hat{T}_{QR}^{(k)} - T^{(k)}] = 0,$$

where $\hat{T}_{QR}^{(k)}$ is computed by using $\mathbf{s}_k$ and $\hat{T}_{QR}$, as described above, and $E_p[\cdot]$ is the expectation operator with respect to the sampling plan p, with inclusion probabilities in $\mathbf{\Pi}$.

Define the population quantity

(9.2.1) $$T_{QR} = \mathbf{1}'_N \mathbf{X}\mathbf{B} + \mathbf{1}'_s \mathbf{R}_s \epsilon_s,$$

where $\epsilon_s = \mathbf{y}_s - \mathbf{X}_s \mathbf{B}$ and

(9.2.2) $$\mathbf{B} = (\mathbf{X}'\mathbf{Q}^*\mathbf{X})^{-1}\mathbf{X}'\mathbf{Q}^*\mathbf{y}.$$

We can write

$$\hat{T}_{QR} - T_{QR} = (\mathbf{1}'_N \mathbf{X} - \mathbf{1}'_s \mathbf{R}_s \mathbf{X}_s)(\hat{\beta}_Q - \mathbf{B}).$$

some conditions on the second-order inclusion probabilities (Isaki and Fuller, 1982; Montanari, 1987) ensure that

(9.2.3) $$\hat{\beta}_{Qk} - \mathbf{B}_k \overset{P[p]}{\to} \mathbf{0},$$

and

(9.2.4) $$\lim_{k\to\infty} E_p[\hat{T}_{QR}^{(k)} - T_{QR}^{(k)}] = 0.$$

The following theorem presents a sufficient condition for the ADU property in the QR class under the asymptotic framework described above. The proof is based on geometrical arguments.

Theorem 9.2.1. *A sufficient condition for predictor $\hat{T}_{QR}$ to be ADU is that*

$$\mathbf{\Pi}^{-1}\mathbf{1}_N = \mathbf{R}\mathbf{1}_N + \mathbf{Q}^*\mathbf{X}\boldsymbol{\delta}, \tag{9.2.5}$$

for some vector $\boldsymbol{\delta}$.

Proof. Assuming that Eq. (9.2.4) holds, a sufficient condition for $\hat{T}_{QR}$ to be ADU is that

$$E_p[T_{QR} - T] = 0, \text{ for all } \mathbf{y},$$

that is, T_{QR} is exactly design unbiased. Thus, since, conditional on $\mathbf{y}$, $\mathbf{B}$ is nonrandom,

$$\begin{aligned} T - E_p[T_{QR}] &= \mathbf{1}'_N(\mathbf{y} - \mathbf{XB} - \mathbf{R\Pi}\boldsymbol{\epsilon}) \\ &= \mathbf{1}'_N(\mathbf{I}_N - \mathbf{R\Pi})\boldsymbol{\epsilon}, \end{aligned}$$

where $\boldsymbol{\epsilon} = \mathbf{y} - \mathbf{XB}$. Now, the last expression is equal to zero if and only if

$$\mathbf{Q}^{*^{-1}}(\mathbf{I} - \mathbf{R\Pi})\mathbf{1}_N \in \mathcal{M}(\mathbf{X}),$$

which holds, if and only if, $\mathbf{\Pi}^{-1}\mathbf{1}_N = \mathbf{R}\mathbf{1}_N + \mathbf{QX}\boldsymbol{\delta}$, concluding the proof. $\mathcal{M}(\mathbf{X})$ denotes the subspace of E^N generated by the columns of the matrix $\mathbf{X}$.

(Q.E.D.)

Example 9.2.1. Consider the ψ-model of Example 9.1.1. It is easily shown that $\hat{T}_{HTR}$, $\hat{T}_{CR}$, $\hat{T}_{GREG}$, and $\hat{T}_{BR}$ are always ADU, for all sampling designs and all x_j. Now, using Theorem 9.2.1, $\hat{T}_{BLU}$ is ADU if, and only if, $v_j(\pi_j^{-1} - 1)$ is a multiple of x_j, that is, there is a constant $\delta > 0$, such that

$$\pi_j = \frac{1}{1 + v_j^{-1}x_j\delta} = \frac{\alpha}{\alpha + v_j^{-1}x_j},$$

for $\alpha = 1/\delta$ and $j = 1, \ldots, N$. If $v_j = \sigma^2 x_j$, then $\hat{T}_{BLU} = \hat{T}_R$, is ADU if, and only if,

$$\pi_j = \frac{n}{N}, \ j = 1, \ldots, N.$$

We thus conclude that the simple random sampling design protects the ratio estimator against model misspecification in large samples. This result agrees with the results in Example 5.1.3 since large samples obtained from the simple random sampling design are typically well balanced.

Example 9.2.2. Consider the GREG predictor defined in Eq. (9.1.6). Since $\mathbf{R} = \mathbf{\Pi}^{-1}$, Theorem 9.2.1 with $\boldsymbol{\delta} = \mathbf{0}$ implies that the GREG predictor fulfills the ADU condition under the regression model $\psi = \psi(\boldsymbol{\beta}, \mathbf{V})$.

We state the next result as a corollary of Theorem 9.2.1 due to its importance. Notice that if $\mathbf{R} = \mathbf{I}_N$ and $\mathbf{Q}_s = \mathbf{V}_s^{-1}$, then $\hat{T}_{QR} = \hat{T}_{BLU}$.

Corollary 9.2.1. *$\hat{T}_{BLU}$ is ADU if*

$$\boldsymbol{\Pi}^{-1}\mathbf{1}_N = \mathbf{1}_N + \mathbf{V}^{-1}\mathbf{X}\boldsymbol{\delta}, \tag{9.2.6}$$

for some vector $\boldsymbol{\delta}$.

The next result specializes the form of the design for the case where $\mathbf{R} = \mathbf{I}_N$. The following corollary generalizes some results of Tam (1988a).

Corollary 9.2.2. *Consider the QR predictor with $\mathbf{R} = \mathbf{I}_N$.*

(i) Let

$$\mathbf{X}\boldsymbol{\delta} = (\frac{n}{N} - 1)\mathbf{Q}^{-1}\mathbf{1}_N,$$

for some vector $\boldsymbol{\delta}$. Then $\hat{T}_{QR}$ is ADU if, and only if,

$$\boldsymbol{\Pi}\mathbf{1}_N = \frac{n}{N}\mathbf{1}_N,$$

which corresponds to the simple random sampling design.

(ii) Let

$$\mathbf{X}\boldsymbol{\delta} = k\mathbf{Q}^{-1/2}\mathbf{1}_N - \mathbf{Q}^{-1}\mathbf{1}_N,$$

for some vector δ. Then $\hat{T}_{QR}$ is ADU if, and only if,

$$\boldsymbol{\Pi}\mathbf{1}_N = k^{-1}\mathbf{Q}^{-1/2}\mathbf{1}_N$$

and $k = \mathbf{1}'_N\mathbf{Q}^{-1/2}\mathbf{1}_N/n$. In the particular case where $\mathbf{Q}_s = \mathbf{V}_s^{-1}$, a characterization of the ADU property for $\hat{T}_{BLU}$ follows by taking $\pi_j = v_j^{-1/2}/k$, where $k = \sum_{j=1}^N v_j^{1/2}/n$.

Notice that if $\mathbf{Q}_s^{-1} = \mathbf{V}_s$ then (i) is equivalent to the condition L. We explore now the possibility of creating a simple and intuitive form for $\hat{T}_{QR}$ with the ADU property, by extending Theorem 2.1.2. Let $\hat{T}_{QSP} = \mathbf{1}'_N\mathbf{X}\hat{\boldsymbol{\beta}}_Q$. Notice that if $\hat{\boldsymbol{\beta}}_Q = \hat{\boldsymbol{\beta}}_s$ then $\hat{T}_{QSP} = \hat{T}_{SP}$.

Theorem 9.2.2. *$\hat{T}_{QSP}$ is ADU if, and only if,*

$$\mathbf{Q}_s^{-1}\mathbf{R}_s\mathbf{1}_s = \mathbf{X}_s\boldsymbol{\delta},$$

for some $\boldsymbol{\delta}$ and $\pi_j = n/N$, $j = 1, \ldots, N$.

Proof. Notice that $\hat{T}_{QR} = \hat{T}_{QSP}$ if, and only if,

$$\mathbf{e}'_s\mathbf{Q}_s^{-1}\mathbf{R}_s\mathbf{1}_s = 0. \tag{9.2.7}$$

The result follows by combining Eqs. (9.2.5) and (9.2.7).

(Q.E.D.)

The following result shows that any QR predictor which is also ADU is equivalent to a GREG predictor. Thus, when looking for QR predictors with the ADU property one should consider the GREG class of predictors.

Lemma 9.2.1. *Let $\hat{T}_{QR}$ be any QR predictor that is also ADU and $\hat{T}_{GREG}$ the GREG predictor that uses the same matrix* $\mathbf{Q}$ *as $\hat{T}_{QR}$. Then $\hat{T}_{QR} = \hat{T}_{GREG}$ for all $\mathbf{y} \in E^N$ and all samples* $\mathbf{s}$.

Proof. The proof follows directly from Theorem 9.2.1 and from the fact that

$$\hat{T}_{GREG} - \hat{T}_{QR} = \sum_{j \in \mathbf{s}} (\pi_j^{-1} - r_j)\hat{e}_j / N. \tag{9.2.8}$$

(Q.E.D.)

9.3. Optimal ADU Predictors

A criterion of asymptotic optimality and conditions for a QR predictor of T to be asymptotically optimal under the design–model approach are established in the present section. It is assumed that $\hat{T}_{QR}$ satisfy the ADU condition, so that, according to Lemma 9.2.1, it is equivalent to consider the GREG class of predictors. As mentioned above, the sampling design p is characterized by the matrix $\mathbf{\Pi}$.

Now, corresponding to population $\mathcal{P}_k$, define the population quantity $T_*^{(k)}$, which corresponds to

$$T_* = \mathbf{1}'_N \mathbf{X}\boldsymbol{\beta} + \sum_{j \in \mathbf{s}} \frac{e_j}{\pi_j}. \tag{9.3.1}$$

Some conditions on the second–order inclusion probabilities (Isaki and Fuller 1982, Montanari, 1987) ensure that,

$$\lim_{k \to \infty} E_p E_\psi [T_*^{(k)} - \hat{T}_{GREG}^{(k)}]^2 = 0. \tag{9.3.2}$$

Definition 9.3.1. *The asymptotic expected prediction variance of $\hat{T}_{QR}$ is*

$$AV_{p\psi}[\hat{T}_{QR}] = \lim_{k \to \infty} E_\psi E_p [T_*^{(k)} - \hat{T}_{GREG}^{(k)}]^2.$$

Definition 9.3.2. *An ADU predictor $\hat{T}_{QR}^*$ is said to be asymptotically optimal under the regression model ψ, if*

$$AV_{p\psi}[\hat{T}_{QR}^*] \leq AV_{p\psi}[\hat{T}_{QR}],$$

for any ADU predictor $\hat{T}_{QR}$.

Theorem 9.3.1. *Let $\hat{T}_{QR}$ be a QR–predictor corresponding to model $\psi = \psi(\boldsymbol{\beta}, \mathbf{V})$, where $\mathbf{V} = \sigma^2 \text{diag}\{v_1, \ldots, v_N\}$, which is ADU. Then,*

$$AV_{p\psi}[\hat{T}_{QR}] \geq \{\frac{1}{n}(\sum_{i=1}^{N} v_i^{1/2})^2 - \sum_{i=1}^{N} v_i\}\sigma^2. \tag{9.3.3}$$

Equality is attained by a predictor $\hat{T}^*_{QR}$ with inclusion probabilities given by

$$\pi_j = n\frac{v_j^{1/2}}{\sum_{i=1}^{N} v_i^{1/2}}, \tag{9.3.4}$$

$j = 1, \ldots, N$.

Proof. We may write (suppressing the superscript k)

$$T_* - T = \sum_{i \in \mathbf{s}} \frac{e_i}{\pi_i} - \sum_{i=1}^{N} e_i,$$

so that, by using Exercise 1.1,

$$E_p[T_* - T]^2 = \frac{1}{2}\sum_{i=1}^{N}\sum_{i \neq j}(\pi_i\pi_j - \pi_{ij})(\frac{e_i}{\pi_i} - \frac{e_j}{\pi_j})^2.$$

Applying the identities

$$E_\psi[\frac{e_i}{\pi_i} - \frac{e_j}{\pi_j}]^2 = (\frac{v_i}{\pi_i^2} + \frac{v_j}{\pi_j^2})\sigma^2, \tag{9.3.5}$$

$$\sum_{i \neq j}^{N} \pi_i = n - \pi_j, \text{ and } \sum_{i \neq j}^{N} \pi_{ij} = (n-1)\pi_j, \tag{9.3.6}$$

(see Exercise 9.3), we obtain

$$\begin{aligned}
E_\psi E_p[T_* - T]^2 &= \sum_{i=1}^{N}\sum_{i \neq j}^{N}(\pi_i\pi_j - \pi_{ij})\frac{v_i\sigma^2}{\pi_i^2} \\
&= \sum_{i=1}^{N}[\pi_i(n - \pi_i) - (n-1)\pi_i]\frac{v_i\sigma^2}{\pi_i^2} \\
&= \sum_{i=1}^{N}(\pi_i^{-1} - 1)v_i\sigma^2.
\end{aligned}$$

Under model ψ, for a fixed $n = \sum_{i=1}^{N} \pi_i$, Cauchy–Schwartz inequality implies that

$$(\sum_{i=1}^{N} v_i)^2 \leq (\sum_{i=1}^{N} \pi_i)(\sum_{i=1}^{N} \pi_i^{-1} v_i),$$

with equality if, and only if,

$$\pi_i = n \frac{v_i^{1/2}}{\sum_{i=1}^{N} v_i^{1/2}}.$$

Thus,

$$AV_{p\psi}[\hat{T}_{QR}] \geq \{\frac{1}{n}(\sum_{i=1}^{N} v_i^{1/2})^2 - \sum_{i=1}^{N} v_i\}\sigma^2, \tag{9.3.7}$$

with equality if, and only if, Eq. (9.3.4) holds. This proves the theorem. (Q.E.D.)

Notice that the right side of Eq. (9.3.7) was first derived as a lower bound of the expected variance of design unbiased predictors by Godambe and Joshi (1965).

Example 9.3.1. Consider the superpopulation model of Example 9.1.1. It follows from Theorems 9.2.1 and 9.3.1 that the ADU and asymptotically optimal predictor should be based on

$$\pi_j = n \frac{v_j^{1/2}}{\sum_{j=1}^{N} v_j^{1/2}} \text{ and } q_j = \frac{(\pi_j^{-1} - 1)}{x_j},$$

$j = 1, \ldots, N$. The resulting asymptotically optimal predictor is therefore

$$\hat{T}_{QR}^{*} = n\bar{y}_s + \hat{\beta}_{BR} \sum_{j \notin s} x_j. \tag{9.3.8}$$

Notice that predictor (9.3.8) is Brewer's (1979) predictor, defined in Eq. (9.1.5). Thus, $\hat{T}_R$ is ADU but not asymptotically optimal according to the optimality criterion of Definition 9.3.2.

According to Theorem 9.3.1, the GREG predictor

$$\hat{T}_{GREG} = \mathbf{1}'_N \mathbf{X} \hat{\beta} + \mathbf{1}'_s \mathbf{\Pi}_s^{-1} \hat{\mathbf{e}}_s, \tag{9.3.9}$$

with inclusion probabilities given in Eq. (9.3.4) is asymptotically optimal. This GREG optimal predictor is denoted by $\hat{T}_{GREG}^{*}$. The next result, which appears in Tam (1988b), provides a set of sufficient conditions for any linear predictor $\mathbf{h}'_s \mathbf{y}_s$ to be asymptotically optimal in the sense of Definition 9.3.2. Let $k = \sum_{i=1}^{N} v_i^{1/2}/n$.

Theorem 9.3.2. *The sufficient condition for any linear predictor $\hat{T} = \mathbf{h}_s'\mathbf{y}_s$ with inclusion probabilities given in Eq. (9.3.4) to be asymptotically optimal is that*

(i) $\mathbf{h}_s'\mathbf{X}_s = \mathbf{1}_N'\mathbf{X}$ *and*
(ii) $\mathbf{Q}_s^{-1}(\mathbf{h}_s - k\mathbf{V}_s^{-1/2}\mathbf{1}_s) \in \mathcal{M}(\mathbf{X}_s)$,

for any sample $\mathbf{s}$ *with* $p(\mathbf{s}) > 0$, *where* $\mathbf{Q}_s$ *satisfy the ADU condition of Theorem 8.2.2 and* $\mathcal{M}(\mathbf{X}_s)$ *is as in Section 2.1.*

Proof. From (ii), there exists a $n \times 1$ vector $\boldsymbol{\lambda}$, such that

$$\mathbf{h}_s' = \mathbf{1}_s'\boldsymbol{\Pi}_s^{-1} + \boldsymbol{\lambda}'\mathbf{X}_s'\mathbf{Q}_s, \tag{9.3.10}$$

where

$$\boldsymbol{\Pi}_s = k^{-1}\text{diag}\{v_1^{1/2}, \ldots, v_N^{1/2}\}.$$

Using (i) and solving for $\boldsymbol{\lambda}$ in Eq. (9.3.10), it follows that

$$\boldsymbol{\lambda}' = (\mathbf{1}_N'\mathbf{X} - \mathbf{1}_s'\boldsymbol{\Pi}_s^{-1}\mathbf{X}_s)\mathbf{X}_s'\mathbf{Q}_s\mathbf{X}_s. \tag{9.3.11}$$

Substituting Eq. (9.3.11) into Eq. (9.3.10), we obtain

$$\mathbf{h}_s'\mathbf{y}_s = \hat{T}_{GREG}^*.$$

(Q.E.D.)

Tam (1988) considered robustness of sampling strategies (design-based model-based approach) from the point of view of attaining asymptotically the Godambe–Joshi lower bound. Asymptotically designed unbiased, although a necessary condition for robustness in the sense considered above is not sufficient. Notice that the conditions of Theorem 9.3.2 are similar to those of Lemma 5.1.2 for robustness of model-based predictors.

Following Tam (1988b), we consider these definitions:

Definition 9.3.3. *A linear predictor* $\mathbf{h}_s'\mathbf{y}_s$ *of* T *is said to be robust against covariance matrix misspecification (or weakly robust) if, and only if, condition (ii) of Theorem 9.3.2 is satisfied for all samples* $\mathbf{s}$ *with* $p(\mathbf{s}) > 0$.

Definition 9.3.4. *A linear predictor* $\mathbf{h}_s'\mathbf{y}_s$ *of* T *is said to be robust against covariance matrix and design matrix misspecification (or strongly robust) if conditions (i) and (ii) of Theorem 9.3.2 hold for all samples* $\mathbf{s}$ *with* $p(\mathbf{s}) > 0$.

The working model can fail in two ways: (i) the design matrix is misspecified and (ii) the covariance matrix is misspecified. For type (i) model misspecification, robustness of the predictor is provided by condition (i) of Theorem 9.3.2 if important auxiliary variables are omitted from $\mathbf{X}$ [see Exercise 9.4(i)]; or condition (ii) if unimportant auxiliary variables are included in $\mathbf{X}$ [see Exercise 9.4(ii)]. For type two model misspecification, robustness is provided by condition (ii) of Theorem 9.3.2 (see Exercise 9.5).

The next example illustrates the applicability of Theorem 9.3.2. See also Exercise 9.6.

Example 9.3.2. Consider the model–based predictor

$$\hat{T}_{BLU} = \mathbf{h}_s^{*'}\mathbf{y}_s = \mathbf{1}'_N\mathbf{X}\hat{\boldsymbol{\beta}}_s + \mathbf{1}'_s(\mathbf{y}_s - \mathbf{X}_s\hat{\boldsymbol{\beta}}_s),$$

considered in Section 2.1 specialized for the case of $\mathbf{V}_{sr} = \mathbf{0}$. Since $\hat{T}_{BLU}$ is the ψ-BLUP of T, it is model–unbiased and according to Lemma 5.1.2 it follows that

$$\mathbf{V}_s\mathbf{h}_s^* - \mathbf{V}_s\mathbf{1}_s \in \mathcal{M}(\mathbf{X}_s).$$

Thus,

$$(\mathbf{V}_s\mathbf{h}_s^* - k\mathbf{V}_s^{1/2}\mathbf{1}_s) = (\mathbf{V}_s\mathbf{h}_s^* - \mathbf{V}_s\mathbf{1}_s) + (\mathbf{V}_s\mathbf{1}_s - k\mathbf{V}_s^{1/2})\mathbf{1}_s,$$

where $k = \sum_{i=1}^N v_i^{1/2}/n$. Therefore, according to Theorem 9.3.2, a sufficient condition for $\hat{T}_{BLU}$ to be strongly robust is that

$$(\mathbf{V}_s - k\mathbf{V}_s^{1/2})\mathbf{1}_s \in \mathcal{M}(\mathbf{X}_s).$$

9.4. Exercises

[9.1] Verify expressions (9.2.3) and (9.2.4).
[9.2] Verify expressions (9.2.7) and (9.2.8).
[9.3] Verify expressions (9.3.5) and (9.3.6).
[9.4] (Montanari, 1987) Consider the regression model $\psi = \psi(\boldsymbol{\beta}, \mathbf{V})$, where $\mathbf{V}$ is a general positive definite covariance matrix. Let

$$\mathbf{Q} = \begin{pmatrix} \mathbf{Q}_s & \mathbf{Q}_{sr} \\ \mathbf{Q}_{rs} & \mathbf{Q}_r \end{pmatrix} \text{ and } \mathbf{Q}^* = \begin{pmatrix} \pi_1 q_{11} & \cdots & \pi_{1N} q_{1N} \\ \vdots & \ddots & \vdots \\ \pi_{N1} q_{N1} & \cdots & \pi_N q_{NN} \end{pmatrix},$$

where π_{ij} (π_i) denote the second- (first-) order inclusion probabilities and $\mathbf{Q}$ is a matrix with entries $q_{ij} \geq 0$. Define the QR predictor in this more general model ($\mathbf{V}$ not necessarily diagonal) as in Definition 9.1.1. Let

$$\mathbf{B} = (\mathbf{X}'\mathbf{Q}^*\mathbf{X})^{-1}\mathbf{X}'\mathbf{Q}^*\mathbf{y}.$$

(i) Prove that $\hat{T}_{QR}$ is ADU provided

$$\hat{\boldsymbol{\beta}}_Q \xrightarrow{P[p]} \mathbf{B}, \text{ as } n, N \to \infty,$$

and

$$(\mathbf{I} - \mathbf{R\Pi})\mathbf{1}_N \in \mathcal{M}(\mathbf{Q}^*\mathbf{X}).$$

(ii) As in Section 9.1, the GREG predictor is defined as

$$\hat{T}_{GREG} = \mathbf{1}'_N\mathbf{X}\hat{\beta}_Q + \mathbf{1}'_s\mathbf{\Pi}_s^{-1}\hat{\mathbf{e}}_s.$$

Find sufficient conditions under which any $\hat{T}_{QR} = \hat{T}_{GREG}$.
(iii) Find sufficient conditions under which any QR predictor may be written as a simple projection predictor.
(iv) Find sufficient conditions under which the BLUP of T given in (2.1.5) is ADU.

[9.5] (Tam, 1988b) Let $\mathbf{W} = (\mathbf{X}, \mathbf{Q})$ and $\mathbf{X} = (\mathbf{P}, \tilde{\mathbf{X}})$, where $\mathbf{Q}$ is an $N \times q$, $\mathbf{P}$ an $N \times r$ and $\tilde{\mathbf{X}}$ an $N \times m$ matrix such that $r + m = p$. Define

$$\mathbf{W} = (\mathbf{W}_s, \mathbf{W}_r) \text{ and } \mathbf{P} = (\mathbf{P}_s, \mathbf{P}_r).$$

Consider the two predictors

$$\hat{T}^*(\mathbf{H}_s) = \mathbf{1}'_s\mathbf{\Pi}_s^{-1}\mathbf{y}_s + (\mathbf{1}'\mathbf{H} - \mathbf{1}'_s\mathbf{\Pi}_s^{-1}\mathbf{H}_s)(\mathbf{H}'_s\mathbf{V}_s^{-1}\mathbf{H}_s)^{-1}\mathbf{H}'_s\mathbf{V}_s^{-1}\mathbf{y}_s,$$

with inclusion probabilities given by Eq. (9.3.4) and where $\mathbf{H} = \mathbf{P}$ or $\mathbf{W}$. Show that

(i) $\hat{T}^*(\mathbf{P}_s)$ is robust if condition (i) of Theorem 9.3.2 holds;
(ii) $\hat{T}^*(\mathbf{W}_s)$ is robust if condition (ii) of Theorem 9.3.2 is satisfied.

[9.6] (Tam, 1988b) Suppose that $\mathbf{V}^* = \sigma^2\text{diag}\{g_1 \ldots, g_N\}$ is adopted instead of the model $\psi = \psi(\boldsymbol{\beta}, \mathbf{V})$, where $\mathbf{V} = \sigma^2\text{diag}\{v_1, \ldots, v_N\}$. Let $\mathbf{h}'_s\mathbf{y}_s$ be a linear predictor of T such that

$$\text{(i) } \mathbf{h}'_s\mathbf{X}_s = \mathbf{1}'_N\mathbf{X} \text{ and (ii) } \mathbf{Q}_s^{-1}(\mathbf{h}_s - \mathbf{\Pi}_s^*\mathbf{1}_s) \in \mathcal{M}(\mathbf{X}_s).$$

Then, for any *ADU* predictor $\hat{T}_{QR}$ of T,

$$E_{p^*}E_\psi[\hat{T}_{QR} - T]^2 \geq E_{p^*}E_\psi[\mathbf{h}'_s\mathbf{y}_s - T]^2$$

$$= \sum_{j=1}^{N}(k^*g_i^{1/2} - 1)v_i\sigma^2,$$

and p^* is the optimal design corresponding to the alternative model. In this case, $k = \sum_{i=1} g_i^{1/2}/n$.

[9.7] Prove that a sufficient condition for $\hat{T}_{SP} = \mathbf{1}'_N\mathbf{X}\hat{\boldsymbol{\beta}}_s$ to be strongly robust, according to the Definition 9.3.4, is that

$$\mathbf{V}_s^{1/2}\mathbf{1}_s \in \mathcal{M}(\mathbf{X}_s).$$

Glossary of Predictors

I. Predictors of the Population Total T

Predictor	Formula	Name of Predictor (Estimator)
$\hat{T}$	(1.1.5)	Horwitz–Thompson
$\hat{T}_E$	(1.3.4)	Expansion
$\hat{T}_R$	(1.3.5)	Ratio
$\hat{T}_L$	(2.1.1)	General Linear
$\hat{T}_{BLU}$	(1.5.3)–(2.1.5)	Best Linear Unbiased, general
$\hat{T}_{BLU}$	(2.1.9)	Best Linear Unbiased, SM6
$\hat{T}_{SR}$	(1.5.6)	Stratified Ratio
$\hat{T}_{SP}$	(2.1.11)	Simple Projection
$\hat{T}_{RE}$	(2.2.17)	Regression
$\hat{T}_{MRE}$	(2.3.14)	Minimum Risk Equivariant (MRE), Location Model
$\hat{T}_{MRE}$	(2.3.24)	MRE, Scale Model
$\hat{T}_{MRE}$	(2.3.30)	MRE, Location–Scale Model
$\hat{T}_c$	(2.4.7)	Shrinkage Predictor
$\hat{T}_B$	(3.1.15)	Bayes, Normal Model
$\hat{T}_{GS}$	(3.2.2)	Linear Bayes, Smouse
$\hat{T}_{GO}$	(3.2.3)	Linear Bayes, O'Haggan
$\hat{T}_B(a, b)$	(3.3.1)	Bayes, Binomial Model
$\hat{T}_{DBt}$	(3.4.8)	Dynamic (Recursive)
$\hat{T}_{DBt}$	(3.4.12)	Dynamic (Recursive), SM1
$\hat{T}_{BE}$	(4.1.7)	Best Estimative, Exponential
$\hat{T}_{EML}$	(4.2.2)	Estimative Maximum Likelihood
$\hat{T}_{MPL}$	(4.2.5)	Maximum Profile Likelihood
$\hat{T}_{MLHP}$	(4.2.8)	Maximum Lauritzen–Hinkley

$\hat{T}_{EE}$	(7.1.3)	Estimative Predictor
$\hat{T}_{RML}$	(4.2.20)	Royall MLP
$\hat{T}_{ML}$	(4.2.23)	Royall MLP, SM3
$\hat{T}_{GRE}$	(7.1.11)	General Regression, Errors
$\hat{T}_{REE}$	(7.1.12)	Regression, Errors
$\hat{T}_{REU}$	(7.1.23)	Regression, Errors
$\hat{T}_{REC}$	(7.1.30)	Conditional, Regression Errors
$\hat{T}_{BE}$	(7.2.3)	Bayes, Errors
$\hat{T}_{BE}$	(7.2.8)–(7.2.9)	Bayes, Errors, SM6
$\hat{T}_{BE}$	(7.2.28)	Bayes, Regress., Errors
$\hat{T}_{QR}$	(9.1.1)	QR-type
$\hat{T}_{HTR}$	(9.1.3)	Horvitz–Thompson, SM2
$\hat{T}_{CR}$	(9.1.4)	Combined Regression, SM2
$\hat{T}_{BR}$	(9.1.5)	Brewer, SM2
$\hat{T}_{GREG}$	(9.1.6)	Generalized Regression

II. Predictors of the Population Variance

Predictor	Formula	Name
S_y^2	(1.3.2)	Population Variance
s_y^2	(1.1.7)	Sample Variance
$\hat{S}_y^2$	(1.1.8)	p–Unbiased
$\hat{\sigma}_s^2$	(2.2.5)	Estimator of σ^2
$\hat{\sigma}_s^2$	(2.2.11)	Estimator of σ^2, SM2
$\hat{S}_{BU}^2$	(2.2.7)	Best Unbiased, General Regression
$\hat{S}_{BU}^2$	(2.2.8)	Best Unbiased, SM1 (σ known)
$\hat{S}_{BU}^2$	(2.2.9)	Best Unbiased, SM1
$\hat{S}_{BU}^2$	(2.2.10)	Best Unbiased, SM2
$\hat{S}_{MRE}^2$	(2.3.37)	Minimum Risk Equivariant (MRE), Location–Scale Model
$\hat{S}_{MRE}^2$	(2.3.40)	MRE, Normal Model
$\hat{S}_B^2$	(3.1.25)	Bayes
$\hat{S}_{DBt}^2$	(3.4.14)	Dynamic Bayes
$\hat{S}_{DBt}^2$	(3.4.15)	Dynamic Bayes, SM1
$\hat{S}_{EML}^2$	(4.3.1)	Estimative Maximum–Likelihood
$\hat{S}_{BE}^2$	(4.3.2)	Best Estimative, Normal Regression Model
$\hat{S}_{MLH}^2$	(4.3.10)	Maximum–Likelihood, Lauritzen–Hinkley
$\hat{S}_{BM}^2$	(6.5.6)	Bayes Modeling
$\hat{S}_{BM}^2$	(6.5.14)	Bayes Modeling, SM1–SM2
$\hat{S}_{BE}^2$	(7.2.5)	Bayes, SM1, Errors
$\hat{S}_{BE}^2$	(7.2.20)	Bayes, SM3, Errors

III. Predictors of the Population (Model) Regression Coefficients

Predictor	Formula	Name
β	(1.2.1)	Regression Coefficients
β_N	(1.3.9)	Population Regression Coefficients
$\hat{\beta}_s$	(1.3.14)	Weighted Least Squares
$\hat{\beta}_{BU}$	(2.2.20)	Best Unbiased, β_N
$\hat{\beta}_{LN}$	(2.2.22)	General Linear, β_N
$\hat{\beta}_B$	(3.1.5)	Bayes, β
$\hat{\beta}_{BN}$	(3.1.21)	Bayes, β_N
$\hat{\beta}_{EML}$	(4.1.6)	Estimative ML Predictor, Exponential Model
$\hat{\beta}_{MLH}$	(4.4.1)	Maximum L–H Predictor
$\hat{\beta}_{1E}$	(7.1.13)	Least Squares, Errors, β
$\hat{\beta}_{cs}$	(8.3.4)	Errors, β, β_N
$\hat{\beta}^*_{cs}$	(8.3.5)	Errors, β, β_N
$\hat{\beta}_Q$	(9.1.2)	QR-type, β

Bibliography

1. Aitchison, J. and Dusmore, I.R. (1975), *Statistical Prediction Analysis*, Cambridge University Press, Cambridge.
2. Arnold, S.B. (1981), *The Theory of Linear Models*, Wiley, New York.
3. Basu, D. (1955), On statistics independent of a complete and sufficient statistics, *Sankhyā*, A15, 377-380.
4. Basu, D. (1971), An essay on the logical foundations of survey sampling in *Foundations of Statistical Inference*, V.P. Godambe and D.A. Sprott, eds., Holt, Rinehart and Winston, Toronto.
5. Basu, D. (1975), Statistical information and likelihood, *Sankhyā*, A, 37, 1-71.
6. Bellhouse, D. (1987), Model based inference in finite population sampling, *The American Statistician*, 41, 4, 260-262.
7. Berger, J. (1980), *Statistical Decision Theory, Foundations, Concepts and Methods*, Springer–Verlag, New York.
8. Bjørnstad, J.F. (1990), Prediction likelihood: A review, *Statistical Science*, 5, 242-265.
9. Blight, B.G.N. and Scott, A.J. (1973), A stochastic model for repeated surveys, *Journal of the Royal Statistical Society*, B35, 61-66.
10. Blyth, C.R. (1951), On minimax statistical decision procedures and their admissibility, *Annals of Mathematical Statistics*, 22, 22-42.
11. Bock, R.D. and Peterson, A.C. (1975), A multivariate correction for attenuation, *Biometrika*, 62, 673-678.
12. Bolfarine, H. (1987), Minimax prediction in finite populations, *Communications in Statistics, Theory and Methods*, 16, 12, 3683-3700.
13. Bolfarine, H. (1988a), Finite population prediction under dynamic generalized linear superpopulation models, *Communications in Statistics, Computation and Simulation*, 17, 1, 187-208.

14. Bolfarine, H. (1988b), Bayesian modelling in finite populations, *South African Statistical Journal*, 23, 157-166.
15. Bolfarine, H. (1989a), Population variance prediction under normal dynamic superpopulation models, *Statistics and Probability Letters*, 8, 35-39.
16. Bolfarine, H. (1989b), Equivariant prediction in finite populations, *Communications in Statistics, Theory and Methods*, 18, 3, 927-942.
17. Bolfarine H. (1989c), Finite population prediction under assymetric loss functions, *Communications in Statistics, Theory and Methods*, 18, 5, 1863-1870.
18. Bolfarine, H. (1990), Bayesian linear prediction in finite populations. *Annals of the Institute of Statistical Mathematics*, 42, 3, 435-444.
19. Bolfarine, H. (1991), Finite population prediction under error in variables superpopulation models, *Canadian Journal of Statistics*, 19, 2, 191-207.
20. Bolfarine, H., Pereira, C.A.B. and Rodrigues, J. (1987), Robust linear prediction in finite populations, a Bayesian perspective, *Sankhyā*, B, 49, 23-35.
21. Bolfarine, H. and Rodrigues, J. (1988), On the simple projection predictor in finite populations, *Australian Journal of Statistics*, 3, 338-341.
22. Bolfarine, H. and Rodrigues, J. (1990), Finite population prediction under a linear functional superpopulation model, a Bayesian perspective, *Communication in Statistics, Theory and Methods*, 19, 7, 2577-2595.
23. Bolfarine, H. and Sandoval, M.C. (1990), Finite population prediction under error in variables models with known reliability ratio, Technical Report, University of São Paulo, Department of Statistics.
24. Bolfarine, H. and Sandoval, M.C. (1990), Prediction of the finite population distribution function under Gaussian superpopulation models, Technical Report, University of São Paulo, Department of Statistics.
25. Bolfarine, H. and Zacks, S. (1991), Bayes and minimax prediction in finite populations, *Journal of Statistical Planning and Inference*, 28, 3, 139-151.
26. Bolfarine, H., Elian, S.N., Rodrigues, J. and Zacks, S. (1991), Optimal prediction of the regression coefficient in finite populations, Technical Report, University of São Paulo, Department of Statistics.
27. Brewer, K.R.W. (1979), A class of robust sampling designs for large scale surveys, *Journal of the American Statistical Association*, 74, 368, 911-915.
28. Butler, R.W. (1986), Predictive likelihood inference with applications, *Journal of the Royal Statistical Society*, B48, 1-38.
29. Cassel, C.M., Särndal, C.E. and Wretman, J.H. (1977), *Foundation of Inference in Sample Surveys*, Wiley, New York.
30. Chandhook, P.K. (1982), A study of the effects of measurement error in survey sampling, Ph.D. dissertation, Iowa State University, Ames, Iowa.
31. Cochran, W.G. (1968), Errors of measurements in statistics, *Technometrics*, 10, 637-666.

32. Cochran, W.G. (1977), *Sampling Techniques*, 3rd edition, Wiley, New York.
33. Cooley, W.W. and Lohnes, P.R. (1971), *Multivariate Data Analysis*, Wiley, New York.
34. Cox, D.R. and Reid, N. (1987), Parameter orthogonality and approximate conditional inference (with discussion). *Journal of the Royal Statistical Society*, B49, 1-49.
35. Cumberland, W.G. and Royall, R.M. (1981), Prediction models in unequal probability sampling, *Journal of the Royal Statistical Society*, B43, 353-367.
36. Dempster, A.P., Laird, N.M. and Rubin, D.B. (1977), Maximum likelihood from incomplete data via the EM–algorithm, *Journal of the Royal Statistical Society*, B39, 1-38.
37. Ericson, W.A. (1969), Subjective Bayesian models in sampling finite populations, *Journal of the Royal Statistical Society*, B31, 195-224.
38. Ericson, W.A. (1988), Bayesian inference in finite populations in *Handbook of Statistics*, Vol. 6, 213-246. P.R. Krishnaiah and C.R. Rao, eds., North Holland, New York.
39. Ferguson, T.S. (1967), *Mathematical Statistics, A Decision Theoretic Approach*, Academic Press, New York.
40. Fuller, W.A. (1975a), Regression analysis for sample surveys, *Sankhyā*, C37, 117-132.
41. Fuller, W.A. (1975b), *The Statistical Analysis of Time Series*, Wiley, New York.
42. Fuller, W.A. (1987), *Measurement Error Models*, Wiley, New York.
43. Ghosh, M. and Meeden, G. (1986), Empirical Bayes estimation in finite populations sampling, *Journal of the American Statistical Association*, 81, 1058-1062.
44. Ghosh, M. and Lahiri, P. (1987), Robust Empirical Bayes estimation of means from stratified samples, *Journal of the American Statistical Association*, 82, 1153-1162.
45. Godambe, V.P. and Joshi, V.M. (1965), Admissibility of Bayes estimation in finite populations, *Annals of Mathematical Statistics*, 36, 1707-1722.
46. Guttman, I. (1970), *Statistical Tolerance Regions: Classical and Bayesian*, Griffin, London.
47. Hájek, J. (1971), Comments on the essay by D. Basu, in *Foundations of Statistical Inference*, V.P. Godambe and D.A. Sprott, eds., Holt, Rinehart and Winston, Toronto.
48. Hartley, H.O. and Sielken, R.L. Jr. (1975), A superpopulation viewpoint to finite population sampling, *Biometrics*, 31, 411-422.
49. Herson, J. (1976), An investigation of the relative efficiency of least–squares prediction to conventional probability sampling plans, *Journal of the American Statistical Association*, 71, 700-703.

50. Hidiroglou, M. (1974), Estimation of regression parameters for finite populations, Ph.D. thesis, Iowa State University, Ames.
51. Hinkley, D.V. (1979). Predictive likelihood, *Annals of Statistics*, 7, 718-728.
52. Isaki, C.T. and Fuller, W.A. (1982), Survey design under a regression superpopulation model, *Journal of the American Statistical Association*, 77, 89-96.
53. Johnson, N.L. and Kotz, S. (1970), *Distribution in Statistics: Continuous Univariate Distributions*, Houghton and Mifflin, Boston.
54. Jones, R.G. (1980), Best linear unbiased estimators for repeated sampling, *Journal of the Royal Statistical Society*, B42, 221-226.
55. Konijn, H.S. (1962), Regression analysis in sample surveys, *Journal of the American Statistical Association*, 57, 590-606.
56. Kagan, A.M., Linik, Y.V. and Rao, C.R. (1965), On the characterization of the normal law based on a property of the sample average, *Sankhyā*, A27, 405-406.
57. Lehmann, E.L. (1984), *Theory of Point Estimation*, Wiley, New York.
58. Lindley, D.V. (1966), Discussion on a generalized least squares approach to linear functional relationship, *Journal of the Royal Statistical Society*, B28, 279-297.
59. Lindley, D.V. and El Sayad, G. (1968), The Bayesian estimation of a linear functional relationship, *Journal of the Royal Statistical Society*, B30, 190-202.
60. Montanari, G.M. (1987), Post sampling efficient QR prediction in large sample surveys, *International Statistic Review*, 55, 191-202.
61. O'Haggan, A. (1986), Bayes linear estimators for finite populations, Technical Report No. 58, University of Warwick.
62. Orchard, T., Woodbury, M. (1972), A missing value information principle, *Proceedings of the 6th Berkeley Symposium on Mathematical Statistics and Probability*, 1, 697-715.
63. Pereira, C.A.B., and Rodrigues, J. (1983), Robust linear prediction in finite populations, *International Statistical Review*, 51, 293-300.
64. Pfeffermann, D. (1984), Note on large sample properties of balanced sample, *Journal of the Royal Statistic Society*, B46, 38-41.
65. Rao, C.R. (1973), *Linear Statistical Inference and its Applications*, Wiley, New York.
66. Reilly, P. and Patino–Leal, H. (1981), A Bayesian study of the error in variable models, *Technometrics*, 23, 3, 221-231.
67. Robbins, H. (1955), The empirical Bayes approach to statistics, Proceedings of the 3rd Berkeley Symposium on Mathematical Statistics and Probability, 1, 157-164.
68. Rodrigues, J. (1987), Some shrinkage predictors in finite populations with a multinormal superpopulation model, *Statistics and Probability Letters*, 5, 347-351.

69. Rodrigues, J. (1988), Some results on restricted Bayes least squares predictors for finite populations, *South African Statistics Journal*, 22, 45-53.
70. Rodrigues, J. and Bolfarine, H. (1984), *Teoria da Previsão em Populacoes Finitas*, VI Brasilian Symposium in Probability and Statistics. (In Portuguese)
71. Rodrigues, J., Bolfarine, H. and Rogakto, A. (1985), A general theory of prediction in finite populations, *International Statistical Review*, 53, 239-254.
72. Rodrigues, J. and Bolfarine, H. (1987), A Kalman filter model for single and two stage repeated surveys, *Statistics and Probability Letters*, 5, 299-303.
73. Rodrigues, J. and Bolfarine, H. (1988), Some asymptotic results on generalized regression predictors in survey sampling, *Pakistan Journal of Statistics*, 4, 129-138.
74. Rodrigues, J. and Bolfarine, H. (1989), A note on asymptotically design unbiased designs in survey sampling, Technical Report, University of São Paulo, Department of Statistics.
75. Rodrigues, J. and Elian, S.N. (1989), The coordinate free estimation in finite population sampling, *Statistics and Probability Letters*, 7, 293-295.
76. Royall, R.M. (1970), On Finite Population Sampling Theory Under Certain Linear Regression Models, *Biometrika*, 57, 377-387.
77. Royall, R.M. (1976a), The linear least squares prediction approach to two stage sampling, *Journal of the American Statistical Association*, 71, 657-664.
78. Royall, R.M. (1976b), Likelihood function in finite population sampling theory, *Biometrika*, 63, 605-614.
79. Royall, R.M. (1988), The prediction approach to sampling theory, in *Handbook of Statistics*, 6, 399-413, P.R. Krishnaiah and C.R. Rao, eds., North Holland, New York.
80. Royall, R.M. and Cumberland, W.G. (1978), Variance estimation in finite population sampling, *Journal of the American Statistical Association*, 73, 351-358.
81. Royall, R.M. and Cumberland, W.G. (1981a), An empirical study of the ratio estimator and estimators of its variance, *Journal of the American Statistical Association*, 76, 66-77.
82. Royall, R.M. and Cumberland, W.G. (1981b), The finite population linear regression estimator and estimators of its variance, an empirical study, *Journal of the American Statistical Association*, 76, 924-930.
83. Royall, R.M. and Eberhardt, K.R. (1975), Variance estimates of the ratio estimator, *Sankhyā* C37, 43-52.
84. Royall, R.M. and Herson, J. (1973a), Robust estimation in finite populations I, *Journal of the American Statistical Association*, 68, 880-889.
85. Royall, R.M. and Herson, J. (1973b), Robust estimation in finite populations II, Stratification on a size variable, *Journal of the American Statistical Association*, 68, 890-893.

86. Royall, R.M. and Pfeffermann, D. (1982), Balanced samples and Robust Bayesian inference in finite population sampling, *Biometrika*, 69, 401-409.
87. Särndal, C.E. (1980), On π–inverse weighting versus best linear unbiased weighting in probability sampling, *Biometrika*, 67, 641-650.
88. Särndal, C.E. (1982), Implications of survey design for generalized regression estimation of linear functions, *Journal of Statistical Planning and Inference*, 19, 155-170.
89. Särndal, C.E. and Wright, R.L. (1984), Cosmetic forms of estimators in survey sampling, *Scandinavian Journal of Statistics*, 11, 146-156.
90. Scott, A.J., Brewer, K.W. and Ho, E.W. (1978), Finite population sampling and robust estimation, *Journal of the American Statistical Association*, 73, 359-361.
91. Scott, A.J. and Smith, T.M.F. (1969), Estimation in multi–stage surveys, *Journal of the American Statistical Association*, 64, 830-840.
92. Scott, A.J. and Smith, T.F.M. (1974), Analysis of repeated surveys using time series methods, *Journal of the American Statistical Association*, 69, 674-678.
93. Searle, S.R. (1971), *Linear Models*, Wiley, New York.
94. Seber, G.H.F. (1977), *Linear Regression Analysis*, Wiley, New York.
95. Serfling, R.J. (1980), *Approximation Theorems in Mathematical Statistics*, Wiley, New York.
96. Shah, B.V., Holt, M.M. and Folsom, R.E. (1977), Inference about regression models from sample survey data, *Bulletin of the International Statistical Institute*, 47, 3, 43-57.
97. Skiener, C.J. (1983), Multivariate prediction from selected samples, *Biometrika*, 70, 289-292.
98. Smith, A.F.M. (1986), Some Bayesian thoughts on modelling and model choice, *The Statistician*, 35, 97-102.
99. Smouse, E. (1984), A note on Bayesian least squares inference for finite population models, *Journal of the American Statistical Association*, 79, 386, 390-392.
100. Sprent, P. (1966), A generalized least squares approach to linear functional relationships, *Journal of the Royal Statistical Society*, B28, 279-297.
101. Srivastava, M.S. and Khatri, C.G. (1979), *An Introduction to Multivariate Statistics*, North Holland, New York.
102. Tallis, G.M. (1978), Note on robust estimation in finite populations, *Sankhyā*, C40, 136-138.
103. Tam, S.M. (1986), Characterization of best model–based predictors in survey sampling, *Biometrika*, 74, 659-660.
104. Tam, S.M. (1987a), Optimality of Royall's predictor under a Gaussian superpopulation model, *Biometrika*, 74, 659-660.
105. Tam, S.M. (1987b), Analysis of repeated surveys using a dynamic linear model, *International Statistics Review*, 55, 63-73.

106. Tam, S.M. (1988a), Asymptotically design–unbiased predictors in survey sampling, *Biometrika*, 75, 175-177.
107. Tam, S.M. (1988b), Some results on robust estimation in finite population sampling, *Journal of the American Statistical Association*, 83, 242.
108. Thomsen, I. (1981), The use of Markov chain models in sampling from finite populations, *Scandinavian Journal of Statistics*, 8, 1-9.
109. Thomsen, I. and Tesfu, D. (1988), On the use of models in sampling from finite populations, in *Handbook of Statistics*, 6, 369-398, P.R. Krishnaiah and C.R. Rao, eds., North Holland, New York.
110. Thompson, J.R. (1968), Some shrinkage techniques for estimating the mean, *Journal of the American Statistical Association*, 63, 113-123.
111. Tierney, L. and Kadane, J. (1986), Accurate approximations for posterior moments and marginal densities, *Journal of the American Statistical Association*, 81, 82-86.
112. Wright, R.L. (1983), Finite population sampling with multivariate auxiliary information, *Journal of the American Statistical Association*, 78, 879-884.
113. Zacks, S. (1971), *The Theory of Statistical Inference*, Wiley, New York.
114. Zacks, S. (1981), *Parametric Statistical Inference: Basic Theory and Modern Approaches*, Pergamon Press, Oxford.
115. Zacks, S. (1981), Bayes equivariant estimators of the variance of a finite population for exponential priors, *Communications in Statistics, Theory and Methods*, 10, 427-437.
116. Zacks, S. and Bolfarine, H. (1991), Equivariant prediction of the population variance under location–scale superpopulation models, *Sankhyā*, B53.
117. Zacks, S. and Bolfarine, H. (1990), Maximum likelihood prediction of finite population quantities, Technical Report, University of São Paulo, Department of Statistics.
118. Zacks, S. and Rodrigues, J. (1986), A note on the missing value principle and the EM–algorithm for estimation and prediction in sampling from finite populations with a multi–normal superpopulation model, *Statistics and Probability Letters*, 4, 35-37.
119. Zacks, S. and Solomon, H. (1981), Bayes equivariant estimators of the variance of a finite population: Part I, simple random sampling, *Communications in Statistics, Theory and Methods*, 10, 407-426.
120. Zellner, A. (1971), *An Introduction to Bayesian Inference in Econometrics*, Wiley, New York.
121. Zellner, A. (1986), Bayes estimation and prediction using asymmetric loss functions, *Journal of the American Statistical Association*, 81, 446-451.
122. Zyskind, G. (1967), On canonical forms, nonnegative covariance matrices and best and simple least squares linear estimators in linear models, *Annals of Mathematical Statistics*, 38, 1092-1109.

Author Index

Subject Index

Springer Series in Statistics

(continued from p. ii)

Siegmund: Sequential Analysis: Tests and Confidence Intervals.
Tong: The Multivariate Normal Distribution.
Vapnik: Estimation of Dependences Based on Empirical Data.
West/Harrison: Bayesian Forecasting and Dynamic Models.
Wolter: Introduction to Variance Estimation.
Yaglom: Correlation Theory of Stationary and Related Random Functions I: Basic Results.
Yaglom: Correlation Theory of Stationary and Related Random Functions II: Supplementary Notes and References.